Metabolic Flux Analysis in Eukaryotic Cells

Methods and Protocols

Edited by

Deepak Nagrath

Department of Biomedical Engineering, University of Michigan, Ann Arbor, MI, USA

Humana Press

Editor
Deepak Nagrath
Department of Biomedical Engineering
University of Michigan
Ann Arbor, MI, USA

ISSN 1064-3745 ISSN 1940-6029 (electronic)
Methods in Molecular Biology
ISBN 978-1-0716-0158-7 ISBN 978-1-0716-0159-4 (eBook)
https://doi.org/10.1007/978-1-0716-0159-4

This Humana imprint is published by the registered company Springer Science+Business Media, LLC, part of Springer Nature.
The registered company address is: 233 Spring Street, New York, NY 10013, U.S.A.

Preface

Recent advancements in metabolic flux analysis (MFA) are instrumental in revealing mechanistic underpinnings of cellular metabolism. Especially in the identification of metabolically vulnerable targets and the dissection of crosstalk between diseased cells and their microenvironment, MFA has been the backbone of new advancements in targeting several diseases. The chapters in this special issue feature state-of-the-art metabolic flux techniques that span analyses of cellular, organ-level, and whole-body metabolism.

Stable isotopes are commonly used as labels for analysis of metabolism of living cells. The uptake and utilization of isotopically enriched nutrients leads to the formation of enriched intracellular metabolites and lipids. The enrichment data obtained from mass spectrometry (MS) techniques is incorporated in metabolic models for metabolic flux estimation. Mairinger and Hann present the accurate and precise analysis of isotopologue and tandem mass isotopologue ratios in heavy stable isotope labeling experiments in the presence of measurement uncertainty. Dudek et al. describe a data processing workflow in non-targeted stable isotope labeling experiments to generate metabolite levels, mass isotopomer distribution, and similarity and variability analysis of metabolites. Damini et al. provide a detailed overview of methods for polar metabolite analysis in reverse phase ion pairing and hydrophilic interaction chromatography for ^{13}C MFA. To interrogate intracellular compartments, ^{2}H (deuterium) tracing approaches have gained popularity. Here, Lim et al. provide a detailed description of ^{2}H tracing applications for the interrogation of mitochondrial versus cytosolic NAD(P)H metabolism in mammalian cells. To broaden the scope of MFA from cells to whole body, ^{13}C-based in vivo flux analysis can be used. However, the complexity of handling mice in disease models had discouraged early in vivo tracing analysis. Recent advances in MS techniques and increased instrument sensitivity have however encouraged development of these methodologies. Altea-Manzano et al. fill an important gap by presenting a methodology for understanding the metabolism of metastases in vivo.

Biomass evaluation is an important component in MFA, and in the past generic data has been used to model this important flux. Széliová et al. present a detailed method for the determination of biomass flux from Chinese hamster ovary cells. Nitric oxide involvement in cancer and several other diseases has recently been uncovered. Still, there is a lack of methodologies for the measurement of NO flux. Sivaloganathan presents a detailed protocol, which includes experimental measurements and computational modeling, to estimate the NO flux distributions. To increase the accuracy and quantification of low-abundance metabolites, an accurate modeling workflow is needed. Jaiswal and Wangikar present a methodology called sequential windowed acquisition of all theoretical fragment ion mass spectra, which allows quantification of isotopic ^{13}C enrichment in a large number of cellular metabolites and fragments. To increase the scope of flux analysis to heterogeneous cellular systems, a methodology which can capture metabolic communication between different cell types is required. Achreja et al. present an integrated empirical and computational platform to quantify metabolic crosstalk between source and recipient cells. Their platform allows the estimation of contribution of source cell-derived extracellular vesicles to recipient cells. Garrity et al. describe a method to combine mRNA and metabolomics data in a genome-scale metabolic model to curate a biologically feasible model for constrained MFA. Selivanov et al. present a software supporting a workflow of analysis of stable isotope-resolved metabolomics data obtained with MS and their use in a kinetic model based on ordinary differential equations for isotopomers of metabolites of the corresponding biochemical network to estimate a dynamic flux map. Campit and Chandrasekaran present a genome-scale modeling approach that uses time-course metabolomics to predict dynamic flux rewiring during transitions between metabolic states.

Metabolic flux approaches have also been garnering attention as a cost-effective platform to develop and test drugs for their efficacy toward specific metabolic targets. Rawls et al. present the application of metabolic flux approaches for drug development. Unraveling heterogeneity is difficult with respect to metabolic changes in living cells. Filippo et al. present a new computational framework called single-cell Flux Balance Analysis that aims to set up digital metabolic twins that also use laboratory patient cell models to unravel changes in heterogeneous populations. Toit et al. present an interesting application of MFA in autophagy, which is a cellular homeostasis process that maintains cellular nutrients. Peres and Fromion present a protocol for the integration of thermodynamic constraints in metabolic models to eliminate non-physical fluxes or inconsistencies in the metabolic system.

The chapters in this book will be of great interest to both experts in MFA techniques and researchers getting initiated in the role of quantitative studies to unravel the secrets of dysregulated pathways in human diseases. I would like to thank all the contributing authors for their valuable support in presenting their work and advancements toward making MFA an incisive and decisive technique to dissect metabolic states in diseases. Furthermore, this issue would not have been possible without support from my lab members, precious colleagues, and budding scientists, Dr. Abhinav Achreja, Anjali Mittal, Olamide Animasahun, and Noah Meurs. They dedicatedly helped me with editorial corrections and provided their input many times on short notice, and they are themselves looking forward to advancing the field of MFA. I would also like to thank Dr. John Walker and Anna Rakovsky (Springer Nature) for their editorial inputs and assistance.

Ann Arbor, MI, USA ***Deepak Nagrath***

Contents

Contributors

ABHINAV ACHREJA • *Department of Biomedical Engineering, University of Michigan, Ann Arbor, MI, USA; Biointerfaces Institute, University of Michigan, Ann Arbor, MI, USA*

ROBERT AHRENDS • *Leibniz Institut für Analytische Wissenschaften—e.V., Dortmund, Germany*

LILIA ALBERGHINA • *SYSBIO Centre of Systems Biology, Milan, Italy*

PATRICIA ALTEA-MANZANO • *Laboratory of Cellular Metabolism and Metabolic Regulation, VIB-KU Leuven Center for Cancer Biology, VIB, Leuven, Belgium; Laboratory of Cellular Metabolism and Metabolic Regulation, Department of Oncology, KU Leuven and Leuven Cancer Institute (LKI), Leuven, Belgium*

NICOLE BORTH • *Austrian Centre of Industrial Biotechnology, Vienna, Austria; University of Natural Resources and Life Sciences, Vienna, Austria*

DORIEN BROEKAERT • *Laboratory of Cellular Metabolism and Metabolic Regulation, VIB-KU Leuven Center for Cancer Biology, VIB, Leuven, Belgium; Laboratory of Cellular Metabolism and Metabolic Regulation, Department of Oncology, KU Leuven and Leuven Cancer Institute (LKI), Leuven, Belgium*

MARK P. BRYNILDSEN • *Department of Chemical and Biological Engineering, Princeton University, Princeton, NJ, USA*

SCOTT CAMPIT • *Program in Chemical Biology, University of Michigan, Ann Arbor, MI, USA*

MARTA CASCANTE • *Department of Biochemistry and Molecular Biomedicine, Faculty of Biology, Universitat de Barcelona, Barcelona, Spain; Institute of Biomedicine of Universitat de Barcelona (IBUB), Barcelona, Spain; Centro de Investigación Biomédica en Red de Enfermedades Hepáticas y Digestivas (CIBEREHD), Instituto de Salud Carlos III (ISCIII), Madrid, Spain; INB-Bioinformatics Platform Metabolomics Node, Instituto de Salud Carlos III (ISCIII), Madrid, Spain*

SRIRAM CHANDRASEKARAN • *Program in Chemical Biology, University of Michigan, Ann Arbor, MI, USA; Department of Biomedical Engineering, University of Michigan, Ann Arbor, MI, USA*

CRISTINA COMAN • *Leibniz Institut für Analytische Wissenschaften—e.V., Dortmund, Germany*

CHIARA DAMIANI • *SYSBIO Centre of Systems Biology, Milan, Italy; Department of Informatics, Systems and Communication, University of Milano-Bicocca, Milan, Italy; Department of Biotechnology and Biosciences, University of Milano-Bicocca, Milan, Italy*

PEDRO DE ATAURI • *Department of Biochemistry and Molecular Biomedicine, Faculty of Biology, Universitat de Barcelona, Barcelona, Spain; Institute of Biomedicine of Universitat de Barcelona (IBUB), Barcelona, Spain; Centro de Investigación Biomédica en Red de Enfermedades Hepáticas y Digestivas (CIBEREHD), Instituto de Salud Carlos III (ISCIII), Madrid, Spain; INB-Bioinformatics Platform Metabolomics Node, Instituto de Salud Carlos III (ISCIII), Madrid, Spain*

MARZIA DI FILIPPO • *SYSBIO Centre of Systems Biology, Milan, Italy; Department of Informatics, Systems and Communication, University of Milano-Bicocca, Milan, Italy*

BONNIE V. DOUGHERTY • *Department of Biomedical Engineering, University of Virginia, Charlottesville, VA, USA*

JOÃO A. G. DUARTE • *Laboratory of Cellular Metabolism and Metabolic Regulation, VIB-KU Leuven Center for Cancer Biology, VIB, Leuven, Belgium; Laboratory of Cellular Metabolism and Metabolic Regulation, Department of Oncology, KU Leuven and Leuven Cancer Institute (LKI), Leuven, Belgium*

CHRISTIAN-ALEXANDER DUDEK • *Department of Bioinformatics and Biochemistry, Braunschweig Integrated Centre of Systems Biology (BRICS), Technische Universität Braunschweig, Braunschweig, Germany*

ANDRÉ DU TOIT • *Department of Biochemistry, Stellenbosch University, Stellenbosch, South Africa; Department of Physiological Sciences, Stellenbosch University, Stellenbosch, South Africa*

TERESA W.-M. FAN • *Markey Cancer Center, University of Kentucky, Lexington, KY, USA; Center for Environment and Systems Biochemistry and the Resource Center for Stable Isotope Resolved Metabolomics, University of Kentucky, Lexington, KY, USA; Department of Toxicology and Cancer Biology, University of Kentucky, Lexington, KY, USA*

SARAH-MARIA FENDT • *Laboratory of Cellular Metabolism and Metabolic Regulation, VIB-KU Leuven Center for Cancer Biology, VIB, Leuven, Belgium; Laboratory of Cellular Metabolism and Metabolic Regulation, Department of Oncology, KU Leuven and Leuven Cancer Institute (LKI), Leuven, Belgium*

JUAN FERNÁNDEZ-GARCÍA • *Laboratory of Cellular Metabolism and Metabolic Regulation, VIB-KU Leuven Center for Cancer Biology, VIB, Leuven, Belgium; Laboratory of Cellular Metabolism and Metabolic Regulation, Department of Oncology, KU Leuven and Leuven Cancer Institute (LKI), Leuven, Belgium*

VINCENT FROMION • *INRA, UR1404, MaIAGE, Université Paris-Saclay, Jouy-en-Josas, France*

NINA GRANKVIST • *Cardiovascular Medicine Unit, Department of Medicine, Solna, Karolinska Institutet, Stockholm, Sweden; Karolinska University Hospital, Stockholm, Sweden; Center for Molecular Medicine, Karolinska Institutet, Stockholm, Sweden; Department of Medicine, University of California San Diego, La Jolla, CA, USA; Department of Pharmacology, University of California San Diego, La Jolla, CA, USA*

STEPHEN HANN • *Department of Chemistry, University of Natural Resources and Life Sciences—BOKU Vienna, Vienna, Austria; Austrian Center for Industrial Biotechnology, Vienna, Austria*

RICHARD M. HIGASHI • *Markey Cancer Center, University of Kentucky, Lexington, KY, USA; Center for Environment and Systems Biochemistry and the Resource Center for Stable Isotope Resolved Metabolomics, University of Kentucky, Lexington, KY, USA; Department of Toxicology and Cancer Biology, University of Kentucky, Lexington, KY, USA*

KARSTEN HILLER • *Department of Bioinformatics and Biochemistry, Braunschweig Integrated Centre of Systems Biology (BRICS), Technische Universität Braunschweig, Braunschweig, Germany; Computational Biology of Infection Research, Helmholtz Centre for Infection Research, Braunschweig, Germany*

JAN HENDRIK S. HOFMEYR • *Department of Biochemistry, Stellenbosch University, Stellenbosch, South Africa*

MOHIT JAIN • *Department of Medicine, University of California San Diego, La Jolla, CA, USA; Department of Pharmacology, University of California San Diego, La Jolla, CA, USA*

Damini Jaiswal • *Department of Chemical Engineering, Indian Institute of Technology Bombay, Mumbai, India*

Sigurður T. Karvelsson • *Center for Systems Biology, School of Health Sciences, University of Iceland, Reykjavik, Iceland*

Špela Knez • *University of Ljubljana, Ljubljana, Slovenia*

Gunda Koellensperger • *University of Vienna, Vienna, Austria*

Andrew N. Lane • *Markey Cancer Center, University of Kentucky, Lexington, KY, USA; Center for Environment and Systems Biochemistry and the Resource Center for Stable Isotope Resolved Metabolomics, University of Kentucky, Lexington, KY, USA; Department of Toxicology and Cancer Biology, University of Kentucky, Lexington, KY, USA*

Esther W. Lim • *Department of Bioengineering, University of California San Diego, La Jolla, CA, USA*

Ben Loos • *Department of Physiological Sciences, Stellenbosch University, Stellenbosch, South Africa*

Teresa Mairinger • *Department of Chemistry, University of Natural Resources and Life Sciences—BOKU Vienna, Vienna, Austria; Department of Environmental Chemistry, Eawag: Swiss Federal Institute of Aquatic Science and Technology, Duebendorf, Switzerland*

Silvia Marin • *Department of Biochemistry and Molecular Biomedicine, Faculty of Biology, Universitat de Barcelona, Barcelona, Spain; Institute of Biomedicine of Universitat de Barcelona (IBUB), Barcelona, Spain; Centro de Investigación Biomédica en Red de Enfermedades Hepáticas y Digestivas (CIBEREHD), Instituto de Salud Carlos III (ISCIII), Madrid, Spain*

Davide Maspero • *Department of Informatics, Systems and Communication, University of Milano-Bicocca, Milan, Italy; Istituto Nazionale dei Tumori, Milan, Italy*

Giancarlo Mauri • *SYSBIO Centre of Systems Biology, Milan, Italy; Department of Informatics, Systems and Communication, University of Milano-Bicocca, Milan, Italy*

Sarah McGarrity • *School of Science and Engineering, Reykjavik University, Reykjavik, Iceland; Center for Systems Biology, School of Health Sciences, University of Iceland, Reykjavik, Iceland*

Christian M. Metallo • *Department of Bioengineering, University of California San Diego, La Jolla, CA, USA; Moores Cancer Center, University of California San Diego, La Jolla, CA, USA; Diabetes and Endocrinology Research Center, University of California San Diego, La Jolla, CA, USA; Institute of Engineering in Medicine, University of California San Diego, La Jolla, CA, USA*

Noah Meurs • *Department of Biomedical Engineering, University of Michigan, Ann Arbor, MI, USA; Biointerfaces Institute, University of Michigan, Ann Arbor, MI, USA*

Anjali Mittal • *Department of Chemical Engineering, University of Michigan, Ann Arbor, MI, USA*

Deepak Nagrath • *Department of Chemical Engineering, University of Michigan, Ann Arbor, MI, USA; Department of Biomedical Engineering, University of Michigan, Ann Arbor, MI, USA; Biointerfaces Institute, University of Michigan, Ann Arbor, MI, USA*

Roland Nilsson • *Cardiovascular Medicine Unit, Department of Medicine, Solna, Karolinska Institutet, Stockholm, Sweden; Karolinska University Hospital, Stockholm, Sweden; Center for Molecular Medicine, Karolinska Institutet, Stockholm, Sweden*

JASON PAPIN • *Department of Biomedical Engineering, University of Virginia, Charlottesville, VA, USA*
SETH J. PARKER • *Department of Radiation Oncology, Perlmutter Cancer Center, New York University School of Medicine, New York, NY, USA*
SABINE PERES • *LRI, Université Paris-Sud, CNRS, Université Paris-Saclay, Orsay, France; INRA, UR1404, MaIAGE, Université Paris-Saclay, Jouy-en-Josas, France*
DARIO PESCINI • *SYSBIO Centre of Systems Biology, Milan, Italy*
MÉLANIE PLANQUE • *Laboratory of Cellular Metabolism and Metabolic Regulation, VIB-KU Leuven Center for Cancer Biology, VIB, Leuven, Belgium; Laboratory of Cellular Metabolism and Metabolic Regulation, Department of Oncology, KU Leuven and Leuven Cancer Institute (LKI), Leuven, Belgium*
EVELYN RAMPLER • *University of Vienna, Vienna, Austria*
KRISTOPHER RAWLS • *Department of Biomedical Engineering, University of Virginia, Charlottesville, VA, USA*
ÓTTAR ROLFSSON • *Center for Systems Biology, School of Health Sciences, University of Iceland, Reykjavik, Iceland*
DAVID E. RUCKERBAUER • *Austrian Centre of Industrial Biotechnology, Vienna, Austria; University of Natural Resources and Life Sciences, Vienna, Austria*
LISA SCHLICKER • *Department of Bioinformatics and Biochemistry, Braunschweig Integrated Centre of Systems Biology (BRICS), Technische Universität Braunschweig, Braunschweig, Germany*
HARALD SCHOENY • *University of Vienna, Vienna, Austria*
VITALY A. SELIVANOV • *Department of Biochemistry and Molecular Biomedicine, Faculty of Biology, Universitat de Barcelona, Barcelona, Spain; Institute of Biomedicine of Universitat de Barcelona (IBUB), Barcelona, Spain; Centro de Investigación Biomédica en Red de Enfermedades Hepáticas y Digestivas (CIBEREHD), Instituto de Salud Carlos III (ISCIII), Madrid, Spain; INB-Bioinformatics Platform Metabolomics Node, Instituto de Salud Carlos III (ISCIII), Madrid, Spain*
ÓLAFUR E. SIGURJÓNSSON • *School of Science and Engineering, Reykjavik University, Reykjavik, Iceland; Center for Systems Biology, School of Health Sciences, University of Iceland, Reykjavik, Iceland*
DARSHAN M. SIVALOGANATHAN • *Lewis-Sigler Institute for Integrative Genomics, Princeton University, Princeton, NJ, USA*
DIANA SZÉLIOVÁ • *Austrian Centre of Industrial Biotechnology, Vienna, Austria; University of Natural Resources and Life Sciences, Vienna, Austria*
JOSEP TARRAGÓ-CELADA • *Department of Biochemistry and Molecular Biomedicine, Faculty of Biology, Universitat de Barcelona, Barcelona, Spain; Institute of Biomedicine of Universitat de Barcelona (IBUB), Barcelona, Spain*
CHRISTINA TROYER • *University of Natural Resources and Life Sciences, Vienna, Austria*
MARCO VANONI • *SYSBIO Centre of Systems Biology, Milan, Italy; Department of Biotechnology and Biosciences, University of Milano-Bicocca, Milan, Italy*
XUANQING WAN • *Department of Chemical and Biological Engineering, Princeton University, Princeton, NJ, USA*
PRAMOD P. WANGIKAR • *Department of Chemical Engineering, Indian Institute of Technology Bombay, Mumbai, India; DBT-PAN IIT Centre for Bioenergy, Indian Institute of Technology Bombay, Mumbai, India; Wadhwani Research Centre for Bioengineering, Indian Institute of Technology Bombay, Mumbai, India*

Jeramie D. Watrous • *Department of Medicine, University of California San Diego, La Jolla, CA, USA; Department of Pharmacology, University of California San Diego, La Jolla, CA, USA*
Jürgen Zanghellini • *Austrian Biotech University of Applied Sciences, Tulln, Austria; Austrian Centre of Industrial Biotechnology, Vienna, Austria; University of Natural Resources and Life Sciences, Vienna, Austria*

Chapter 1

Determination of Isotopologue and Tandem Mass Isotopologue Ratios Using Gas Chromatography Chemical Ionization Time of Flight Mass Spectrometry - Methodology and Uncertainty of Measurement

Teresa Mairinger and Stephen Hann

Abstract

The accurate and precise analysis of isotopologue and tandem mass isotopologue ratios in heavy stable isotope labeling experiments is a critical part of assessing absolute intracellular metabolic fluxes. Resulting from feeding the organism of interest with a specifically isotope-labeled substrate, the principal characteristics of these labeling experiments are the metabolites' non-naturally distributed isotopologue patterns. For the purpose of inferring metabolic rates by maximizing the fit between a priori simulated and experimentally obtained labeling patterns, ^{13}C is the preferred stable isotope of use.

The analysis of the obtained labeling patterns can be approached by different mass spectrometric approaches. Gas chromatography (GC) features broad metabolite coverage and excellent separation efficiency of biologically relevant isomers. These advantages compensate for laborious derivatization steps and the resulting need for interference correction for natural abundant isotopes.

Here, we describe a workflow based on GC-high resolution mass spectrometry with chemical ionization for the analysis of carbon-isotopologue distributions and some positional labeling information of primary metabolites. To study the associated measurement uncertainty of the resulting ^{13}C labeling patterns, guidance to uncertainty estimation according to the EURACHEM guidelines with Monte-Carlo simulation is provided.

Key words Gas chromatography, Chemical ionization, Isotopologue distribution, Tandem mass isotopologue distribution, Primary carbon metabolism, ^{13}C based metabolic flux analysis, Measurement uncertainty

1 Introduction

Information on time-dependent motions within a metabolic network, that is, metabolic fluxes, has become a key in understanding condition-dependent regulation hierarchies that govern the metabolic phenotype of a biological system [1–4]. Stable isotope labeling experiments render the estimation of these in vivo intracellular metabolic rates possible. In brief, a specifically heavy stable

Deepak Nagrath (ed.), *Metabolic Flux Analysis in Eukaryotic Cells: Methods and Protocols*, Methods in Molecular Biology, vol. 2088, https://doi.org/10.1007/978-1-0716-0159-4_1, © Springer Science+Business Media, LLC, part of Springer Nature 2020

isotope-labeled substrate, mostly with ^{13}C, is fed to the organism of interest. This isotope label is incorporated into the downstream metabolite pool, leading to a non-naturally distributed labeling pattern [1]. Together with data on growth, biomass composition, etc., experimentally obtained information on isotopic enrichment can be then implemented into a metabolic network. Maximizing the fit between a-priori simulated and experimental labeling data, allows to infer absolute metabolic fluxes [5]. Since ^{13}C is most frequently used, this chapter focuses on the analysis of labeling experiments involving this specific isotope tracer.

Labeling patterns are commonly analyzed using mass spectrometric (MS) detection coupled with chromatographic separation techniques. Different types of mass analyzers are suitable for the analysis of isotopic enrichments within intracellular metabolites. Depending on the MS instrumentation employed, two types of information can be assessed: either on isotopologue distributions (ID), that is, molecular entities that differ in their isotopic composition [6] (as depicted in Fig. 1) or additional information on the position of the incorporated isotope label within the molecule can be obtained (as indicated in Fig. 2). The latter can be achieved by making use of fragmentation capability in MS/MS instruments (e.g., quadrupole tandem MS systems, quadrupole-time of flight (Q-TOF) systems, or fragmenting ion traps (3D ion traps, linear ion traps or orbitraps)). The general challenge of acquiring positional information via MS is a matter of *time*. This time problem concerns the mass spectrometer, as for the measurement of each mass transition a certain dwell time is needed (*mainly* to get high ion counts) and thereby affecting the cycle time, which will determine the achievable number of points describing a chromatographic peak. Especially for narrow chromatographic peaks and metabolites with a higher number of carbons in the backbone, achieving a reasonable cycle time (resulting from the number of transitions as well as the set dwell time) is quite challenging. Apart from its acquisition speed, here, the dual-stage MS hybrid of Q and TOFMS is regarded advantageous due to its high resolution, thereby facilitating the unambiguous identification of (non)selective fragments and its recording of all fragments of a given labelling state without restriction of time [7–9].

Applying MS/MS, so-called tandem mass isotopologue distributions (TMID) can be calculated. Each tandem mass isotopologue is characterized using a two number code $X.Y$, where X is indicating the number of ^{13}C atoms in the precursor ion and Y is indicating the number of ^{13}C atoms in the product ion. The exact masses of the precursor and product ions can be calculated by substituting each ^{12}C-atom from the carbon backbone with a ^{13}C-atom. When positional information is of interest to the study, selective fragmentation of an intact carbon backbone facilitates mass spectral interpretation. Hence, soft ionization techniques,

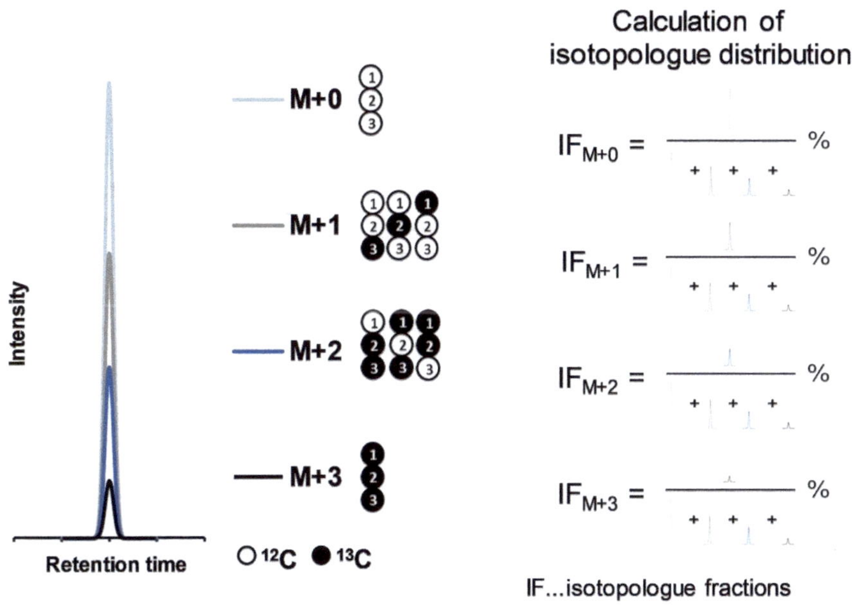

Fig. 1 Schematic overview assessing isotopologue distribution via mass spectrometric detection coupled to chromatographic separation

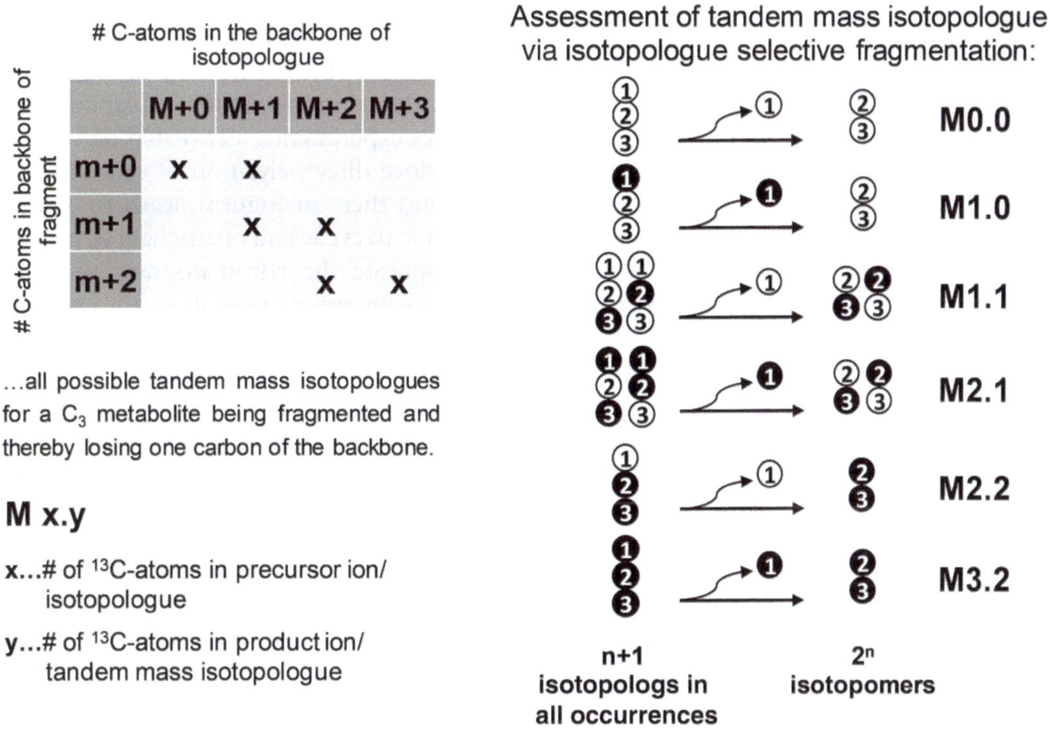

Fig. 2 Schematic overview assessing tandem mass isotopologues, including the definition and the matrix used for evaluating all possible transitions

where a high abundant intact analyte ion is attainable, are ideal. The type of ionization technique employed depends on the front-end separation and here, either liquid (LC) or gas chromatographic (GC) approaches are mainly used for the analysis of metabolites. For LC approaches, most frequently electrospray ionization is employed, whereas in GC based analysis the standard ionization source is electron ionization (EI). Since EI is a harsh ionization technique with extensive in-source fragmentation, it is only under certain circumstances [10] suitable for the analysis of ^{13}C labeling patterns. Chemical ionization (CI) with methane as reaction gas has proven to be a highly suitable; therefore, we regard this approach as fit for purpose [11, 12].

To make polar analytes GC-amenable, a derivatization step and consequently also an extensive natural isotope interference correction [13–15] is required. And yet GC analysis compensates these drawbacks with excellent separation efficiency, including the separation of biologically relevant constitutional isomers, and a wide metabolite coverage within a single analytical run. Besides, it is noteworthy that if a sample preparation robot is available, the derivatization can be easily automated.

Thermal stability and volatility of polar analytes is increased by replacing active hydrogen atoms in various functional groups either with trimethylsilyl (TMS) or with ter-butyldimethylsilyl (TBDMS) groups. Since the TBDMS moiety is bulkier, derivatization of vicinal active hydrogen is often not possible [16], hence often TMS is the derivatization reagent of choice in metabolomics experiments [11, 17–19]. In metabolomics experiments, a two-step derivatization is typically performed, since direct silylation of certain compound classes (e.g., sugars and their analogues), leads to a large number of different isomers due to cyclic and open-chain structures [20] and would hence complicate the chromatogram and data evaluation process. By alkoximation prior to silylation, the carbonyl group is protected and thereby locks sugars and their analogues in the open-chain structure and prevents decarboxylation [17, 20].

Notably, the measured labeling patterns do not reflect the true ^{13}C-isotope distributions and are masked by naturally abundant heavy stable isotopes, either already present in the native molecule itself (e.g., ^{34}S), or introduced by the aforementioned derivatization procedures [13–15, 21]. Only when correcting for these natural abundant isotope interferences, an unbiased ID and TMID is achievable [12, 15, 22]. Depending on the data type, ID or TMID, several (open-source) software packages are available (e.g., [13–15, 23–25]).

Elucidating the method's capability to provide analytical results in the required quality is also here for ^{13}C-based metabolic flux analysis experiments of key importance. Apart from assessing the overall method performance parameters [26], including, for example, precision under repeatability or reproducibility conditions of measurement, trueness (error) and sensitivity, the theoretical concept of measurement uncertainty budgeting can be implemented [27]. The evaluation of measurement uncertainty gives a measure of confidence being subjected to the result and allows to identify and quantify major uncertainty components.

In the following, GC-CI-QTOFMS based method for the analysis of ^{13}C labeling patterns of *free* intracellular metabolites is presented.

2 Materials

2.1 Chemicals for Metabolite Standard Preparation and Derivatization

Metabolite standards of interest (e.g., metabolites covering the central carbon metabolism, including glycolysis, pentose phosphate pathway, and citric acid cycle, as well as proteinogenic amino acids) should be purchased analytical grade (e.g., 6-Phosphogluconic acid (6PGA), 2-phosphoglyceric acid (2PG), 3-phosphoglyceric acid (3PG), dihydroxyacetone phosphate (DHAP), erythrose-4-phosphate (E4P), fructose-6-phosphate (F6P), glucose-6-phosphate (G6P), glyceraldehyde-3-phosphate (GAP), manose-6-phosphate (M6P), ribulose-5-phosphate (Rul5P), ribose-5-phosphate (R5P), sedoheptulose-7-phosphate (S7P); cis-aconitic acid (Aco), α-ketoglutaric acid (AKG), citric acid (Cit), isocitric acid (I-Cit), fumaric acid (Fum), malic acid (Mal), succinic acid (Suc); L-alanine (Ala), L-asparagine (Asn), L-aspartic acid (Asp), L-homoserine (H-Ser), L-glutamine (Gln), L-glutamic acid (Glu), L-isoleucine (Ile), L-glycine (Gly), L-leucine (Leu), L-lysine (Lys), L-phenylalanine (Phe), L-proline (Pro), L-serine (Ser), L-threonine (Thr), L-tyrosine(Tyr), L-valine (Val)).

For quality control, in addition to metabolite standards, it is highly recommended to use a cell extract with a defined non-naturally distributed C-isotope pattern, as it was proposed by Millard et al. [28] or if not available at least a representative cell extract from a cultivation with non-isotopically labeled substrate.

Solvents (LC-MS grade water, water-free pyridine for derivatization and pyridine for syringe cleaning) are purchased analytical grade. For the two-step derivatization procedure, ethoxyamine hydrochloride (EtOx) and *N*-methyl-*N*-(trimethylsilyl) trifluoroacetamide (MSTFA) with 1% trimethylchlorosilane (TMCS) are used. Please consult material safety data sheets before starting to work with the derivatization reagents.

2.2 Equipment for Derivatization and GC-HR-MS/MS Analysis

Samples are dried and subsequently derivatized in 2 mL crimp-top vials with flat-bottom glass inserts. For evaporation of the sample and metabolite standard solutions to complete dryness prior derivatization a centrifugal solvent evaporator, operating at low pressure (≤2 mbar) is employed.

Analysis is conducted on a high-resolution quadrupole-time of flight (Q-TOF) mass spectrometer coupled to a gas chromatograph (GC). For sample injection and to automate derivatization, the GC should be equipped with a dual head sample preparation robot with two quasi-independently operating towers (syringe volumes: 10 μL for sample injection and 100 μL for adding derivatization reagents, both towers are also used for transportation of vials), and a temperature-controlled agitator for sample incubation. For sample vaporization either a programmable temperature vaporization (PTV), or a *standard* split/splitless inlet can be employed. Depending on the inlet, a deactivated single tapered splitless liner, or a deactivated baffled liner, in both cases without glass wool, is used. Helium (purity 5.0) is used as carrier gas. Derivatives are separated on a 60 m nonpolar 100% dimethylpolysiloxane stationary phase analytical column (Macherey-Nagel, Germany, Optima 1MS Accent 60 m × 0.25 mm i.d., 0.25 μm film thickness, is recommended). For protection of the analytical column a deactivated nonpolar guard column is employed. Ionization is performed using chemical ionization, with methane (purity 5.0) as reaction gas. Nitrogen (purity 5.0) is used for collision induced dissociation (CID) when employing MS2 functionality of the instrument.

3 Methods

3.1 Metabolite Standard and Sample Preparation

1. Stock standard solutions of all metabolites are prepared by dissolving an appropriate amount of solid standard in LC-MS-grade water or 0.1 M HCl. A standard mixture solution of all single analytes with a concentration of 50 μM is prepared in LC-MS-grade water by appropriate dilution. Single stock standards and analyte mixture solutions kept at −80 °C and are stable for a minimum of 4 weeks.
2. For ^{13}C based MFA of *Pichia pastoris*, quenching and metabolite extraction procedure is applied according to [29]: In brief, cells are rapidly sampled into 60:40 methanol–water at −30 °C, filtered through cellulose acetate filters (0.45 μm) using a vacuum pump. The cell pellet on the filter is washed once with cold quenching solvent before it is transferred to a precooled tube and stored at −80 °C until extraction. The intracellular metabolites are extracted by adding 4 mL of 75% ethanol (v/v) at 85 °C to the frozen cell pellets, vortexing and incubating it for 3 min at 85 °C in a water bath. After rapid cooling-down on dry-ice, the cell extract is separated by centrifugation (4000 × *g* for 10 min at −20 °C). The supernatant, that is, ethanolic cell extract, is decanted into precooled tube and stored at −80 °C. For *P. pastoris*, a cell dry weight of approximately 20 mg was found to be a suitable amount to be extracted [12].

3. The final sample injection volume, that is, the fraction of initially extracted cell dry weight injected on column, was approximately 1/1600. It is worth mentioning that depending on the scope on analytes preconcentration can be easily achieved by adjusting the volume of ethanolic cell extract to be evaporated.

Even though analyzing labeling patterns in ^{13}C based MFA is not performed in an absolute quantitative manner, special care needs to be taken during the quenching step [30], since the isotopic enrichment must remain unaltered during sample preparation. Hence, for sampling of intracellular metabolites of eukaryotic organisms other than *Pichia pastoris*, it is of utmost importance to adapt especially quenching, but also extraction conditions (e.g., [31]). Moreover, reproducibility of the experiment has to be assured via accurate control and documentation of fermentation conditions and sampling time. *See* **Notes** Sect. 4.1 for further information.

3.2 Derivatization

For automated just-in-time online derivatization, a two-step reaction was performed, using EtOx in the first step and MSTFA with 1% TMCS in the second step. The procedure is in line with a variation of that described by Koek et al. [32]. *See* **Notes** Sect. 4.2 for further information.

1. As a prerequisite of the employed derivatization procedure, the samples have to be completely dry. To protect the ketogroups already during this evaporation step [33], 10 μL of freshly prepared ethoxyamine hydrochloride solution (c (EtOx) = 19 mg mL^{-1} in water-free pyridine) is added to all samples (including metabolite standards).
2. The dried samples will be reconstituted in 60 μL of derivatization reagent; hence, if a preconcentration step is required, the sample volume can be adjusted accordingly.
3. After, evaporation to complete dryness in a vacuum centrifuge, the sample vials are crimped, and stored at −80 °C until analysis. Dried samples are stable for a minimum of 8 weeks. To ensure complete dryness, the samples are dried again for 0.5 h before analysis.
4. For ethoximation, the dried sample is reconstituted in 18 μL EtOx (c(EtOx) = 19 mg mL^{-1} in water-free pyridine) and incubated at 40 °C for 90 min.
5. In the second step, the sample is silylated by adding 42 μL of MSTFA with 1% TMCS and incubation for 50 min at 40 °C.
6. For sample cool down, the derivatized samples are then put at 4 °C for approximately 4 min and are then injected in the GC-inlet.

3.3 GC-HR-MS/MS Analysis

1. Injection of 1.0 μL aliquots of derivatized sample is performed by either applying PTV (70 °C for 0.6 min, 12 °C min^{-1} to 260 °C, (hold 1 min), 12 °C min^{-1} to 280 °C, 5 min hold) [12] or using a standard split/splitless inlet, set at 250 °C .

   The 10-μL syringe used for injection is washed with pyridine before and after injection, each time twice.

2. The ethoximated and trimethylsilylated samples are separated on a 60 m × 0.25 mm i.d analytical column with 100% dimethylpolysiloxane as stationary phase (0.25 μm film thickness) and a 5 m nonpolar guard column. The carrier gas Helium is set to constant flow, 1.3 mL min^{-1}.GC temperature programmed followed the conditions optimized by Troyer et al. [11]: 70 °C for 1 min, 20 °C min^{-1} to 190 °C, 5 °C min^{-1} to 225 °C, 3 °C min^{-1} to 260 °C, and 20 °C min^{-1} to 310 °C (hold for 1 min).
3. Positive chemical ionization was performed according to the optimized parameters described in [11] using methane as reagent gas. The parameters are as following: the ion source temperature is set to 150 °C, for the electron energy the tuned value is taken, the emission current is kept at 10 μA. 40% methane gas flow (equal to 2 mL min^{-1}) was employed. (*It is noteworthy that parameters were optimized on an Agilent 7200B GC-QTOFMS system; hence, parameters might change depending on the manufacturer.*)
4. The scan speed, that is, the number for TOF spectra per second, was set to 3.3 Hz. In order to extend the linear dynamic range, the detector was operated in 2 GHz-extended dynamic range mode.
5. For assessment of tandem mass isotopologue distributions, the MS2 functionality using CID is employed—optimized parameters for the analysis of over 40 metabolites covering the central carbon metabolism can be found in [12]. For three exemplary compounds, namely, the amino acid alanine (Ala), the organic acid malic acid (Mal) and the sugar phosphate ribulose-5-phosphate (Rul5P), collision energies, acquisition times, proposed chemical structure of precursor and product ions, and the respective retention time, are shown in Table 1.

Chromatographic separation of over 40 metabolites, covering the central carbon metabolism, including glycolysis, pentose phosphate pathway and citric acid cycle, as well as proteinogenic amino acids is depicted in Fig. 3. *See* **Notes** Sect. 4.3 for further information.

3.4 Data Evaluation

The isotopologue distribution (ID), representing the metabolite's labeling pattern, that is, incorporation of the stable isotope label, consists of the respective isotopologue fractions (IF) and is calculated according to Eq. 1:

Table 1
GC-CI-QTOFMS method: acquisition parameters, structural information as well as exact masses for exemplary metabolites [12]

Compound name/ derivative	RT (min)	CE	Acquisition time (ms /spectrum)	Precursor ion structure and sum formula	Product ion structure and sum formula	Transitions
Alanine 2TMS	7.65	15	50	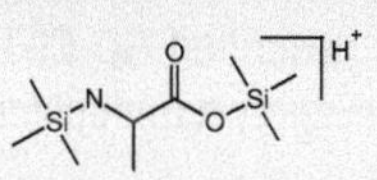 $C_9H_{24}NO_2Si_2^+$	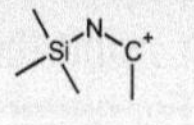 $C_5H_{14}NSi^+$	Ala_M0.0: 234.1340 → 116.0890 Ala_M1.0: 235.1374 → 116.0890 Ala_M1.1: 235.1374 → 117.0924 Ala_M2.1: 236.1407 → 117.0924 Ala_M2.2: 236.1407 → 118.0957 Ala_M3.2: 237.1441 → 118.0957
Malate 3TMS	10.50	10	50	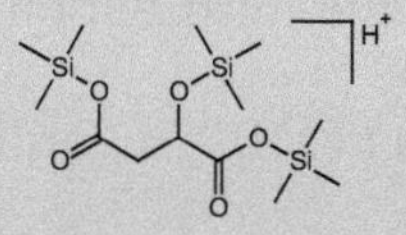 $C_{13}H_{31}O_5Si_3^+$	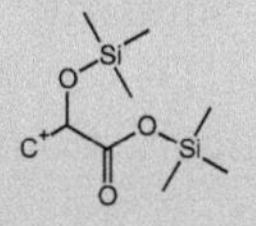 $C_9H_{21}O_3Si_2^+$	Mal_M0.0: 351.1474 → 233.1024 Mal_M1.0: 352.1507 → 233.1024 Mal_M1.1: 352.1507 → 234.1057 Mal_M2.1: 353.1541 → 234.1057 Mal_M2.2: 353.1541 → 235.1091 Mal_M3.2: 354.1574 → 235.1091 Mal_M3.3: 354.1574 → 236.1124 Mal_M4.3: 355.1608 → 236.1124
Ribulose-5-phosphate 5TMS 1EtOx	20.20	25	50	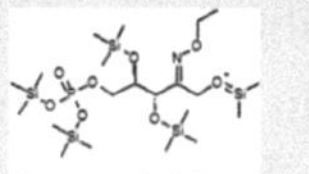 $C_{21}H_{53}NO_8PSi_5^+$	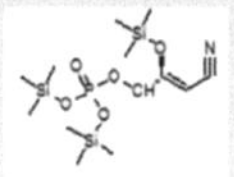 $C_{13}H_{29}NO_5PSi_3^+$	Rul5P_M0.0: 618.235 → 394.1086 Rul5P_M1.0: 619.2383 → 394.1086 Rul5P_M1.1: 619.2383 → 395.1119 Rul5P_M2.1: 620.2417 → 395.1119 Rul5P_M2.2: 620.2417 → 396.1153 Rul5P_M3.2: 621.2450 → 396.1153 Rul5P_M3.3: 621.2450 → 397.1186 Rul5P_M4.3: 622.2484 → 397.1186 Rul5P_M4.4: 622.2484 → 398.122 Rul5P_M5.4: 623.2517 → 398.122

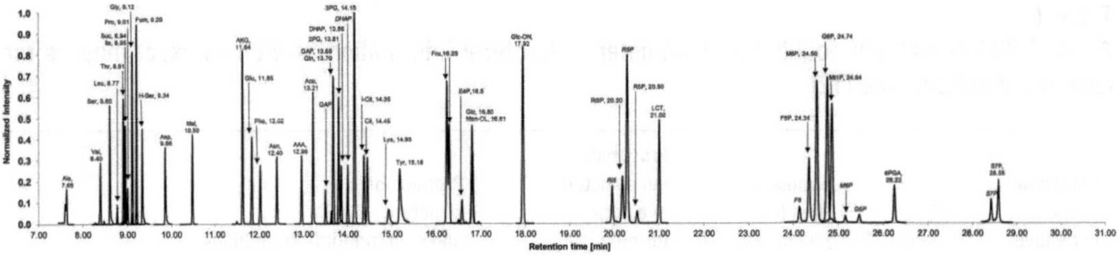

Fig. 3 Representative chromatogram using the described GC-CI-TOFMS method [12], analyzing a standard mixture in a concentration of 50 μM. For better visibility a multiplication factor (0.5–20) was applied to the low and very high abundant signals

$$IF_i = \frac{A_i}{\sum_{i=0}^{n} A_i} \quad (1)$$

where IF_i represents the measured isotopologue fraction i and n is the number of carbon atoms in the metabolite backbone. A_i is the peak area of the respective isotopologue i, obtained by peak area integration of the extracted ion chromatograms.

Similarly, the distribution of tandem mass isotopologues (TMID) can be assessed, by calculating the tandem mass isotopologue fractions (TMIF), according to Eq. 2:

$$TMIF_i = \frac{a_i}{\sum_{i=0}^{n} a_i} \quad (2)$$

where $TMIF_i$ is the measured tandem mass isotopologue (TMI) fraction i and n is the number of the remaining carbon atoms in the metabolite fragment, and a_i corresponds to the peak area of the respective TMI obtained by isotopologue selective fragmentation and subsequent integration of the extracted ion chromatograms.

1. After manual inspection and reintegration if necessary, the ID and TMID of naturally distributed samples are calculated for reasons of quality control.
2. The trueness in terms of error, can be determined by comparing the IDs to the theoretical fraction of the naturally abundant isotope distribution of the molecule with the experimental value. In case of TMID, the peak areas of transitions leading to the same product ion, though having different precursor ions (isotopologues) are summed up and compared with the natural isotope distribution of the fragment.
3. Data is subsequently corrected for interferences stemming from the contributions of naturally distributed heavy stable isotopes (^{29}Si, ^{30}Si, ^{13}C, ^{2}H, ^{15}N, ^{18}O, ^{34}S) using the open-source software tool ICT (isotope correction tool box) [15]. ICT is written in the multiplatform programming language Perl and is capable of correcting isotopologue as well as tandem mass isotopologue distributions for natural isotope interferences. A screenshot how to operate ICT is shown in Fig. 4.

Fig. 4 Screenshot of isotope correction toolbox [15] and the three necessary input files, namely definition of the compounds' elemental composition, including information on the carbon backbone ("−c"), natural isotope abundances ("−i") and the measured labeling patterns to be interference corrected ("−m"). Additionally, a file name for the output needs to be defined

Apart from the natural isotope abundances [34] ("−i") used for the interference correction, information on the number of carbons in the backbone as well as elements and number of atoms to be corrected for needs to be defined per metabolite ("−c"). Finally, data of the measured IDs following the compound's abbreviation, defined in "−c" and "$\_M_{0-n}$; or TMID, following the compounds abbreviation and previously mentioned notation of "$M_{x.y}$" is used as input. After installing Perl, the "ict.pl" script can be run by simply using, for example, the command line interface [15]. After heavy stable isotope interference correction, the labeling pattern of the carbon backbone is calculated. Together with data on growth, biomass composition as well as uptake and secretion rates, these C-isotopologue and/or tandem mass isotopologue distributions can be then implemented, into biochemical network [5]. *See* **Notes** Sect. 4.4 for further information.

3.5 Accuracy and Uncertainty of Measurement of Isotopologue Fractions

3.5.1 Accuracy

The assessment of the accuracy of measured isotopologue fractions can be directly performed using quality control data obtained within the experiment. It is recommended to determine the measurement error, that is, the difference between the measured value and a reference value using either natural material or material with a defined isotopologue distribution (see above). The obtained error is usually expressed as the relative difference between the two values.

3.5.2 Precision

Precision of isotopologue fractions can be obtained under repeatability or reproducibility conditions of measurement utilizing procedural replicates.

Precision under repeatability conditions can be obtained from one single fermentation using replicate samples from a very short sampling interval, or several aliquots from one single sampling event. At least 3, but better 6, replicates should be available. Precision is then reported as the (relative) standard deviation of n independently measured replicates.

Procedural precision obtained under repeatability conditions should always be compared with instrumental precision under repeatability conditions. Instrumental precision is obtained via the repeated injection of one and the same sample from the same sample vial. (*Typically, no absolute quantification but rather analysis of (TM)ID is performed in isotope labeling experiments and hence potential quantitative changes in concentration (due to stability of the derivative) can be neglected over these 6 repeated injections.*) Evidently, the difference between instrumental and procedural precision gives information on the impact of the sampling, quenching and extraction process.

Precision obtained under reproducibility conditions of measurement can be considered as the (relative) standard deviation obtained for isotopologue fractions of samples from replicate fermentations. As already mentioned, in this case precision is strongly depending on the reproducibility of the fermentation process and the timing of sampling.

3.5.3 Uncertainty of Measurement

The measurement of isotopologue fraction can be affected by numerous sources of uncertainty, which are not covered by the above-mentioned precision data. For both better understanding and effective optimization of an experiment and the metrological process we strongly suggest assessing the uncertainty of measurement following the relevant documentation and the principles outlined in state-of-the-art documents as the ISO Guide for the

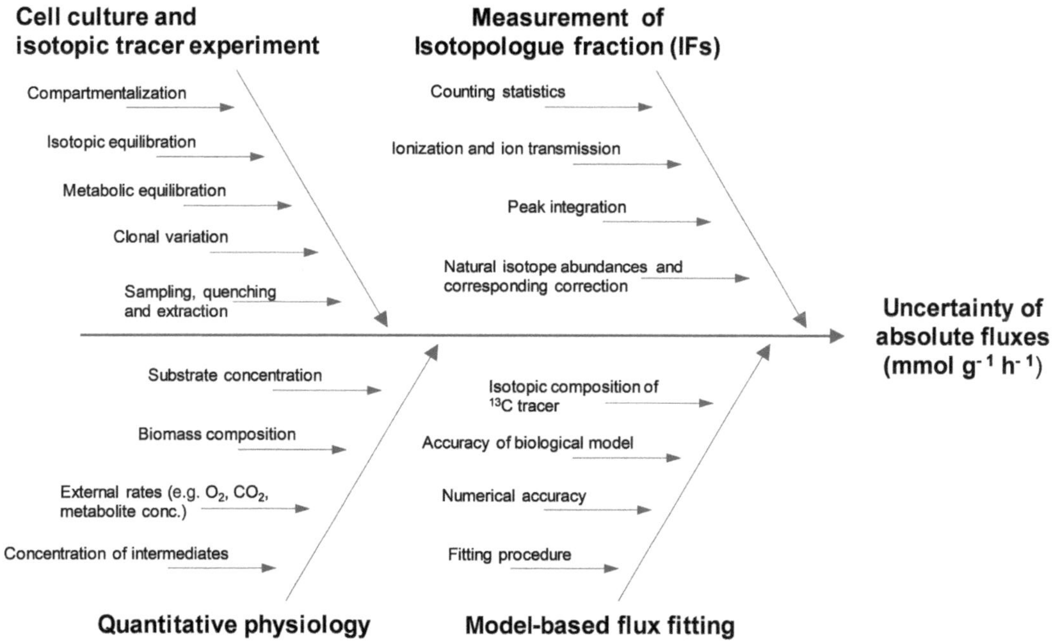

Fig. 5 Ishikawa diagram (also known as a "cause-and-effect-diagram") for identification of possible sources of measurement uncertainty of absolute flux values showing the most critical contributors to the uncertainty of absolute fluxes [22]

Expression of Uncertainty in Measurement and QUAM [26, 27]. In brief, uncertainty of measurement is estimated by (1) specification of the measurand (e.g., the isotopologue fractions of alanine in cell extracts), (2) identification of uncertainty sources (*see* Fig. 5), (3) setup of a model equation containing all uncertainty components with their uncertainty distribution, (4) propagation of the uncertainties for calculation of the standard uncertainty and expanded uncertainty. It is noteworthy that an important benefit of this procedure lies in the fact that the relative contribution of each uncertainty component is quantified and can be used for further process optimization. The described procedure and calculations are available in Mairinger et al. where uncertainty estimation has been performed for a ^{13}C labeling experiment studying glycolysis and the pentose phosphate pathway in a yeast cell factory [22]. *See* **Notes** Sect. 4.5 for further information.

4 Notes

4.1. Metabolite standard and sample preparation:

- For reasons of quality control, it is recommended to analyze also metabolite standard mixture within the measurement sequence.

4.2. Derivatization:

- During the incubation steps of the derivatization procedure, the samples are shaken (approx. 250 rpm in a temperature-controlled agitator on the sample preparation robot).
- For automated just-in-time derivatization: after each addition step, the syringe is washed twice, first with pyridine and then with water-free pyridine.

4.3. GC-HR-MS/MS analysis:

- Prior to analysis and every approximately 4–5 h within a measurement sequence, the mass axis of the high resolution mass spectrometer is calibrated automatically with an internal calibrant solution applying the mass calibration protocol of the instrument.
- The septa and liner are exchanged after approximately 100 injections.
- The system is conditioned by injecting MSTFA and derivatized samples.

4.4. Data evaluation:

- Since the exact mass in an isotope labeling experiment is unknown and the mass resolution of the employed TOFMS instrument does not allow for resolving isotopic fine structure, a mass extraction window of either ±50 ppm or ±100 ppm, is recommended for isotopologue analysis. This depends on the number of derivatized groups and possible chromatographic coelution/interferences. In case of tandem mass isotopologues a mass extraction window of ±50 ppm is used, as generally lower sensitivity and as expectedly lower mass accuracy in QTOFMS mode is observed.

4.5. Accuracy and uncertainty of measurement of isotopologue fractions:

- For uncertainty propagation we recommend the use of either the spread sheet approach outlined by Kragten [35], or Monte Carlo simulation utilizing fit-for-purpose tools such as the free tool from NIST (NIST Uncertainty Machine, uncertainty.nist.gov) or commercial software (e.g., @RISK, Palisade). We would like to emphasize that, in our opinion, the Monte-Carlo approach is advantageous as it facilitates operating with asymmetric uncertainty distributions as well as correlated uncertainty components.

References

1. Sauer U (2006) Metabolic networks in motion: ^{13}C-based flux analysis. Mol Syst Biol 2:62. https://doi.org/10.1038/msb4100109
2. Sauer U, Lasko DR, Fiaux J, Hochuli M, Glaser R, Szyperski T, Wuthrich K, Bailey JE (1999) Metabolic flux ratio analysis of genetic and environmental modulations of *Escherichia coli* central carbon metabolism. J Bacteriol 181:6679–6688
3. Wiechert W (2001) ^{13}C metabolic flux analysis. Metab Eng 3:195–206. https://doi.org/10.1006/mben.2001.0187
4. Zamboni N (2011) ^{13}C metabolic flux analysis in complex systems. Curr Opin Biotechnol 22:103–108. https://doi.org/10.1016/j.copbio.2010.08.009
5. Antoniewicz MR (2013) ^{13}C metabolic flux analysis: optimal design of isotopic labeling experiments. Curr Opin Biotechnol 24:1116–1121. https://doi.org/10.1016/j.copbio.2013.02.003
6. McNaught AD, Wilkinson A (1997) IUPAC. Compendium of chemical terminology, 2nd edn. Blackwell Scientific Publications, Oxford
7. Kappelmann J, Klein B, Geilenkirchen P, Noack S (2017) Comprehensive and accurate tracking of carbon origin of LC-tandem mass spectrometry collisional fragments for ^{13}C-MFA. Anal Bioanal Chem 409 (9):2309–2326. https://doi.org/10.1007/s00216-016-0174-9
8. McCloskey D, Young JD, Xu S, Palsson BO, Feist AM (2016) MID max: LC–MS/MS method for measuring the precursor and product mass Isotopomer distributions of metabolic intermediates and cofactors for metabolic flux analysis applications. Anal Chem 88:1362–1370. https://doi.org/10.1021/acs.analchem.5b03887
9. Mairinger T, Hann S (2017) Implementation of data-dependent isotopologue fragmentation in ^{13}C-based metabolic flux analysis. Anal Bioanal Chem 409:3713–3718. https://doi.org/10.1007/s00216-017-0339-1
10. Zamboni N, Fischer E, Sauer U (2005) FiatFlux—a software for metabolic flux analysis from ^{13}C-glucose experiments. BMC Bioinformatics 6:209. https://doi.org/10.1186/1471-2105-6-209
11. Chu DB, Troyer C, Mairinger T, Ortmayr K, Neubauer S, Koellensperger G, Hann S (2015) Isotopologue analysis of sugar phosphates in yeast cell extracts by gas chromatography chemical ionization time-of-flight mass spectrometry. Anal Bioanal Chem 407:2865–2875. https://doi.org/10.1007/s00216-015-8521-9
12. Mairinger T, Steiger M, Nocon J, Mattanovich D, Koellensperger G, Hann S (2015) Gas chromatography-Quadrupole time-of-flight mass spectrometry-based determination of Isotopologue and tandem mass Isotopomer fractions of primary metabolites for ^{13}C-metabolic flux analysis. Anal Chem 87:11792–11802. https://doi.org/10.1021/acs.analchem.5b03173
13. Van Winden WA, Wittmann C, Heinzle E, Heijnen JJ (2002) Correcting mass isotopomer distributions for naturally occurring isotopes. Biotechnol Bioeng 80:477–479. https://doi.org/10.1002/bit.10393
14. Millard P, Letisse F, Sokol S, Portais J-C (2012) IsoCor: correcting MS data in isotope labeling experiments. Bioinformatics 28:1294–1296. https://doi.org/10.1093/bioinformatics/bts127
15. Jungreuthmayer C, Neubauer S, Mairinger T, Zanghellini J, Hann S (2015) ICT: isotope correction toolbox. Bioinformatics 32 (1):154–156. https://doi.org/10.1093/bioinformatics/btv514
16. Halket JM, Waterman D, Przyborowska AM, Patel RKP, Fraser PD, Bramley PM (2005) Chemical derivatization and mass spectral libraries in metabolic profiling by GC/MS and LC/MS/MS. J Exp Bot 56:219–243. https://doi.org/10.1093/jxb/eri069
17. Kind T, Wohlgemuth G, Lee DY, Lu Y, Palazoglu M, Shahbaz S, Fiehn O (2009) FiehnLib: mass spectral and retention index libraries for metabolomics based on Quadrupole and time-of-flight gas chromatography/mass spectrometry. Anal Chem 81:10038–10048. https://doi.org/10.1021/ac9019522
18. Koek MM, Jellema RH, van der Greef J, Tas AC, Hankemeier T (2011) Quantitative metabolomics based on gas chromatography mass spectrometry: status and perspectives. Metabolomics 7:307–328. https://doi.org/10.1007/s11306-010-0254-3
19. Cipollina C, ten Pierick A, Canelas AB, Seifar RM, van Maris AJA, van Dam JC, Heijnen JJ (2009) A comprehensive method for the quantification of the non-oxidative pentose phosphate pathway intermediates in *Saccharomyces cerevisiae* by GC–IDMS. J Chromatogr B 877:3231–3236. https://doi.org/10.1016/j.jchromb.2009.07.019
20. Harvey D, Horning M (1973) Characterization of the trimethylsilyl derivatives of sugar

phosphates and related ompounds by gas chromatography and gas chromatography-mass spectrometry. J Chromatogr 76(1):51–62
21. Zamboni N, Fendt S-M, Rühl M, Sauer U (2009) ^{13}C-based metabolic flux analysis. Nat Protoc 4:878–892. https://doi.org/10.1038/nprot.2009.58
22. Mairinger T, Wegscheider W, Peña DA, Steiger MG, Koellensperger G, Zanghellini J, Hann S (2018) Comprehensive assessment of measurement uncertainty in ^{13}C-based metabolic flux experiments. Anal Bioanal Chem 410:3337–3348. https://doi.org/10.1007/s00216-018-1017-7
23. Wahl SA, Dauner M, Wiechert W (2004) New tools for mass isotopomer data evaluation in ^{13}C flux analysis: mass isotope correction, data consistency checking, and precursor relationships. Biotechnol Bioeng 85:259–268. https://doi.org/10.1002/bit.10909
24. Moseley HN (2010) Correcting for the effects of natural abundance in stable isotope resolved metabolomics experiments involving ultra-high resolution mass spectrometry. BMC Bioinformatics 11:139. https://doi.org/10.1186/1471-2105-11-139
25. Niedenführ S, ten Pierick A, van Dam PTN, Suarez-Mendez CA, Nöh K, Wahl SA (2016) Natural isotope correction of MS/MS measurements for metabolomics and ^{13}C fluxomics. Biotechnol Bioeng 113:1137–1147. https://doi.org/10.1002/bit.25859
26. BIPM (2012) International vocabulary of metrology—basic and general concepts and associated terms (VIM). BIPM, Sèvres
27. Ellison S, Williams A (2012) Eurachem/CITAC guide: quantifying uncertainty in analytical measurement, 3rd edn. LGC (Teddington) Limited, Teddington
28. Millard P, Massou S, Portais J-C, Létisse F (2014) Isotopic studies of metabolic systems by mass spectrometry: using Pascal's triangle to produce biological standards with fully controlled labeling patterns. Anal Chem 86:10288–10295. https://doi.org/10.1021/ac502490g
29. Russmayer H, Troyer C, Neubauer S, Steiger MG, Gasser B, Hann S, Koellensperger G, Sauer M, Mattanovich D (2015) Metabolomics sampling of Pichia pastoris revisited: rapid filtration prevents metabolite loss during quenching. FEMS Yeast Res 15:fov049. https://doi.org/10.1093/femsyr/fov049
30. van Gulik WM (2010) Fast sampling for quantitative microbial metabolomics. Curr Opin Biotechnol 21:27–34. https://doi.org/10.1016/j.copbio.2010.01.008
31. Hernández Bort JA, Shanmukam V, Pabst M, Windwarder M, Neumann L, Alchalabi A, Krebiehl G, Koellensperger G, Hann S, Sonntag D, Altmann F, Heel C, Borth N (2014) Reduced quenching and extraction time for mammalian cells using filtration and syringe extraction. J Biotechnol 182–183 (1):97–103
32. Koek MM, Muilwijk B, van der Werf MJ, Hankemeier T (2006) Microbial metabolomics with gas chromatography/mass spectrometry. Anal Chem 78:1272–1281
33. Vielhauer O, Zakhartsev M, Horn T, Takors R, Reuss M (2011) Simplified absolute metabolite quantification by gas chromatography–isotope dilution mass spectrometry on the basis of commercially available source material. J Chromatogr B 879:3859–3870. https://doi.org/10.1016/j.jchromb.2011.10.036
34. Berglund M, Wieser ME (2011) Isotopic compositions of the elements 2009 (IUPAC technical report). Pure Appl Chem 83:397–410. https://doi.org/10.1351/PAC-REP-10-06-02
35. Kragten J (1994) Tutorial review. Calculating standard deviations and confidence intervals with a universally applicable spreadsheet technique. Analyst 119:2161–2165. https://doi.org/10.1039/AN9941902161

Chapter 2

Non-Targeted Mass Isotopolome Analysis Using Stable Isotope Patterns to Identify Metabolic Changes

Christian-Alexander Dudek, Lisa Schlicker, and Karsten Hiller

Abstract

Gas chromatography coupled with mass spectrometry can provide an extensive overview of the metabolic state of a biological system. Analysis of raw mass spectrometry data requires powerful data processing software to generate interpretable results. Here we describe a data processing workflow to generate metabolite levels, mass isotopomer distribution, similarity and variability analysis of metabolites in a nontargeted manner, using stable isotope labeling. Using our data analysis software, no bioinformatic or programming background is needed to generate results from raw mass spectrometry data.

Key words Gas chromatography, Mass spectrometry, GCMS, Data analysis, Metabolism, Mass isotopomer distribution, Stable isotope labeling, Nontargeted metabolomics

1 Introduction

Gas chromatography (GC) or liquid chromatography (LC) coupled to mass spectrometry (MS) is widely used to measure metabolites in biological samples [1]. While classic metabolomics approaches only provide a static view of metabolite concentrations, metabolic flux analysis investigates dynamic metabolite conversion rates [2]. In this regard, flux balance analysis (FBA) relies on stoichiometric reaction models and predicts a solution space of feasible combinations of metabolic fluxes within the modeled network. This solution space can be reduced by thermodynamic constraints and by measured extracellular metabolic fluxes. On the other hand, ^{13}C metabolic flux analysis (MFA) provides exact metabolic fluxes for specific experimental conditions. It employs a combination of experimental data from stable-isotope labeling experiments and mathematical modeling of atom transitions. Both methods have in common that they rely on extensive prior knowledge of the metabolic reaction network and underlying stoichiometries [3] and often such a detailed information is not available a priori. Due to limitations of our biochemical knowledge, many

Deepak Nagrath (ed.), *Metabolic Flux Analysis in Eukaryotic Cells: Methods and Protocols*, Methods in Molecular Biology, vol. 2088, https://doi.org/10.1007/978-1-0716-0159-4_2,

metabolites have not yet been connected to a metabolic network or cannot even been identified after mass spectrometry. To overcome some of these limitations, we established a non-targeted mass isotopolome analysis, which allows for the identification of metabolic fluxes without a priori knowledge of the underlying metabolic network [4–7].

The initial part of such a study is a stable-isotope labeling experiment. For this purpose, a ^{13}C, ^{15}N or any other stable-isotope tracer is applied to the biological system and tracer derived isotopes are incorporated into downstream metabolites. After metabolite extraction and mass-spectrometry driven measurement, all detectable known and unknown isotopically enriched metabolites can be identified with algorithms such as nontargeted tracer fate detection (NTFD), MetExtract or X^{13} CMS [8–11]. In a next step, mass isotopomer distributions (MIDs) for all detected and thus labeled metabolites are determined based solely on the mass-spectrometric data of the nonlabeled counterparts as no structural information is available at this point [7, 8, 12]. All these MIDs are determined by metabolic fluxes through the underlying biochemical reaction network and can be applied to reveal both, the network structure and metabolic flux changes [7, 9].

In the following we will describe in detail two options for a nontargeted mass isotopolome analysis (Fig. 1).

2 Materials

The following software packages are required to perform the described nontargeted analysis of GC-MS data recorded in the context of a stable-isotope labeling experiment.

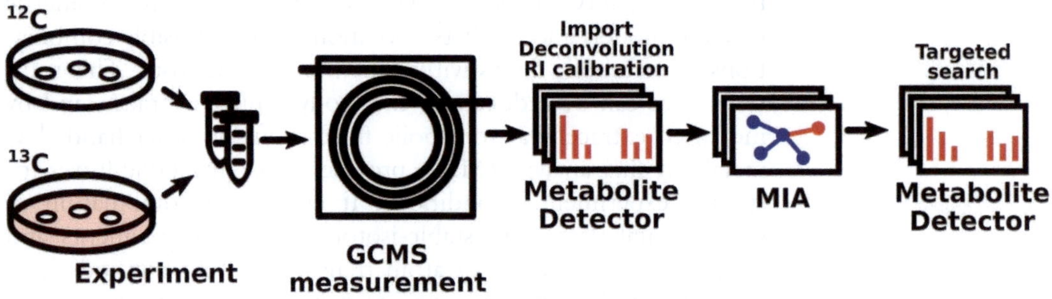

Fig. 1 The presented workflow of a nontargeted experiment and data analysis. The workflow starts with the cell culture experiment and the preparation of metabolite extracts followed by gas chromatography and mass spectrometry (GCMS) measurement. The raw data in netCDF format needs to be imported, calibrated and deconvoluted in MetaboliteDetector. The similarity and variability analysis to create a network based on MID similarity is performed with MIA. For the subsequent targeted data analysis, MetaboliteDetector can be used to obtain more sensitive MID determination and quantification data

2.1 MetaboliteDetector

MetaboliteDetector is a software designed for the semiautomatic analysis of GC-MS raw data. The software provides a graphical user interface and can be applied for targeted and nontargeted approaches. For that, MetaboliteDetector implements an extensive set of different GC-MS data analysis algorithms, including retention time correction, ion-chromatographic deconvolution, metabolite detection, metabolite identification, metabolite quantification, and first statistical analyses. Moreover, this program can be applied to determine MIDs for target compounds and correct these for natural occurring isotopes. Finally, a wide variety of data visualization options are available, like single- and total ion chromatograms, box plots and MID bar plots [13].

MetaboliteDetector is freely available for Linux operating systems at http://metabolitedetector.tu-bs.de.

2.2 Mass Isotopolome Analyzer

The Mass Isotopolome Analyzer (MIA) is a tool to visualize labeled metabolites in a MID similarity-based network. Additionally, a variability analysis of the different experimental conditions can reveal metabolic flux changes. MIA first detects all isotopically enriched metabolites present in a GC-MS data set [8] and then contextualizes MIDs of labeled compounds in a network visualization. A graph is created which connects metabolites with similar MID patterns based on dynamic programming and distance calculation. Different cutoffs for distance measurement and inter-experiment MID variability can be applied to analyze the datasets [7]. If a reference compound library is supplied, detected metabolites can be identified.

MIA is freely available for Windows and Linux operating systems at http://mia.bioinfo.nat.tu-bs.de.

2.3 Optional: Metabolite Spectra Reference Libraries

While the approaches described below are all of a nontargeted manner, the biological interpretation needs an incorporation of the nontargeted data into a known metabolic context. Therefore, identification of as many metabolites as possible is of advantage.

Both, MetaboliteDetector and MIA can import and export reference libraries in MSL format and as such the Golm Metabolome Database (GMD) can be applied for compound identification. It is freely available [14] and currently contains 26,590 spectra of 3568 derivatized analytes of 2222 different metabolites and additionally 3488 spectra of purchased reference substances from GC-MS measurements. *See* **Note 1** on how to convert a library in MSL format into MetaboliteDetector format.

3 Methods

3.1 Experimental Prerequisites

Any nontargeted detection of isotopic labeled metabolites starts with an experimental setup applying a stable isotope labeled metabolite as a tracer. In general, any stable isotope, for example carbon (^{13}C) or nitrogen (^{15}N), can be employed. For every condition, two experiments are needed, one using the isotopic tracer and the other using the same substrate, but without enrichment. To get the best results, both experiments should be performed simultaneously under identical experimental conditions. It is also advised to perform biological replicates to validate the results of mass isotopomer calculations [7, 8].

As the retention time of analytes varies between instruments and different GC-MS columns, it is recommended to measure a mix of *n*-alkanes (usually 1 μL of a mixture of C_{10} to C_{40}, splitless injection) along with the samples. This enables the calculation of retention indices for improved metabolite identification in addition to spectrum similarity alone.

The detailed experimental procedure for cultivation and metabolite extraction highly depends on the metabolites and organism of interest. Different experimental protocols are described elsewhere [9, 15–20]. After GCMS measurement, the raw data need to be exported into the common netCDF format, which can usually be done with the instrument software.

3.2 Data Import

The protocol starts with the raw data in netCDF format. If not otherwise stated, the default settings are a good starting point. First, all netCDF files are imported with MetaboliteDetector's netCDF import function (Fig. 2). Depending on the computational power, data import can take some time. During the import process, files with extension .bin and .idx are created in the same directory as the netCDF files, if no other output directory was selected.

3.3 RI Calibration and Deconvolution

As mentioned before, the retention index calibration is a highly recommended step because the retention time of metabolites can shift slightly, especially when measuring many samples.

MetaboliteDetector provides the RI-Calibration Wizard, which guides through the calculation of retention indices and ion-chromatographic deconvolution. Measured *n*-alkanes need to be selected from the reference spectra library (Fig. 3, left) and after detection of the alkanes, the calibration table lists the detected compounds, detected retention time and the corresponding retention index provided with the library (Fig. 3, right). At this point it is crucial to validate that the peak-alkane assignment is correct and needs to be corrected otherwise. In the next step, deconvolution settings for compound detection need to be defined and should be applied according to your experimental setup (*see* **Note 2**).

After deconvolution and retention index calculation, the data is ready to be further analyzed in MetaboliteDetector or MIA. Since this step is crucial for all downstream analyses, it is highly recommended to verify the quality of the results of the deconvolution and RI calibration process (*see* **Note 4**).

Fig. 2 The netCDF import dialog. Files for import can be selected directly (1) or searched for recursively in a folder (2). The selected files will appear on the left (3). Optionally, a different output directory can be selected (4)

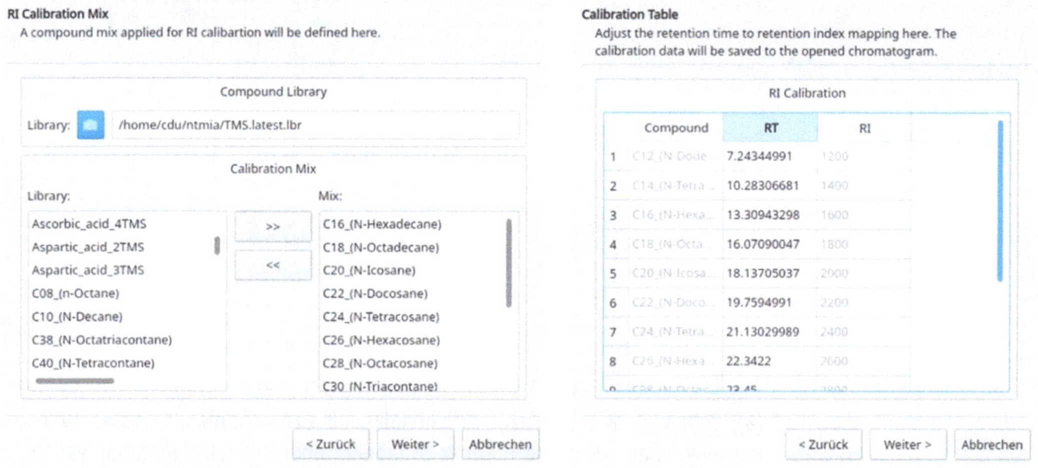

Fig. 3 RI-Calibration Wizard configuration. Left: Selection of alkanes from the library. Right: Calibration table with alkanes, detected retention times and corresponding retention index from the library

3.4 Similarity Analysis

For the interpretation of nontargeted metabolomics data, it is of advantage to first set (known) parts of the obtained results into a biological context and map unidentified labeled metabolites based on MID similarity to this established frame. In this regard, MIA provides networks based on MID similarities and can highlight variability of MIDs and thus underlying metabolic fluxes between different conditions (e.g., control and treatment) within this network.

A MIA analysis starts with the import of labeled and unlabeled chromatograms preprocessed by MetaboliteDetector (see previous paragraph) for every experiment (Fig. 4a). Next, some settings need to be defined: For the compound identification settings, the *RI Tolerance* setting is the maximum retention index tolerance for

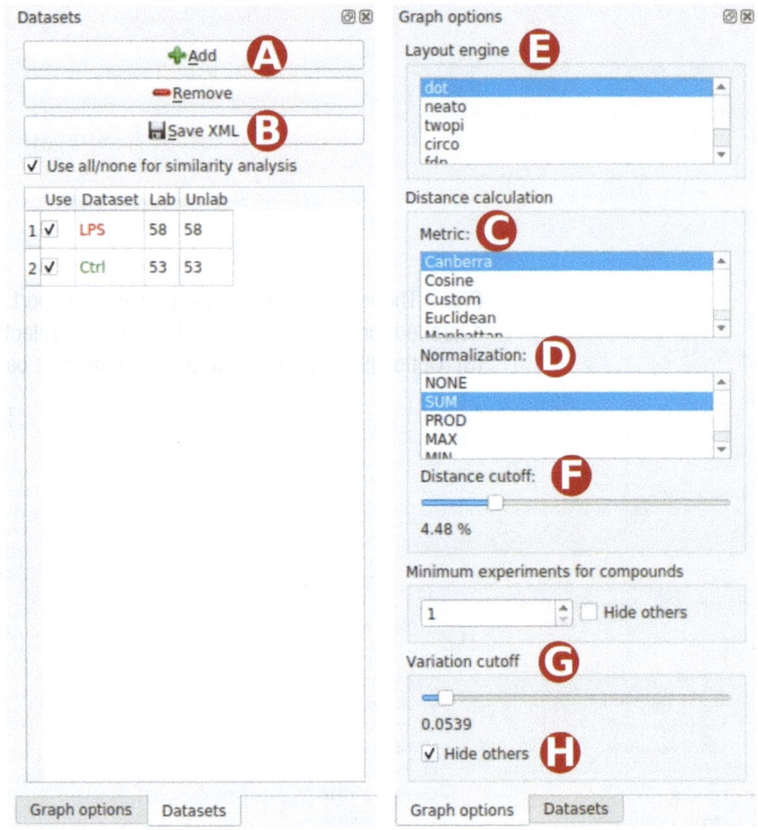

Fig. 4 MIA option tabs. Left: In the datasets tab new experiments can be added (**a**), removed or saved (**b**). Additionally, all experiments are listed with the corresponding color and number of labeled and unlabeled metabolites. Right: In the graph options the layout engine (**e**), distance metric (**c**), and normalization (**d**) can be selected. Using the distance cutoff slider (**f**) the minimum MID similarity distance for node connection can be adjusted. The variability slider (**g**) can be used to only show connections for nodes with MID variation between experiments. If Hide others is checked (**h**), all other nodes will not be shown

library matching (smaller values require stricter matching), the *Compound identification cutoff score* is the minimum score needed for compound identification (1 – perfect match, 0 – no match) and *Show top n hits* is the number of library matches for potential compound identification and name assignment. In the label detection section, *Maximum fragment deviation* is the maximum deviation of the summed MIs from 1.0. The summed fractions of all mass isotopomers should be 1.0 because each MI is typically given in percental values, *Minimum number of labeled fragments* is the minimum number of fragments per compound which need to be detected as labeled in order to get recognized as labeled compound, *Required amount of isotopic enrichment* is the minimum amount of labeling needed, *Minimum* R^2 is the minimum coefficient of determination (R^2) from MID determination, *Minimum* M_0 *abd.* is the minimum abundance of the unlabeled mass isotopomer (M_0) needed and *Ignore compounds with M_n|with n >* ... is the number of mass isotopomers from which a compound gets excluded. Additionally the *Gap penalty for the Needleman-Wunsch-Scoring* for MID alignment and the initial *Distance cutoff for graph edges* can be defined. It is highly advisable to select the same settings for every experiment to achieve better comparability across experiments. The settings for every experiment can be saved in XML format for future runs of the program (Fig. 4b). After starting the isotope detection and MID profiling, every labeled metabolite will be represented by a node depicting the MID in the main window of MIA. In addition, all labeled metabolites will be listed in the left side of the window. Up till now the complete procedure is nontargeted and does not require any structural information on the detected metabolites. However, a compound library for identification can be applied at this stage to bring parts of the data into a biochemical context.

With the experiments and identifications in place, various options for graph layout, distance calculation (Fig. 4c) and normalization (Fig. 4d) can be applied. In most cases the default values (Canberra distance and normalization by sum) should be sufficient. The layout engine defines the way the nodes will be arranged in the network and should be selected based on the amount of nodes. While the "*dot*" layout is a general purpose engine, "*circo*" of "*twopi*" are suitable for fewer connected nodes and small clusters (Fig. 4e). The *distance cutoff* sets the minimum similarity that two nodes need to have to get connected (Fig. 4f). While increasing this cutoff, the nodes will start to rearrange and connect to each other based on their MID similarities. The edge color indicates the corresponding experiment. Unconnected metabolites can be hidden from the visualization through the *options* menu. Connected metabolites are similar based on their MIDs and in many cases be metabolically related.

3.5 Variability Analysis

Metabolism is a highly dynamic but regulated system and can switch to different states depending on the environmental conditions. In many cases, changes in metabolic fluxes are not directly linked to changes of the metabolome or metabolite levels and are thus hidden in metabolomics analyses. However, stable isotope labeling has the power to reveal changes in underlying fluxes and for this purpose, MIA provides an MID variability analysis to detect flux changes between experiments. Using the previously created graph of metabolites with similar MIDs, the *variation cutoff* slider can be increased slowly (Fig. 4g). While increasing the cutoff, nodes will start to disconnect, if the variability between experiments does not match the given cutoff. Additionally, if *Hide others* (Fig. 4h) is enabled, the nodes will not only disconnect, but also disappear from the graph. All remaining connected nodes will have at least the given MID variation between experiments as defined by the cutoff slider.

3.6 Library Export

All detected MIDs can be exported in CSV format for further processing with other programs. Additionally, MIA can generate and export a MetaboliteDetector library containing all detected labeled metabolites. This library can then be opened in MetaboliteDetector to perform additional sensitive profiling such as quantification or a more sensitive MID determination in a targeted manner (Subheadings 3.7 and 3.8).

3.7 Targeted Search for Quantification

While MIA is a powerful tool to detect labeling in metabolites and create networks based on MID similarity, it lacks features for metabolite quantification and statistical analysis. Furthermore, MIAs sensitivity for MID determination is lower due to the non-targeted approach. To overcome these shortcomings, it is possible to apply MetaboliteDetector and perform a targeted search for all metabolites detected by MIA. Because MIA provides exact mass spectra and retention times for all revealed known and unknown labeled compounds, these metabolites can be further profiled in MetaboliteDetector even if their structure is not known.

To obtain semiquantitative metabolite levels for MIA compounds with MetaboliteDetector, the metabolite library exported by MIA can be used (Tools—Batch quantification...). The batch quantification wizard will guide through the settings (Fig. 5). First, the files need to be selected as replicate groups (*see* **Note 3**). Next, as analysis type, *Targeted Analysis* needs to be selected, because only the previously detected metabolites from MIA are of interest for this analysis. The last panel of the quantification wizard contains different settings:

- *Ref. library* refers, in this case, to the library of labeled compound spectra supplied by MIA.
- The retention index difference (ΔRI) is the maximum difference of retention indices of two compounds to get matched. A small

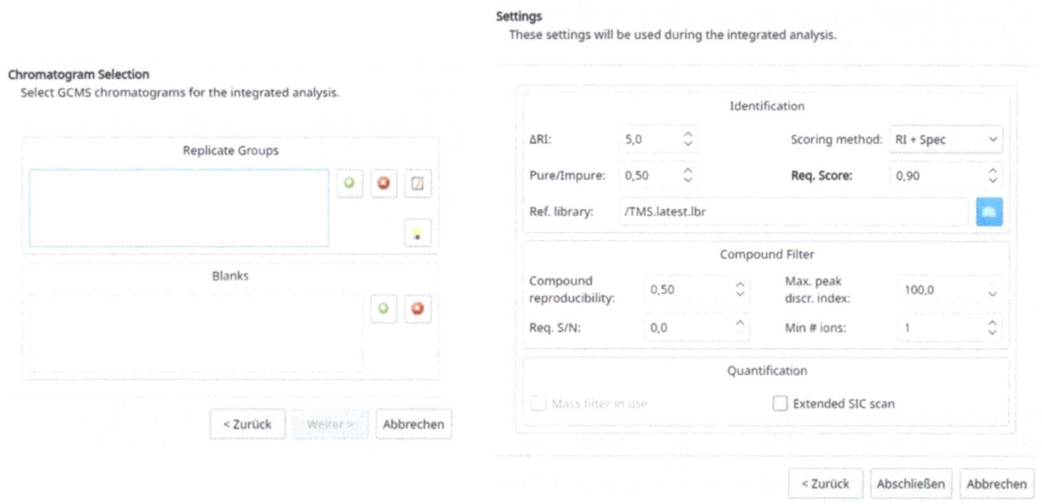

Fig. 5 Batch quantification wizard. The files for replicates (left) can either be select manually, using the plus button, or alternatively using the lightbulb button (*see* **Note 3**) for regular expression based replicate selection. The settings window (right) contains the settings for metabolite identification and additional compound filtering

value like 5.0 should be appropriate, but highly depends on the calibration quality (*see* **Note 4**).

- The identification score (*Req. Score*) defines the strictness of mapping to compounds together. Higher values require higher similarities in terms of spectrum and retention time to be matched. To avoid matching of compound with different structures, this value should be set to a higher value (e.g., 0.9).

After finalization of calculation, a new tab (Batch Quantification) appears in the bottom of the window and contains the quantification results. Each row represents one of the metabolites previously detected with MIA and is named accordingly. The data can be normalized by the summed sample signal ($\frac{1}{\sum}$) or by the peak integral of a selected metabolite, in most cases an internal standard ($\frac{1}{A}$). The statistics subtab presents the mean (and normalized if selected previously) intensities of the replicates. Additionally, the p-value (ANOVA) is calculated as a measurement for the statistical difference between the groups. Selected rows of the statistics subtab can also be displayed as a box plot. From the quantification subtab and the statistics subtab, the data can be exported in CSV format.

3.8 Targeted Search for MIDs

Another option is to determine MIDs for all MIA compounds. This can be useful if MIDs could not be detected for some conditions within MIA or if additional chromatograms should be processed. However, for classical MID determination the sum formula needs

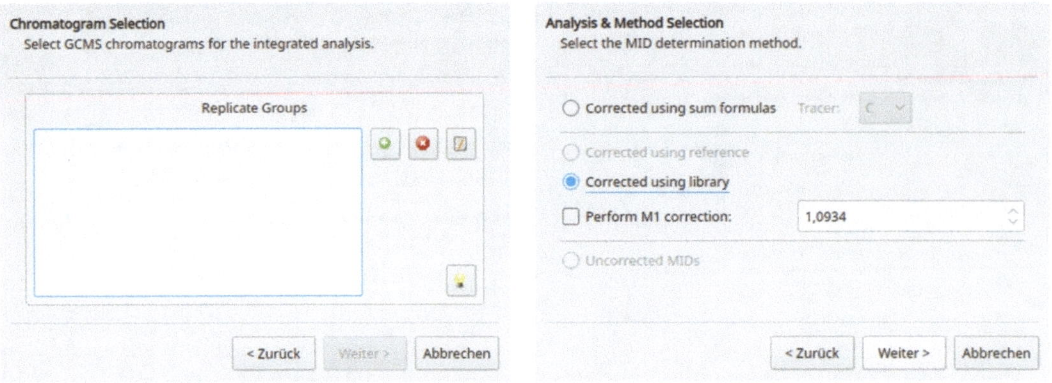

Fig. 6 The MID wizard. The files for replicates (left) can either be select manually, using the plus button, or alternatively using the lightbulb button (*see* **Note 3**) for regular expression based replicate selection. In the analysis and methods section (right) it is important to select "Corrected using library" to correct for natural isotopes abundance using unlabeled reference spectra

to be supplied for each target compound to correct for natural occurring isotopes. Since this information is not available due to the nontargeted character of the MIA analysis, MetaboliteDetector implements an algorithm to correct MIDs for natural isotopes just based on the unlabeled compound spectrum that was provided at the beginning of the MIA analysis [12].

To perform such an analysis, the library exported by MIA needs to be loaded first (Tools—Settings...—Identification—Compound Lib). Then, the MID wizard (Tools—MID Wizard...) has to be opened, which will guide through all required settings (Fig. 6). In the *analysis & method* panel, *Corrected using library* needs to be selected, because sum formulas for compounds are not available in this case (see above). Next, all files from the tracer experiment need to be selected as replicate groups. The last section of the MID wizard contains the same settings as described in Subheading 3.7. Here it is important to select the previously created library containing the unlabeled spectra from MIA.

The MIDs tab will appear in the bottom of the window when the analysis is finished. In the MIDs sub-tab, selected metabolite MIDs can be visualized as bar-plots. Additionally, the confidence interval (CI) and R^2 of the MID determination for every file is shown for quality control. The MID results should be revised carefully, as the nontargeted approach may return wrong ions, (*see* **Note 5**). The statistics subtab shows the means and standard error of the replicates. The MIDs can be exported in CSV format from the MID sub-tab. The data from both tabs will be exported in one file with different formats. This should be kept in mind for further automatic data processing.

3.9 Biological Interpretation

In the previous sections, different data have been created, that represent different aspects of the metabolic state of the analyzed sample. On one hand, the metabolite levels represent the total abundances of all (nontargeted) detected metabolites. On the other hand, the MID data represent the flux changes between metabolites, which can either correlate with the metabolite levels, or can even reveal flux changes, which are not visible in the metabolome dataset alone. Additionally, MIA provides tools to create metabolic connections based on MID similarity and variability analysis between different experiments. MID similarity based networks can reveal metabolite connections between known and unknown metabolites. The variability analysis can highlight connections where metabolic fluxes change between experiments or on the other side, show parts of a metabolic network which do not change between different experiments. Bringing these data into known metabolic context can reveal new metabolic interconnections and thus provide a better understanding of the observed phenotype.

4 Notes

1. MSL library import and manipulation.

   The MSL file format is a text format for mass spectrometric data, which can be imported into MetaboliteDetector using the library import function. The selected library in MSL will get scanned and the number of metabolites in the file will be displayed prior conversion. After the import, a library file in LBR format will be created. After conversion the new library can be automatically loaded or opened later through the settings dialog.

   Libraries in MetaboliteDetector format can be opened within MetaboliteDetector using the Library Editor. All library compound data, like spectrum, meta data, ions for quantification and MID determination can be reviewed and revised if needed (Fig. 7).

2. Deconvolution settings.

   Ion-chromatographic deconvolution is essential for the detection and separation of mass spectra of coeluting metabolites. Therefore, the adjustment of the deconvolution settings is of high importance. The specific value depends on the GCMS instrument used for measurement. Below is a brief description of the parameters.

   (a) *Peak threshold* must be exceeded by the first derivative of the peak function. Peaks below the threshold will be discarded. A lower value increases the sensitivity.

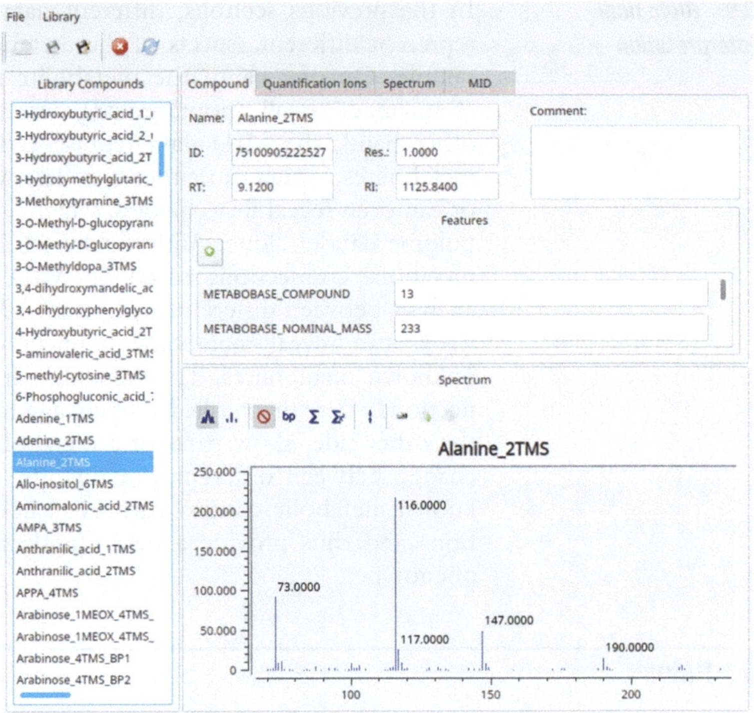

Fig. 7 The library editor window of MetaboliteDetector. All library compounds are listed on the left. On the right, the stored data, including name, retention time, retention index, ions for quantification and MID determination and the compound spectrum, can be reviewed and revised

(b) *Minimum peak height* above the baseline. Peaks below this value will be discarded. A lower value increases the sensitivity.

(c) *Bins/scan*: The number of bins created for every scan before deconvolution. A lower value decreases the ability to separate coeluting compounds, a too high value falsely splits compounds. Normally a value of 10 should be optimal. For detailed description see Stein et al. 1999 [21].

(d) *Deconvolution width* is the maximum number of scans by which a chromatographic peak can differ to still be considered to belong to the same mass spectrum. If the value is too low, coeluting compounds will not be separated, if the value is too high, peaks belonging to one compound will be falsely split in two or more.

(e) *Required intensity (% of base peak)* for peak detection. Peaks assigned to a compound below this value will be discarded. Should be 0 in most cases.

Table 1
Deconvolution settings for different GCMS instruments

Instrument	Peak threshold	Min. peak height	Bin/Scan	Deconvolution width
Agilent 6890 GC	2–10	2–10	10	7.0–8.0
Jeol AccuTOF	5–20	5–20	10	1.0
Trace	5–100	5–20	10	3.0
LECO GC/TOF	20–50	20–50	1	40.0

(f) *Required number of peaks* for a metabolite to get detected. Deconvoluted metabolite spectra with less ions will not be displayed as detected metabolite.

Table 1 lists different settings for some GCMS instruments as reference. Additional fine tuning of the parameters is needed depending of the measured samples. However, the settings are fairly robust when using the same instrument and sample type.

3. Replicate file grouping.

Defining replicate groups in MetaboliteDetector can be tedious especially for large datasets. MetaboliteDetector provides a smart replicate group function (lightbulb symbol in Batch quantification or MID wizard) which is based on regular expressions. If replicate file names were consecutively numbered, the smart selection function automatically defines replicate groups for further processing and subsequent statistical analysis. For example, if the replicates were named like "expA_1", "expA_2", . . .and "expB_1", "expB_2", the regular expression "`(.*)_`", will group all corresponding files in group "expA" and "expB," respectively.

4. Spectrum quality verification.

The deconvolution and retention index calibration have a high impact on quality of the extracted chromatograms and matching of metabolites across samples. Therefore, it is important to revise the results of Subheading 3.3 before proceeding. The first step is to evaluate the quality of the total ion chromatogram. Overloaded peaks that exceed the detector limit have a negative impact on the retention time of following compounds and will not be deconvoluted correctly. If too many peaks are overloaded, it may be advisable to repeat the measurement with diluted extracts or a higher split.

Retention indices of the same metabolite should not differ too much between different samples. In some cases the ΔRI can be adjusted to allow more RI deviation, but a small RI difference is preferable. If the retention indices differ between samples, the retention index calibration should be repeated.

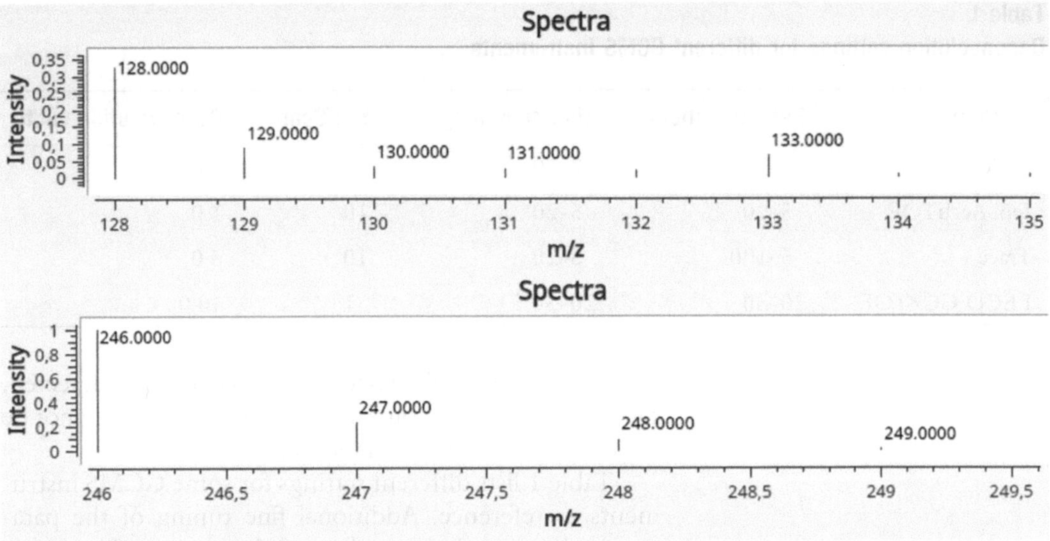

Fig. 8 Spectrum of two fragments with mass 128 (upper spectrum) and 246 (lower spectrum) of an unlabeled metabolite. The upper spectrum has an unexpected high peak at M + 5 (133) isotopomer, which suggests an artifact from insufficient deconvolution. In contrast, the lower spectrum has decreasing intensities for the mass isotopomers, as expected from an unlabeled sample

Especially the automatic detection of the alkanes should be revised carefully.

The single ion chromatogram in MetaboliteDetector can be used to quality control the results of the deconvolution. The different fragments (ions) of a detected metabolite should show the characteristic pattern with decreasing intensities for the mass isotopomers (Fig. 8), which depend on the natural abundance of isotopes (mainly ^{13}C) and the number of atoms. Peaks which do differ from this decreasing pattern may originate from overlapping signals which have not been correctly deconvoluted. While these artifacts cannot be excluded completely, the deconvolution settings should be adjusted to minimize those errors, especially for identified metabolites and metabolites of interest.

5. MID quality verification.

The nontargeted approach will calculate MIDs for every ion which has been previously detected as labeled. This may lead to many different ions for one metabolite with different MID results. Due to fragmentation during the ionization process, the number of detected isotopomers can vary between the different ions. Here it is important to evaluate if the number of mass isotopomers matches with the detected metabolite. For example, with ^{13}C tracer, a metabolite with six detected mass isotopomers, which was been identified as lactic acid (maximum 3 labeled carbon atoms) does not match. Here it is possible, that only one particular ion was detected with the wrong amount of mass isotopomers. In this case it would be possible to exclude that ion from the generated library using MetaboliteDetectors library editor. If all ions are consistent and do not match the expected number of mass isotopomers, the identification may be wrong.

Another critical point is to evaluate the MID pattern based on knowledge of biochemical pathways. Using a specifically labeled tracer, one would expect a specific pattern for metabolites downstream of the tracer. These expectations should be evaluated with corresponding metabolites which have been identified.

References

1. Birkemeyer C, Luedemann A, Wagner C, Erban A, Kopka J (2005) Metabolome analysis: the potential of in vivo labeling with stable isotopes for metabolite profiling. Trends Biotechnol 23(1):28–33
2. Zaitsu K, Hayashi Y, Kusano M, Tsuchihashi H, Ishii A (2016) Application of metabolomics to toxicology of drugs of abuse: a mini review of metabolomics approach to acute and chronic toxicity studies. Drug Metab Pharmacokinet 31(1):21–26
3. Klapa MI, Stephanopoulos G (2000) Metabolic flux analysis. In: Bioreaction engineering. Springer Berlin Heidelberg, Berlin, Heidelberg, pp 106–124
4. Creek DJ, Chokkathukalam A, Jankevics A, Burgess KEV, Breitling R, Barrett MP (2012) Stable isotope-assisted metabolomics for network-wide metabolic pathway elucidation. Anal Chem 84(20):8442–8447
5. Hiller K, Metallo C, Stephanopoulos G (2011) Elucidation of cellular metabolism via metabolomics and stable-isotope assisted metabolomics. Curr Pharm Biotechnol 12 (7):1075–1086
6. Hiller K, Metallo CM, Kelleher JK, Stephanopoulos G (2010) Nontargeted elucidation of metabolic pathways using stable-isotope tracers and mass spectrometry. Anal Chem 82 (15):6621–6628
7. Weindl D, Wegner A, Hiller K (2016) MIA: non-targeted mass isotopolome analysis. Bioinformatics 32(18):2875–2876
8. Hiller K et al (2013) NTFD--a stand-alone application for the non-targeted detection of stable isotope-labeled compounds in GC/MS data. Bioinformatics 29(9):1226–1228
9. Weindl D et al (2016) Bridging the gap between non-targeted stable isotope labeling and metabolic flux analysis. Cancer Metab 4 (1):10
10. Huang X, Chen Y-J, Cho K, Nikolskiy I, Crawford PA, Patti GJ (2014) X ^{13} CMS: global tracking of isotopic labels in untargeted metabolomics. Anal Chem 86(3):1632–1639
11. Bueschl C et al (2017) MetExtract II: a software suite for stable isotope-assisted untargeted metabolomics. Anal Chem 89 (17):9518–9526
12. Jennings ME, Matthews DE (2005) Determination of complex Isotopomer patterns in Isotopically labeled compounds by mass spectrometry. Anal Chem 77(19):6435–6444
13. Hiller K, Hangebrauk J, Jäger C, Spura J, Schreiber K, Schomburg D (2009) Metabolite detector: comprehensive analysis tool for targeted and nontargeted GC/MS based metabolome analysis. Anal Chem 81(9):3429–3439
14. Hummel J, Strehmel N, Selbig J, Walther D, Kopka J (2010) Decision tree supported substructure prediction of metabolites from GC-MS profiles. Metabolomics 6(2):322–333
15. Dietmair S, Timmins NE, Gray PP, Nielsen LK, Krömer JO (2010) Towards quantitative metabolomics of mammalian cells: development of a metabolite extraction protocol. Anal Biochem 404(2):155–164
16. Gonzalez B, François J, Renaud M (1997) A rapid and reliable method for metabolite extraction in yeast using boiling buffered ethanol. Yeast 13(14):1347–1355
17. Naz S, Moreira dos Santos DC, García A, Barbas C (2014) Analytical protocols based on LC–MS, GC–MS and CE–MS for nontargeted metabolomics of biological tissues. Bioanalysis 6(12):1657–1677
18. Lisec J, Schauer N, Kopka J, Willmitzer L, Fernie AR (2006) Gas chromatography mass

spectrometry–based metabolite profiling in plants. Nat Protoc 1(1):387–396
19. Meiser J, Weindl D, Hiller K (2013) Complexity of dopamine metabolism. Cell Commun Signal 11(1):34
20. Meiser J et al (2016) Pro-inflammatory macrophages sustain pyruvate oxidation through pyruvate dehydrogenase for the synthesis of Itaconate and to enable cytokine expression. J Biol Chem 291(8):3932–3946
21. Stein SE (1999) An integrated method for spectrum extraction and compound identification from gas chromatography/mass spectrometry data. Journal of the American Society for Mass Spectrometry 10(8):770–781

Chapter 3

Liquid Chromatography Methods for Separation of Polar and Charged Intracellular Metabolites for ^{13}C Metabolic Flux Analysis

Damini Jaiswal, Anjali Mittal, Deepak Nagrath, and Pramod P. Wangikar

Abstract

Accurate quantification of mass isotopolog distribution (MID) of intracellular metabolites is a key requirement for ^{13}C metabolic flux analysis (^{13}C–MFA). Liquid chromatography coupled with mass spectrometry (LC/MS) has emerged as a frontrunner technique that combines two orthogonal separation strategies. While metabolomics requires separation of monoisotopic peaks, ^{13}C-MFA imposes additional demands for chromatographic separation as isotopologs of metabolites significantly add to the number of analytes. In this protocol chapter, we discuss two liquid chromatography methods, namely, reverse phase ion-pairing and hydrophilic interaction chromatography (HILIC) that together can separate a wide variety of metabolites that are typically used for ^{13}C metabolic flux analysis.

Key words Sugar phosphates, Nucleotides, Reverse phase ion-pairing, HILIC, Metabolic flux analysis

1 Introduction

Metabolomics is an emerging area of omics that focuses on analysis of all the known and unknown metabolites in a biological sample [1]. Metabolome being the closest to the observed phenotype, studying the metabolome-wide alterations give a comprehensive overview of cellular metabolism under different physiological conditions. Unlike DNA, RNA, and proteins that are comprised of relatively smaller number of building blocks, the metabolites are much more diverse in their structure. With 18,505 small molecules in KEGG compound database presently and over 0.2 million metabolites in the plant kingdom, it is great challenge to develop analytical methods for identification and quantification of all of the metabolites [2]. From a biotechnological perspective, a targeted set of metabolites involved in central carbon metabolism are routinely studied, which serve as important precursors for the production of biofuel(s) or biochemical(s). Studies on the alteration in

Deepak Nagrath (ed.), *Metabolic Flux Analysis in Eukaryotic Cells: Methods and Protocols*, Methods in Molecular Biology, vol. 2088, https://doi.org/10.1007/978-1-0716-0159-4_3,

intracellular levels of these central metabolites or the carbon fluxes for the associated reactions pave the path for the rational design of microbial strains for the production of compounds of interest [3–5].

The key metabolites and cofactors involved in central carbon metabolism include sugar phosphates, phosphocarboxylic acid, carboxylic acids, ketoacids, amino acids, nucleotides, and nucleotide sugars. The intracellular concentrations of these metabolites lie in micromolar range [6]. Further, polarity and hydrophobicity of these compounds vary over a wide range, which necessitates development of different chromatographic methods for their separation. Due to the low in vivo concentrations and polar nature of these metabolites, liquid chromatography coupled to mass spectrometry (LC/MS) is the technique of choice for their detection [7, 8]. LC/MS provides sufficient chromatographic separation of metabolites and the necessary mass resolution to detect and quantify isotopologs of metabolites labeled with ^{13}C or ^{15}N tracers used for metabolic flux analysis.

In this chapter, we provide detailed protocol of the two widely used liquid chromatography methods, namely, reverse phase ion–paring and hydrophilic interaction chromatography (HILIC) coupled to ESI-MS, for separation of the polar and charged intracellular metabolites that are central to ^{13}C metabolic flux analysis (^{13}C-MFA) [9, 10]. We do not aim to provide a comprehensive survey of the various liquid chromatography methods used in metabolomics for which some exhaustive reports are already available [7, 8, 11]. Based on the available literature and drawing from our own experience, we discuss the suitability of these two methods in terms of detection, separation, and quantitation of different categories of central metabolites and cofactors and their isotopologs. While each of the two methods allows quantification a number of cellular metabolites, there are some shortcomings of each method that are overcome by the other. Therefore, we propose that analysis of each sample with two different chromatographic methods will provide sufficient coverage of metabolites desirable for ^{13}C-MFA [8]. The chromatographic run time adds significantly to the overall cost of LC/MS analysis and hence a shorter chromatographic method that provides sufficient peak resolution between analytes is preferred. In some cases, the chromatographic separation of certain analytes cannot be achieved using single method even after making a longer run time. In this chapter, we describe the reverse phase ion-pairing and HILIC methods adapted from literature [7, 12] and optimized in-house for chromatographic run times of 25–30 min. Thus, these two chromatographic runs occupy lesser LC/MS instrument time yet allow quantification of greater number of metabolites than a single method that may have a longer run time. Further, we provide examples of coeluting peaks of confounding masses for samples labeled with ^{13}C where no mass conflict is observed for monoisotopic peaks. We advise that users should watch out for such mass conflicts while quantifiying MIDs for ^{13}C-MFA.

2 Materials

2.1 Metabolism Quenching and Metabolite Extraction

1. The cell culture of cyanobacteria *Synechococcus* sp. PCC 7002 and *Synechococcus elongatus* PCC 11801 at metabolic steady state.
2. Fast filtration apparatus equipped with a vacuum pump and nylon membrane filters (0.8 μm, Whatman).
3. Metabolite extraction solvents: 80:20, v/v methanol–water, 0.2 M ammonium hydroxide solution in deionized water and chloroform. Metabolite extraction protocol was adapted from our previous study [13–15].
4. Vortexer.
5. Centrifuge tubes (5 mL and 15 mL).
6. Benchtop cooling centrifuge.
7. Speed vacuum concentrator or lyophilizer with a cold trap for drying the metabolite extract.
8. −80 °C deep freezer to store the metabolite extraction until LC-MS analysis.

2.2 ^{13}C Labeled Dataset

Our previously reported data from the ^{13}C labeling experiment performed *Synechococcus* sp. PCC 7002 is used in the study as a case study [14]. A sample from *Synechococcus elongatus* PCC 11801 with 10 min ^{13}C label incorporation is also used.

2.3 LCMS Analysis

1. A Triple TOF 5600+ mass spectrometer (Sciex, Framingham, MA) interfaced with Shimadzu ultrahigh-performance-liquid chromatography (UPLC) system (Shimadzu, Nexera LC-30 AD, Singapore) and an electrospray ionization (ESI) source.
2. Synergi Hydro-RP LC column 150 × 2 mm, 4 μm particle size (Phenomenex Inc., Torrance, CA).
3. SeQuant® ZIC®-pHILIC 100 × 2.1 mm, 5 μm particle size (EMD Millipore, Billerica, MA, USA).
4. XBridge BEH Amide column (150 mm × 2.1 mm, 2.5 mm particle size, Waters, Milford, MA, USA).
5. Buffers and Additives: LCMS grade water, ethanol, acetonitrile, tributylamine, acetic acid, ammonium carbonate, ammonium bicarbonate, ammonium hydroxide, and formic acid.

6. LCMS grade metabolite standards of amino acids, nucleotides, sugar phosphates, organic acids, nucleotide sugars were purchased from Sigma-Aldrich (St. Louis, MO). The equal amount of standard solutions was injected to compare the reverse phase ion-pairing and HILIC method.
7. Vials, gases, guard columns, 4 mm nylon syringe filters (Phenomenex Inc., Torrance, CA), and other consumables associated with LCMS.

2.4 LCMS Data Analysis

1. PeakView 2.0 and MasterView 1.0 software (Sciex, Framingham, MA) for manual examination of the chromatographic peak quality and baseline separation.

3 Methods

3.1 Sample Preparation

The analysis of metabolites from cells involves four key steps:

1. *Quenching of cellular metabolism.*

   The intracellular metabolites turn over at fast rates. To capture a snapshot of cellular metabolism, it is important to quench the metabolism effectively. This is especially important while performing a dynamic ^{13}C labeling experiment. Centrifugation followed by quenching takes up a significant amount of time to pellet cells and is not advisable in the context of nonstationary ^{13}C-MFA. Among the various methods employed to quench the metabolism, fast filtration followed by quenching in cold solvent (e.g., methanol) is the most effective method in our experience [16]. Rapid mixing of the cell culture with the cold extraction solvent has also been widely used [17], although we find leaching of metabolites to be a key challenge with this technique. Further, acid or heat denaturation has been reported for the analysis of the metabolites that are heat/acid stable [18].
2. *Extraction or release of metabolites from cells.*

   The release of metabolites from the cells requires the breaking of cells and extraction into solvents. Single phase solvent mixtures of acetonitrile, methanol and water or methanol and water have been widely reported for extraction of polar metabolites [8]. For photosynthetic organisms, we find a biphasic solvent system to be quite effective [13]. In the solvent mixture comprising of chloroform, methanol and water, the polar and charged metabolites get extracted into the aqueous-rich phase, while the hydrophobic compounds including the pigments partition into the chloroform-rich phase. This offers partial sample cleanup during extraction itself. Obtaining colorless aqueous phase and cell debris with greenish chloroform phase is a hallmark of effective metabolite extraction [13]. Upon centrifugation, the cell debris typically floats at the interface of the two solvents. Extraction of negatively charged metabolites such as sugar phosphates can be further improved by adding a weak base such as ammonium hydroxide [13].

3. *Storage of metabolites to prevent degradation before analysis.*

   Appropriate care should be taken to prevent degradation of polar metabolites. The aqueous phase containing the extracted metabolites should be lyophilized and stored at −80 °C deep freezer until ready for LC/MS analysis. We find that storage of up to 2 months does not adversely affect the peak quality. However, the compounds that are redox active should be analyzed quickly [8].

4. *Resuspension of metabolite extract in the appropriate buffer before analysis.*

   Reconstitution of the metabolite extract into a suitable buffer is important for two reasons, that is, to completely dissolve the metabolites before LCMS analysis and to obtain reliable retention times for each injection. For the chromatographic methods described here, reverse phase ion pairing and HILIC we recommend reconstitution of the metabolite extract in 50:50 methanol–water and 80:20 acetonitrile–water, respectively. In case of HILIC method, the sample is typically reconstituted in the solvent mixture that is used at $t = 0$ min in the gradient program for elution. As separation on HILIC column involves different kinds of interaction a minor change in the buffer composition can result in dramatic effects on reproducibility [11].

3.2 Reverse Phase Ion-Pairing Chromatography

Reverse phase (RP) chromatography involves a hydrophobic stationary phase and a polar mobile phase. The hydrophobic analytes are adsorbed on the stationary phase through hydrophobic interaction. In a gradient program of chromatography, concentration of the organic solvent is gradually increased resulting in the elution of bound analytes. Polar analytes are eluted first followed by the nonpolar ones. Reverse phase chromatography has been successfully applied for the separation of nonpolar metabolites in positive ion mode [8, 19]. However, this chromatography technique is not effective in retaining highly polar metabolites like sugar phosphates, and nucleotides as these metabolites tend to elute in the void volume of the column [7, 8]. Therefore, taking into consideration the anionic nature of these analytes, anion exchange chromatography (AEC) has been reported for the separation of these polar and negatively charged metabolites [7]. The utilization of solvents with high ionic strength to elute the analytes from the anion exchange column results in the deposition of these nonvolatile salts at the electrospray ionization (ESI) source [7]. Moreover, the multiple isomeric forms of the phosphorylated sugars that are key for isotopic nonstationary ^{13}C-MFA were not resolved using the AEC.

The use of ion-pairing pairing reagents led to significant improvement in the retention and separation of these anionic metabolites on a reverse phase column [20]. Tetraalkylammonium salts were initially tested to separate the anionic compounds. However, these salts are nonvolatile and incompatible with ESI-MS. Subsequently, a volatile ion-pairing reagent, tributylamine, was deployed to separate a wide variety of negatively charged compounds of interest [7, 8, 21–23]. Since then, the reverse phase ion-pair chromatography chromatographic technique has been widely used for separation of negatively charged metabolites under negative ion mode. The use of tributylamine is not recommended under positive ion mode due to strong ion suppression.

In this work, we discuss separation of a number of intracellular metabolites from the extracts of cyanobacteria *Synechococcus sp.* PCC 7002 and *Synechococcus elongatus* PCC 11801 [24] on a C18 reverse phase column using tributylamine as an ion-pairing reagent. We used Synergi™ 4 μm Hydro-RP LC column 150 × 2 mm (Phenomenex Inc., Torrance, CA) for chromatographic separation of intracellular metabolites using a gradient elution method consisting of eluents 10 mM tributylamine +15 mM acetic acid in water (pH = 4.95) [7] (buffer A) and 100% Methanol (buffer B). The gradient method used is as follows: 0% B (0.01 min), 0% B (2 min), 35% B (8 min), 35% B (10.5 min), 90% B (15.50 min), 90% B (20.5 min), 0% B (22 min), and 0% B (30 min). The column temperature and flow rates were 25 °C and 0.3 mL/min, respectively. A Triple TOF 5600+ mass spectrometer (SCIEX, Framingham, MA) interfaced with Shimadzu Ultra Performance- Liquid Chromatography (UPLC) system (Shimadzu, Nexera LC-30 AD, Singapore) equipped with a binary pump, degasser, column oven, and autosampler was used for the analysis. The curtain gas, gas 1, and gas 2 were kept at 35 psi, 40 psi, and 40 psi respectively. The ion source temperature was 450 °C and the voltage was −4500 V. The data was acquired using information dependent acquisition (IDA) method with MS2 scan triggered for the 10 most abundant precursors in each cycle, cycle time 1 s.

3.3 Hydrophilic Interaction Chromatography (HILIC)

HILIC involves a polar stationary phase and a low polarity water-miscible organic solvent and water as the eluents [25]. The polar analytes are retained on the column and are eluted by a gradual increase in the concentration of water. Although the nature of stationary and mobile phases used in HILIC is similar to normal phase chromatography, the separation mechanism in HILIC is much more complex and may include partitioning, adsorption, hydrogen donor interaction, or electrostatic interaction [11]. It combines the characteristics of normal phase, reverse phase and ion-exchange chromatography [11]. There are already some good reports describing the separation mechanism in HILIC [8, 11, 25, 26].

HILIC uses organic solvents that improve the ionization process in ESI and does not require the use of expensive ion-pairing reagents. The selectivity of separation using HILIC can be dramatically changed by the selection of different stationary and mobile phases [11, 27]. Here, we discuss the separation of metabolites from cyanobacterial extracts using a zwitterionic column, SeQuant® ZIC®-pHILIC (EMD Millipore, Billerica, MA, USA) with sulfobetaine functional group. Unlike traditional silica support, ZIC-pHILIC has polymeric support. Because of its zwitterionic nature, it can separate a wide variety of compounds like acids, bases, anions, cations, carbohydrates, metabolites, peptides, and protein digests [11, 27, 28]. The mobile phase consisted of 20 mM ammonium carbonate in water adjusted to pH = 8.0 with formic acid (solvent A) and acetonitrile (solvent B). The gradient program is as follows: 80% B (0.01 min), 80% B (1.5 min), 15% B (14 min), 5% B (17 min), 5% B (21 min), 80% B (23 min), 80% B (29 min). The pH of buffer A plays an important role in ionization and separation of analytes in this method. The flow rate was maintained at 0.15 mL/min. The other instrument parameters and data acquisition method were as described above.

3.4 Comparison Between Reverse Phase Ion-Pairing Chromatography and HILIC Chromatography

1. *Sugar phosphates and isomers.*

   The phosphorylated sugar compounds exist in multiple isomeric forms inside the cells. For example, glucose-6-phosphate (G6P) and fructose-6-phosphate (F6P) are important intermediates in cellular metabolism. These two isomers appear as separate peaks when ion-pairing chromatography is used but do not get resolved using HILIC method (Fig. 1a and Table 1). A similar trend is observed for ribose-5-phosphate (R5P) and ribulose-5-phosphate that are separated using the ion-pairing method but not the HILIC method (Fig. 1b and Table 1).

2. *Nucleotide, nucleotide sugars, and Acyl CoA.*

   The nucleotide, nucleotide sugars, and acyl CoAs are well retained and elute as well-defined peaks in both ion-pairing and HILIC methods. For example, ADP, ATP, UMP, NADPH, UDP-glucose, and NADPH are compared for their retention in ion-pairing and HILIC method as shown in Fig. 2. However, we observe that the peaks for these compounds are in general sharper in the ion-pairing method compared to the HILIC method.

3. *Amino acids.*

   Amino acids are a class of compounds with varying polarity, hydrophobicity, and hydrophilicity. We compared the ion-pairing and HILIC method for their retention and separation capability of amino acids (Table 2 and Fig. 3). We observed that the negatively charged amino acids like glutamate and

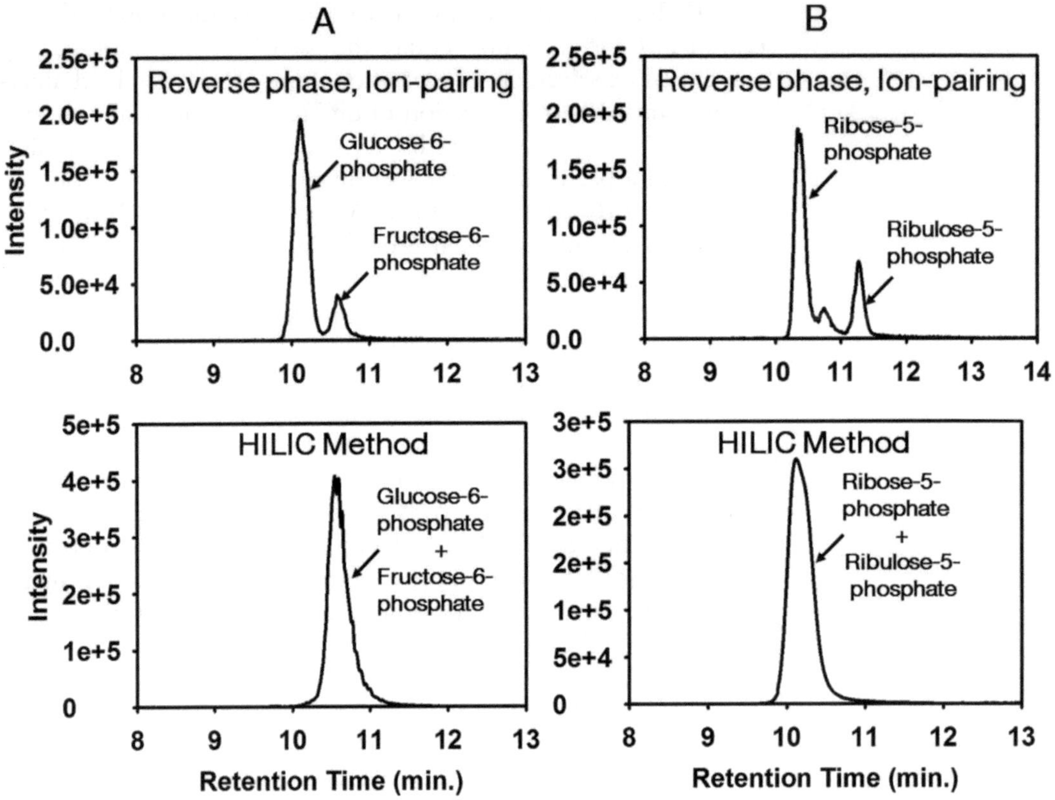

Fig. 1 Comparison between the reverse phase ion-pairing and HILIC methods for peak resolving ability of isomeric compounds like glucose-6-phosphate and fructose-6-phosphate (**a**) and ribose-5-phosphate and ribulose-5-phosphate (**b**)

aspartate are well retained both the methods, although the retention of these amino acids was greater in the HILIC method (Table 2 and Fig. 3). The polar-positively charged amino acids like arginine, histidine, lysine and polar-uncharged amino acids like serine, threonine, asparagine, and glutamine elute early on reverse phase column but are well retained on the HILIC column (Table 2 and Fig. 3). For nonpolar and uncharged amino acids we observed differential retention behavior. The amino acids like glycine, alanine, proline, and valine elute early in the reverse phase method but elute as a well-defined peak in the HILIC method (Table 2). However, other nonpolar amino acids like leucine, isoleucine, and methionine with larger carbon backbone (and thus more hydrophobic) show slightly better retention on the reverse phase column, although the retention of these in HILIC column was far better (Table 2). The aromatic amino acids like phenylalanine, tryptophan, and tyrosine showed much better retention in the reverse phase column compared to other nonpolar uncharged amino acids (Table 2). This might be due to their heavier ring

Table 1
Comparison between reverse phase ion-pairing and HILIC chromatographic methods for retention, and separation of different central metabolites

	Reverse phase ion-pairing				HILIC			
Metabolite	**RT (min)**	**Max. intensity**	**PW50 (min)**	**S/N**	**RT (min)**	**Max. intensity**	**PW50 (min)**	**S/N**
Sugar phosphates								
Glucose-6-phosphate	**10.13**	1.95×10^5	0.24	2.90×10^4	**10.57**[a]	4.07×10^5	0.26[a]	7.92×10^4
Fructose-6-phosphate	**10.6**	3.35×10^4	0.20	5.10×10^3				
Ribose-5-phosphate	**10.36**	1.85×10^5	0.19	2.66×10^4	**10.16**[a]	2.81×10^5	0.19	5.23×10^4
Ribulose-5-phosphate	**10.74**	2.63×10^4	0.2	4.55×10^3				
Ribulose 1,5-bisphosphate	15.18	2.21×10^5	0.11	3.11×10^4	11.52	5.54×10^3	0.38	1.00×10^3
Sedoheptulose-7-phosphate	10.54	1.60×10^5	0.18	2.39×10^4	10.34	3.11×10^5	0.23	5.84×10^4
Nucleotide sugars								
ADP-glucose	14.11	5.38×10^5	0.14	8.46×10^4	9.77	4.82×10^5	0.23	1.13×10^5
UDP-glucose	13.16	3.67×10^5	0.18	6.37×10^4	10.41	4.23×10^5	0.25	1.01×10^5
Carboxylic acids								
Citric acid	15.03	2.33×10^5	0.13	2.37×10^4	11.26	3.65×10^4	0.51	5.70×10^3
Fumaric acid	13.32	2.73×10^4	0.18	4.86×10^3	10.32	2.86×10^4	0.28	5.08×10^3
Malic acid	13.32	1.38×10^5	0.18	1.95×10^4	10.32	1.39×10^5	0.28	2.49×10^4
Succinic acid	13.18	7.29×10^4	0.15	1.21×10^4	9.91	2.83×10^4	0.23	5.78×10^3
Phosphocarboxylic acids								
3-Phoshoglyceric acid	14.61	2.35×10^5	0.11	3.46×10^4	10.55	5.90×10^4	0.27	1.15×10^4
Phosphoenolpyruvic acid	14.99	1.69×10^5	0.10	2.36×10^4	10.72	6.57×10^4	0.18	1.18×10^4
Keto acids								
Alphaketoglutaric acid	13.77	6.12×10^4	0.15	1.01×10^4	10.08	3.02×10^4	0.24	5.45×10^3
Oxaloacetic acid	15.21	2.48×10^4	0.12	4.16×10^3	11.56	2.38×10^3	0.23	4.65×10^2

The major differences in RT between the two methods are highlighted in bold. The m/z corresponds to the molecular mass of the compound minus one (negative ion mode)
RT retention time
PW_{50} peak width at half height
S/N signal-to-noise ratio
[a]These isomers did not separate in the HILIC column when pooled together and injected; hence, the intensity and other variables correspond to the mixed pool of the two isomers

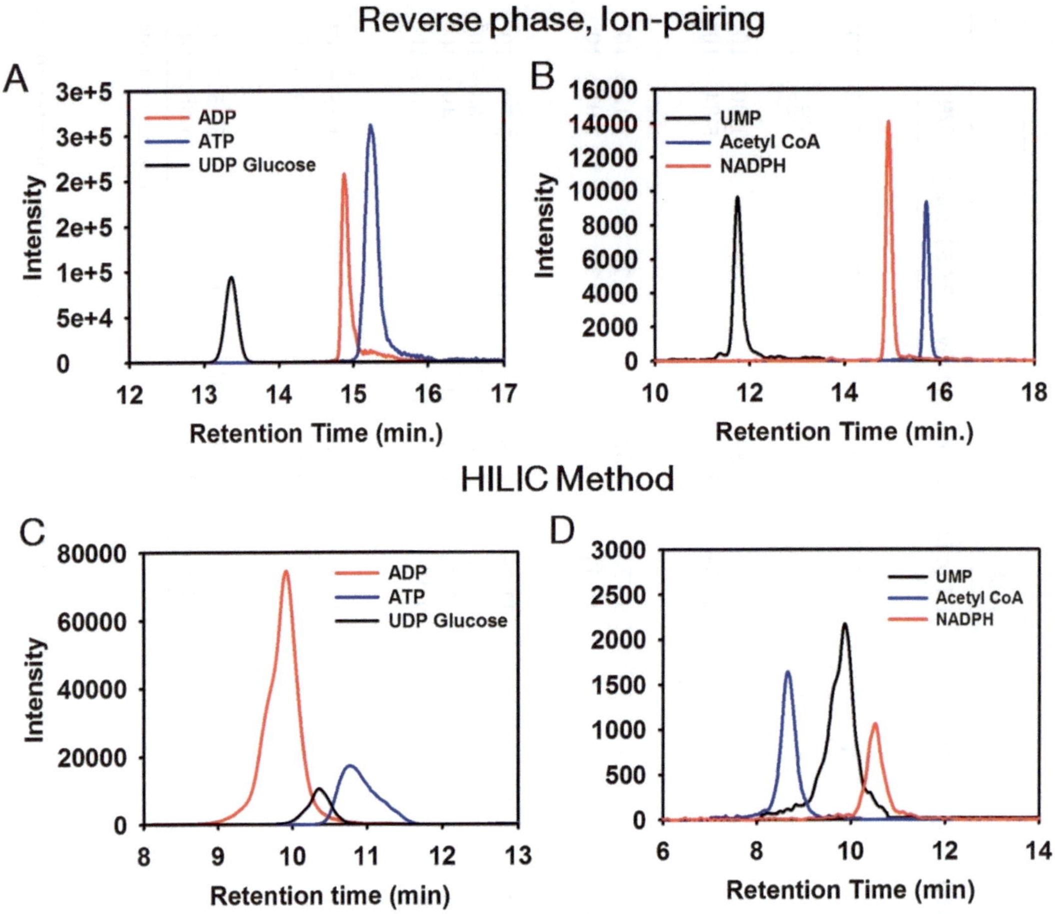

Fig. 2 Comparison between reverse phase ion-pairing and HILIC methods for retention and separation of nucleotides (ADP, ATP, UMP, and NADPH), nucleotide sugar (UDP-glucose) and acyl CoA (acetyl CoA) from the metabolite extract of the cyanobacterium, *Synechococcus elongatus* PCC 11801

structure allowing strong hydrophobic interaction even though the net charge on them in zero. These amino acids are also well retained in the HILIC column.

4. *Carboxylic acids, phosphocarboxylic acids, and keto acids.*

   The carboxylic acids like succinic, citric, fumaric, and malic acid, the phosphocarboxylic acids like 3-phosphoglycerate and phosphoenolpyruvate as well as keto acids like alpha ketoglutaric and oxaloacetic acid all are well retained and elute as good peak in both the methods (Table 1).

To summarize, we observed that compounds like sugar phosphates, nucleotides, nucleotide sugars, carboxylic acids, keto acids, and acyl CoAs are well retained and separated by both ion-pairing and HILIC methods. However, the reverse phase ion-pairing method shows noticeable advantage in resolving the isomers of sugar phosphates [7]. The polar and negatively charged amino

Table 2
Comparison between reverse phase ion pairing and HILIC chromatographic methods for retention and separation of different groups of amino acids

	Reverse phase ion pairing				HILIC			
Metabolite	RT (min)	Max. intensity	PW_{50} (min)	S/N	RT (min)	Max. intensity	PW_{50} (min)	S/N
Polar, positively charged								
Arginine	**1.02**	3.06×10^3	0.07	3.13×10^2	**12.1**	5.90×10^4	0.32	1.10×10^4
Histidine	**1.03**	8.53×10^3	0.07	1.57×10^3	**8.8**	1.18×10^5	0.15	1.50×10^4
Lysine	**1.24**	4.97×10^4	0.08	8.04×10^3	**11.8**	4.01×10^4	0.32	7.30×10^3
Polar, negatively charged								
Aspartate	4.21	3.71×10^4	0.14	6.94×10^3	9.77	7.60×10^4	0.24	1.64E+04
Glutamate	4.02	4.31×10^4	0.085	7.38×10^3	9.66	1.18×10^5	0.24	2.15×10^4
Polar, uncharged								
Serine	**1.18**	9.82×10^3	0.07	2.15×10^3	**9.47**	1.37×10^4	0.23	4.06×10^3
Threonine	**1.25**	1.13×10^4	0.09	2.32×10^3	**8.83**	2.20×10^4	0.25	4.64×10^3
Asparagine	**1.2**	4.87×10^4	0.08	8.04×10^3	**9.36**	4.53×10^4	0.31	8.37×10^3
Glutamine	**1.24**	4.98×10^4	0.08	7.85×10^3	**9.11**	7.33×10^4	0.29	1.34×10^4
Nonpolar, uncharged								
Glycine	**1.18**	3.01×10^3	0.09	6.73×10^2	**9.25**	2.99×10^3	0.19	8.12×10^2
Alanine	**1.2**	2.30×10^3	0.09	9.75×10^2	**9.77**	7.64×10^3	0.23	2.12×10^3
Proline	**1.41**	4.27×10^4	0.07	6.77×10^3	**6.86**	4.05×10^4	0.08	7.27×10^3
Valine	**1.59**	5.89×10^4	0.09	1.03×10^4	**6.74**	6.33×10^4	0.4	1.05×10^4
Leucine	2.8	7.67×10^4	0.13	1.25×10^4	5.05	9.40×10^4	0.2	1.69×10^4
Isoleucine	2.6	6.83×10^4	0.12	1.26×10^4	4.59	9.68×10^4	0.16	1.78×10^4
Methionine	2.19	4.01×10^4	0.10	7.97×10^3	5.72	5.11×10^4	0.59	8.64×10^3
Phenylalanine	7.2	1.98×10^5	0.10	3.35×10^4	4.11	2.55×10^5	0.24	3.93×10^4
Tyrosine	4.43	5.76×10^4	0.14	1.05×10^4	7.34	1.59×10^5	0.43	2.18×10^4
Tryptophan	9.6	1.09×10^5	0.12	1.67×10^4	5.23	2.57×10^5	0.36	4.00×10^4

The major differences in RT between the two methods are highlighted in bold. The m/z corresponds to the molecular mass of the compound minus 1 (negative ion mode)
RT retention time
PW_{50} peak width at half height
S/N signal to noise ratio

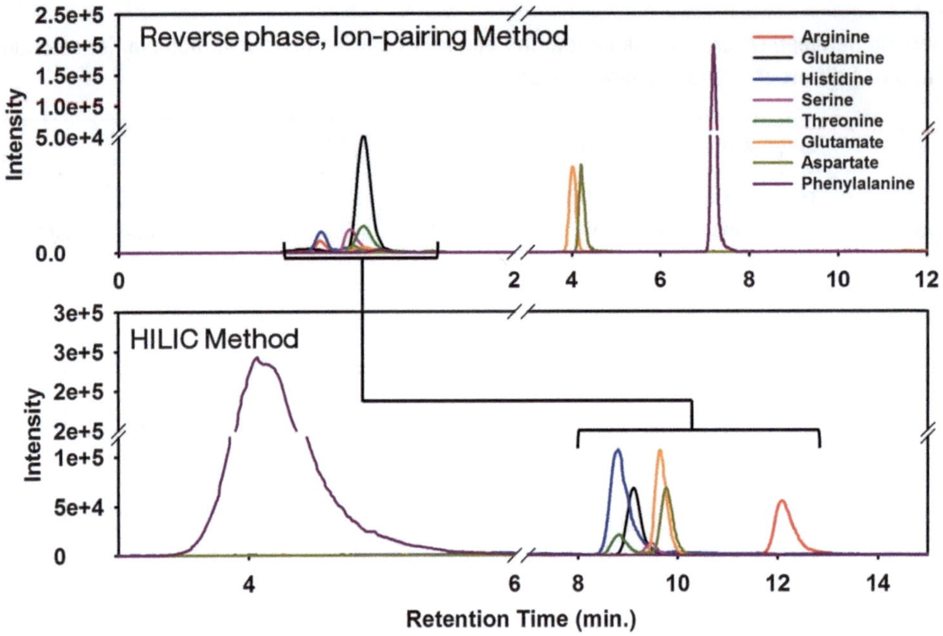

Fig. 3 Comparison between the reverse phase ion-pairing and HILIC methods for retention and separation of amino acids

acids are retained in both the methods. The polar-positively charged and uncharged amino acids are too polar to be retained on the reverse phase column and are separated well in HILIC method. The separation of nonpolar amino acids using the reverse phase ion-pairing method depends on the size of the amino acids. The peak width was generally greater in HILIC method compared to the ion-pairing method (Tables 1 and 2). We suggest the use of these two chromatographic methods as they complement each other very well. The data missed using ion-pairing method can be easily obtained by HILIC method and vice versa.

3.5 HILIC for NAD(+) Metabolome

The NAD(+) metabolome is diverse and comprises metabolites from different classes, including dinucleotides (NAD, nicotinic acid adenine dinucleotide), nucleotides (nicotinamide mono mononucleotide, nicotinic acid mononucleotide), nucleosides (nicotinamide, riboside, nicotinic acid riboside), and nucleobases (nicotinamide, nicotinic acid) [29]. The analysis of these highly hydrophilic compounds by reverse-phase ion-chromatography has low sensitivity. HILIC is a good option for the separation of the metabolites from the NAD(+) metabolome. Amide columns can retain highly polar compounds and provide good retention for these metabolites [30].

Here, we discuss the measurement of the NAD(+) metabolome on an Agilent 1290 Infinity LC system couple to Agilent 6520 accurate-mass Q-TOF mass spectrometer. The chromatography is performed on an XBridge BEH Amide column. Solvent A is 95:5 water–acetonitrile with 20 mM ammonium bicarbonate, pH adjusted to 9.5 with ammonium hydroxide. Solvent B is 90:10 acetonitrile–water with 10 mM ammonium bicarbonate, pH adjusted to 9.5 with ammonium hydroxide. The flow rate is 0.15 mL/min and the column temperature is adjusted to 25 °C. The gradient is 0 min, 85% B; 2 min, 85% B; 3 min, 80% B; 5 min, 80% B; 6 min, 75% B; 7 min, 75% B; 8 min, 70% B; 9 min, 70% B; 10 min, 50% B; 12 min, 50% B; 13 min, 25% B; 16 min, 25% B; 18 min, 0% B; 23 min, 0% B; 24 min, 85% B; 25 min, 85% B, flow rate increased to 0.6 mL/min; 33 min, 85% B, flow rate 0.6 mL/min; 34 min, flow rate decreased to 0.15 mL/min, 35 min, flow rate 0.15 mL/min [30]. The flow rate is increased after 25 min to equilibrate the column in-between runs. The mass spectrometer is set to positive ion mode. The retention time of the metabolites is verified by running standards. Accurate mass and retention time are used to identify the metabolites. The samples are prepared in 85:15 acetonitrile: water, 10 mM ammonium bicarbonate, pH adjusted to 9.5 with ammonium hydroxide. The fraction of water can be increased if the dried sample has low solubility in 80% acetonitrile, but the effect of changing the sample composition on the retention time should be assessed by running standards.

3.6 Identification of Chromatographic Peak in the Sample from LCMS Data

A given sample can have thousands of features representing the metabolites, adducts, contaminants or ghost peaks coming from the plastic wares used during the metabolite extraction. Therefore, it is very necessary to identify the compounds of biological origin that is desirable for a particular study. A simple way to identify targeted set of metabolites in biological samples is to correlate the *m/z*, RT and fragmentation pattern of pure standards injected with the exactly same chromatographic method. Figure 4a shows the example of the correlation of *m/z*, RT and fragmentation pattern of pure UDP-glucose standard and that present in the metabolite extract of *Synechococcus elongatus* PCC 11801. Thus, reliable quantitation of UDP-glucose can be done from the sample. Figure 4b shows another contrasting example where the *m/z* and the RT of pure alpha ketoglutarate match to a feature in the sample. However, the fragmentation pattern of the feature resembling alpha ketoglutarate does not match with that of the standard. This implies it could either an entirely different metabolite or coelution of alpha-ketoglutarate and some other compound in the sample. Thus, as a thumb rule, before quantifying the data of a particular metabolite from the sample, we should make sure that the *m/z*, RT as well the fragmentation pattern are matching with that of the pure standard analyte.

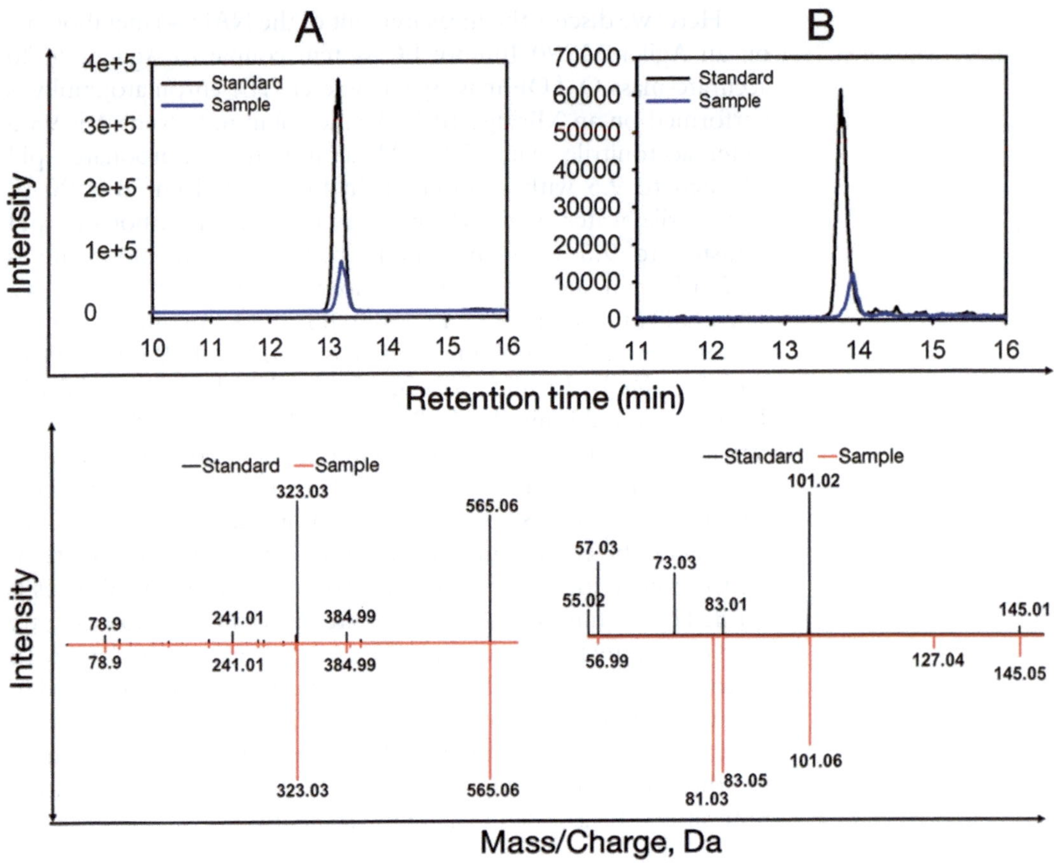

Fig. 4 The XICs of UDP-glucose and alpha ketoglutarate obtained from pure standard injections overlaid with that present in a sample of *Synechococcus elongatus* PCC 11801 (**a**) and (**b**) respectively. The *m/z*, RT and fragmentation pattern of pure standard matches well with that of the sample for UDP-glucose (**a**) but not for alpha-ketoglutarate, where the fragmentation pattern of the sample does not correlate with the pure standard

The metabolites can also be putatively identified by matching the *m/z* and fragmentation pattern in mass spectral databases like METLIN [31], YMDM [32], HMDB [33], Chemspider [34], etc. Metabolite identification tools like MS-DIAL [35] and MetDIA [36] are also helpful. However, for ^{13}C-MFA purpose, it is recommended to confirm the identity of identified metabolites in samples with at least two orthogonal methods.

3.7 Quality Check for Mass Isotopolog Quantitation

The reliable quantitation of the metabolites from any biological sample can be performed by matching the *m/z*, RT, and fragmentation pattern as exemplified in the earlier section. In case of ^{13}C-labeled sample the isotopologs of the metabolites are also identified as distinct features or analytes. Thus, for reliable quantitation of MID of metabolites, an additional criterion should be satisfied (*see* **Note 1**). There should no mass conflict of another metabolite or its isotopologs with any of the isotopologs of a particular metabolite,

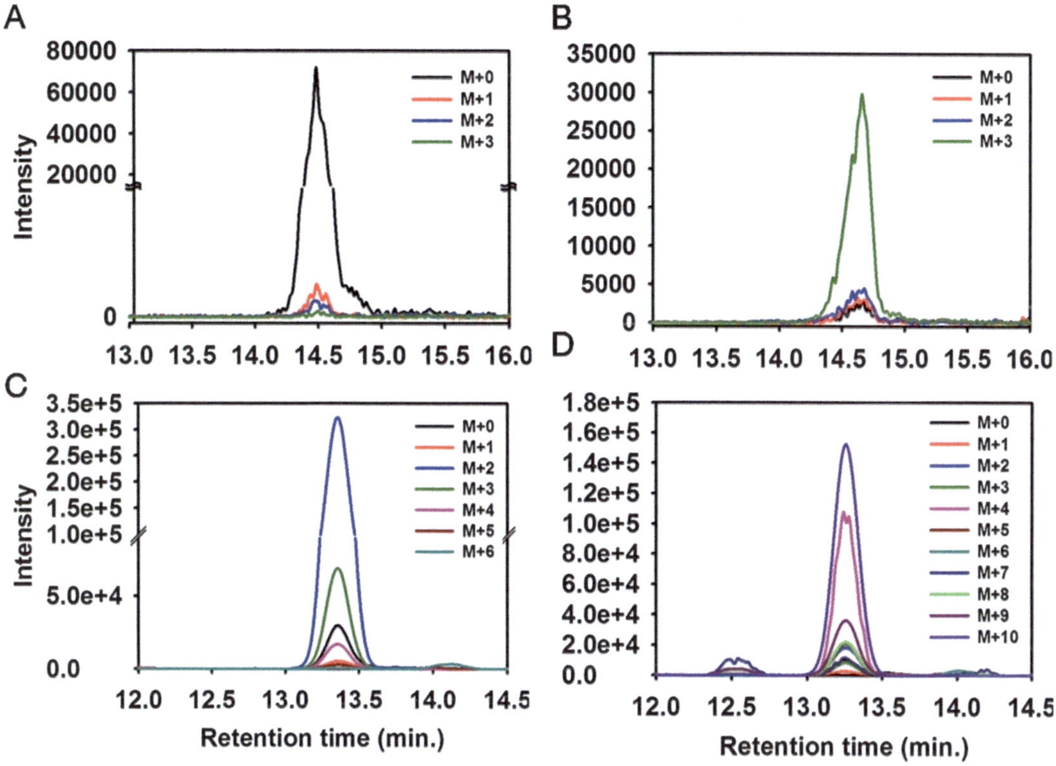

Fig. 5 The mass isotopolog distribution (MID) of 3-phosphoglycerate in an unlabeled sample (**a**) and 10 min after addition of $NaH^{13}CO_3$- (partially labeled sample) (**b**) of *Synechococcus* sp. PCC 7002 showing no mass conflicts in isotopologs. The MID of GDP-mannose in an unlabeled sample (**c**) and a partially labeled sample (**d**) of *Synechococcus elongatus* PCC 11801. The UDP-*N*-acetylglucosamine with *m/z* equal to M + 2 isotopolog of GDP-mannose and its other mass isotopologs shows RT and mass conflict with isotopologs of GDP-mannose

the data for which is desirable for a study. Figure 5a, for example, shows the XICs of all the isotopologs of 3-phosphoglycerate (3C) in an unlabeled sample of *Synechococcus* sp. PCC 7002. There is no mass conflict between 3-phosphoglycerate and its isotopologs with any other peak and follows a suggested natural distribution of ^{13}C carbon. Thus, the isotopologs of 3-phosphoglycerate in a ^{13}C-labeled sample can be reliably quantified (Fig. 5b). We show another example where there are mass and RT conflicts between the isotopologs of GDP-mannose (a C16 compound, *m/z* 604) and UDP-*N*-acetylglucosamine (a C17 compound, *m/z* 606) metabolites in *S. elongatus* PCC 11801. Thus, the M + 0 of UDP-*N*-acetylglucosamine has a conflict with the *m/z* of M + 2 isotopolog of GDP-Mannose (Fig. 5c). In a 10 min ^{13}C-labeled sample of *S. elongatus* PCC 11801, significant labeling is expected in mannose part of GDP-mannose forming M + 6 (*m/z* 610) as the highest isotopolog. But the labeling GDP-mannose is masked by the ^{13}C-labeling of UDP-*N*-

acetylglucosamine, the peak of which is more abundant than GDP-mannose (Fig. 5c, d). Therefore, we see the dominant isotopologs as M + 4 (*m/z* 608) and M + 10 (*m/z* 614) of GDP-mannose that corresponds to the labeling in the acetyl (M + 2, *m/z* 608) and the *N*-acetyl glucosamine (M + 8, *m/z* 614) parts of UDP-*N*-acetylglucosamine (Fig. 5d). The above-mentioned example signifies effect of mass and RT conflict in quantitation of MIDs.

Even after the development of liquid chromatography techniques that suitably resolves some 100 s of metabolites in a biological sample, there could be conflict at isotopolog level as the ^{13}C labeling introduces mass shifts in the metabolites. Therefore, we recommend that mass conflicts arising out of isotopologs need to be checked before quantifying the ^{13}C enrichment. This shows the additional demands imposed on the chromatographic separation for ^{13}C MFA studies.

4 Note

1. The mass isotopolog distribution (MID) is a relative fractional measurement of different isotopologs a metabolite can form upon labeling with a ^{13}C tracer. The number of isotopologs of a metabolite depends on the number of carbons present in it. For example, 3-phosphoglyceric acid that has three carbons and monoisotopic *m/z* of 184.98 (M) can have three isotopologs, 185.98 (M + 1), 186.98 (M + 2), and 187.98 (M + 3). The number of isotopologs in a particular metabolite (M) with *n* carbons will be *n* + 1.

Acknowledgments

This work was supported by a grant from Department of Biotechnology (DBT), Government of India, awarded to PPW toward DBT-Pan IIT Center for Bioenergy (Grant No. BT/EB/PAN IIT/2012).

References

1. Mathew AK, Padmanaban VC (2013) Metabolomics: the apogee of the omics trilogy. Int J Pharm Pharm Sci 5:45–48. https://doi.org/10.1038/nrm3314
2. Fiehn O, Kopka J, Dörmann P et al (2000) Metabolite profiling for plant functional genomics. Nat Biotechnol 18:1157–1161. https://doi.org/10.1038/81137
3. McAtee AG, Jazmin LJ, Young JD (2015) Application of isotope labeling experiments and ^{13}C flux analysis to enable rational pathway engineering. Curr Opin Biotechnol 36:50–56. https://doi.org/10.1016/j.copbio.2015.08.004
4. Orth JD, Thiele I, Palsson BØ (2010) What is flux balance analysis? Nat Biotechnol

28:245–248. https://doi.org/10.1038/nbt.1614
5. Lee SY, Park JM, Kim TY (2011) Application of metabolic flux analysis in metabolic engineering, 1st edn. Elsevier Inc, Amsterdam
6. Ohta E, Dempo Y, Fukusaki E et al (2014) Molar-based targeted metabolic profiling of cyanobacterial strains with potential for biological production. Metabolites 4:499–516. https://doi.org/10.3390/metabo4020499
7. Luo B, Groenke K, Takors R et al (2007) Simultaneous determination of multiple intracellular metabolites in glycolysis, pentose phosphate pathway and tricarboxylic acid cycle by liquid chromatography-mass spectrometry. J Chromatogr A 1147:153–164. https://doi.org/10.1016/j.chroma.2007.02.034
8. Lu W, Bennett BD, Rabinowitz JD (2008) Analytical strategies for LC-MS-based targeted metabolomics. J Chromatogr B Analyt Technol Biomed Life Sci 871:236–242. https://doi.org/10.1016/j.jchromb.2008.04.031
9. Young JD, Shastri AA, Stephanopoulos G, Morgan JA (2011) Mapping photoautotrophic metabolism with isotopically nonstationary ^{13}C flux analysis. Metab Eng 13:656–665. https://doi.org/10.1016/j.ymben.2011.08.002
10. Alagesan S, Gaudana SB, Sinha A, Wangikar PP (2013) Metabolic flux analysis of Cyanothece sp. ATCC 51142 under mixotrophic conditions. Photosynth Res 118:191–198. https://doi.org/10.1007/s11120-013-9911-5
11. Buszewski B, Noga S (2012) Hydrophilic interaction liquid chromatography (HILIC)-a powerful separation technique. Anal Bioanal Chem 402:231–247. https://doi.org/10.1007/s00216-011-5308-5
12. Keasling JD, Adams PD, Benites VT et al (2018) Integrated analysis of isopentenyl pyrophosphate (IPP) toxicity in isoprenoid-producing *Escherichia coli*. Metab Eng 47:60–72. https://doi.org/10.1016/j.ymben.2018.03.004
13. Prasannan CB, Jaiswal D, Davis R, Wangikar PP (2018) An improved method for extraction of polar and charged metabolites from cyanobacteria. PLoS One 13(10):1–16. https://doi.org/10.21228/M8WT20
14. Jaiswal D, Prasannan CB, Hendry JI, Wangikar PP (2018) SWATH tandem mass spectrometry workflow for quantification of mass isotopologue distribution of intracellular metabolites and fragments labeled with isotopic ^{13}C carbon. Anal Chem 90:6486–6493. https://doi.org/10.1021/acs.analchem.7b05329
15. Hendry JI, Prasannan C, Ma F et al (2017) Rerouting of carbon flux in a glycogen mutant of cyanobacteria assessed via isotopically non-stationary ^{13}C metabolic flux analysis. Biotechnol Bioeng 114:2298–2308. https://doi.org/10.1002/bit.26350
16. McCloskey D, Utrilla J, Naviaux RK et al (2014) Fast Swinnex filtration (FSF): a fast and robust sampling and extraction method suitable for metabolomics analysis of cultures grown in complex media. Metabolomics 11:198–209. https://doi.org/10.1007/s11306-014-0686-2
17. Ikeda TP, Shauger AE, Kustu S (1996) Salmonella typhimurium apparently perceives external nitrogen limitation as internal glutamine limitation. J Mol Biol 259:589–607. https://doi.org/10.1006/jmbi.1996.0342
18. Schaub J, Schiesling C, Reuss M, Dauner M (2006) Integrated sampling procedure for metabolome analysis. Biotechnol Prog 22:1434–1442. https://doi.org/10.1021/bp050381q
19. Lu W, Kimball E, Rabinowitz JD (2006) A high-performance liquid chromatography-tandem mass spectrometry method for quantitation of nitrogen-containing intracellular metabolites. J Am Soc Mass Spectrom 17:37–50. https://doi.org/10.1016/j.jasms.2005.09.001
20. Ståhlberg J (1999) Retention models for ions in chromatography. J Chromatogr A 855:3–55. https://doi.org/10.1016/S0021-9673(99)00176-4
21. Qian T, Cai Z, Yang MS (2004) Determination of adenosine nucleotides in cultured cells by ion-pairing liquid chromatography-electrospray ionization mass spectrometry. Anal Biochem 325:77–84. https://doi.org/10.1016/j.ab.2003.10.028
22. Lu W, Clasquin MF, Melamud E et al (2010) Metabolomic analysis via reversed-phase ion--pairing liquid chromatography coupled to a stand alone Orbitrap mass spectrometer. Anal Chem 82:3212–3221
23. Mccloskey D, Gangoiti JA (2015) A pH and solvent optimized reverse-phase ion-paring-LC-MS/MS method that leverages multiple scan-types for targeted absolute quantification of intracellular metabolites. Metabolomics 11:1338–1350. https://doi.org/10.1007/s11306-015-0790-y
24. Jaiswal D, Sengupta A, Sohoni S, Sengupta S (2018) Genome features and biochemical characteristics of a robust , fast growing and naturally transformable Cyanobacterium *Synechococcus elongatus* PCC 11801 isolated

from India. Sci Rep 8:1–13. https://doi.org/10.1038/s41598-018-34872-z

25. Alpert AJ (1990) Hydrophilic-interaction chromatography for the separation of peptides, nucleic acids and other polar compounds. J Chromatogr A 499:177–196. https://doi.org/10.1016/S0021-9673(00)96972-3

26. Navarro-Reig M, Ortiz-Villanueva E, Tauler R, Jaumot J (2017) Modelling of hydrophilic interaction liquid chromatography stationary phases using chemometric approaches. Meta 7:6–9. https://doi.org/10.3390/metabo7040054

27. Kawachi Y, Ikegami T, Takubo H et al (2011) Chromatographic characterization of hydrophilic interaction liquid chromatography stationary phases: Hydrophilicity, charge effects, structural selectivity, and separation efficiency. J Chromatogr A 1218:5903–5919. https://doi.org/10.1016/j.chroma.2011.06.048

28. Creek DJ, Chokkathukalam A, Jankevics A et al (2012) Stable isotope-assisted metabolomics for network-wide metabolic pathway elucidation. Anal Chem 84:8442–8447. https://doi.org/10.1021/ac3018795

29. Trammell SA, Brenner C (2013) Targeted, LCMS-based metabolomics for quantitative measurement of NAD+ metabolites. Comput Struct Biotechnol J 4:e201301012. https://doi.org/10.5936/csbj.201301012

30. Bustamante S, Jayasena T, Richani D et al (2018) Quantifying the cellular NAD+ metabolome using a tandem liquid chromatography mass spectrometry approach. Metabolomics 14:15. https://doi.org/10.1007/s11306-017-1310-z

31. Tautenhahn R, Cho K, Uritboonthai W et al (2012) An accelerated workflow for untargeted metabolomics using the METLIN database. Nat Biotechnol 30:826–828. https://doi.org/10.1038/nbt.2348

32. Jewison T, Knox C, Neveu V et al (2012) YMDB: the yeast Metabolome database. Nucleic Acids Res 40:D815–D820. https://doi.org/10.1093/nar/gkr916

33. Wishart DS, Tzur D, Knox C et al (2007) HMDB: the human metabolome database. Nucleic Acids Res 35:521–526. https://doi.org/10.1093/nar/gkl923

34. Pence HE, Williams A (2010) ChemSpider: an online chemical information resource. J Chem Educ 87:1123–1124. https://doi.org/10.1021/ed100697w

35. Tsugawa H, Cajka T, Kind T et al (2015) MS-DIAL: data-independent MS/MS deconvolution for comprehensive metabolome analysis. Nat Methods 12:523–526. https://doi.org/10.1038/nmeth.3393

36. Li H, Cai Y, Guo Y et al (2016) MetDIA: targeted metabolite extraction of multiplexed MS/MS spectra generated by data-independent acquisition. Anal Chem 88:8757–8764. https://doi.org/10.1021/acs.analchem.6b02122

Chapter 4

Deuterium Tracing to Interrogate Compartment-Specific NAD(P)H Metabolism in Cultured Mammalian Cells

Esther W. Lim, Seth J. Parker, and Christian M. Metallo

Abstract

Oxidation–reduction (redox) reactions are ubiquitous in biology and typically occur in specific subcellular compartments. In cells, the electron transfer between molecules and organelles is commonly facilitated by pyridine nucleotides such as nicotinamide adenine dinucleotide phosphate (NADPH) and nicotinamide adenine dinucleotide (NADH). While often taken for granted, these metabolic reactions are critically important for maintaining redox homeostasis and biochemical potentials across membranes. While ^{13}C tracing and metabolic flux analysis (MFA) have emerged as powerful tools to study intracellular metabolism, this approach is limited when applied to pathways catalyzed in multiple cellular compartments. To address this issue, we and others have applied ^{2}H (deuterium) tracers to observe transfer of labeled hydride anions, which accompanies electron transfer. Furthermore, we have developed a reporter system for indirectly quantifying NADPH enrichment in specific subcellular compartments. Here, we provide a detailed description of ^{2}H tracing applications and the interrogation of mitochondrial versus cytosolic NAD(P)H metabolism in cultured mammalian cells. Specifically, we describe the generation of reporter cell lines that express epitope-tagged R132H-IDH1 or R172K-IDH2 and produce (D)2-hydroxyglutarate in a doxycycline-dependent manner. These tools and methods allow for quantitation of reducing equivalent turnover rates, the directionality of pathways present in multiple compartments, and the estimation of pathway contributions to NADPH pools.

Key words Deuterium, Redox metabolism, Stable isotope tracing, NADH, NADPH, Metabolite extraction, Metabolic flux analysis, Mammalian cell culture, Isotopomer spectral analysis

1 Introduction

Oxidation–reduction (redox) reactions are central to the metabolism of living organisms. In biological systems, most redox reactions utilize cofactors to transfer electrons between metabolites and across membranes. In turn, these reactions play critical roles in supporting reductive biosynthesis, ATP regeneration, signal transduction, and redox homeostasis (i.e., the maintenance of an appropriately reduced environment) [1].

Two important cofactors that facilitate these processes are the pyridine nucleotides nicotinamide adenine dinucleotide (NAD)

Deepak Nagrath (ed.), *Metabolic Flux Analysis in Eukaryotic Cells: Methods and Protocols*, Methods in Molecular Biology, vol. 2088, https://doi.org/10.1007/978-1-0716-0159-4_4,

and nicotinamide adenine dinucleotide phosphate (NADP), which can exist in oxidized (NAD+ or NADP+) or reduced (NADH or NADPH) forms. The ratio of oxidized and reduced forms of these cofactors (NAD+/NADH or NADP+/NADPH) is highly regulated and varies within distinct cellular compartments [2, 3]. In general, the NADH/NAD+ ratio is kept low in the mitochondria (where respiration occurs) and high in the cytoplasm (where glycolysis is catalyzed). The abundance of NAD+ in the mitochondria is consistent with its function as an electron acceptor in catabolic pathways in the mitochondria. In contrast, the NADPH/NADP+ ratio is kept high in the cytosol to facilitate redox buffering by glutathione and maintain a reduced environment beneficial for cell growth [4–8]. Indeed, compartmentalization of cellular reactions is an important feature of eukaryotes and cell metabolism in general.

While ^{13}C tracing and associated MFA studies have greatly improved our understanding of metabolism in higher organisms [9–13], including the de novo synthesis of pyridine nucleotides [14, 15], several recent studies have leveraged the use of ^{2}H tracers to quantify NAD(P)H regeneration and turnover [16–20]. Since electron transfer is accompanied by hydride (H^-) ion transfer in redox reactions, ^{2}H (deuterium) tracers can be used to trace electron flows associated with NAD(P)H-dependent reactions [21, 22]. Furthermore, the segregation of NAD(P)H pools and isozymes within subcellular compartments adds an additional layer of complexity to studying these pathways. Here we describe a protocol that is effective for exploring the metabolic pathways which fuel these compartment-specific redox reactions. Generally, these approaches can be effectively applied to cultured mammalian cells, including cancer cell lines and more functional, differentiated cell types. We first provide an overview of some increasingly used ^{2}H tracers and then describe the generation of reporter cell lines and application of the approach in cancer cells.

1.1 Metabolic Pathways and Deuterium Tracing

1.1.1 Cytosolic NADH Metabolism

[4-^{2}H]glucose effectively traces NADH metabolism through deuterium atom transfer to NAD+ by glyceraldehyde 3-phosphate dehydrogenase (GAPDH) during glycolysis (Fig. 1a). Exchange with water in the aldolase and triose phosphate isomerase reactions, however, decreases the labeling from [4-^{2}H]glucose on glyceraldehyde 3-phosphate (GAP) [16, 19]. Cells maintain glycolysis by regenerating cytosolic NAD+, which is accomplished primarily via lactate dehydrogenase (LDH), malate dehydrogenase (MDH), and glycerol-3-phosphate dehydrogenase (G3PDH). As such, [4-^{2}H] glucose labeling is robustly detected on reduced cytosolic dehydrogenase products such as lactate, glycerol-3-phosphate, and malate (Fig. 1a).

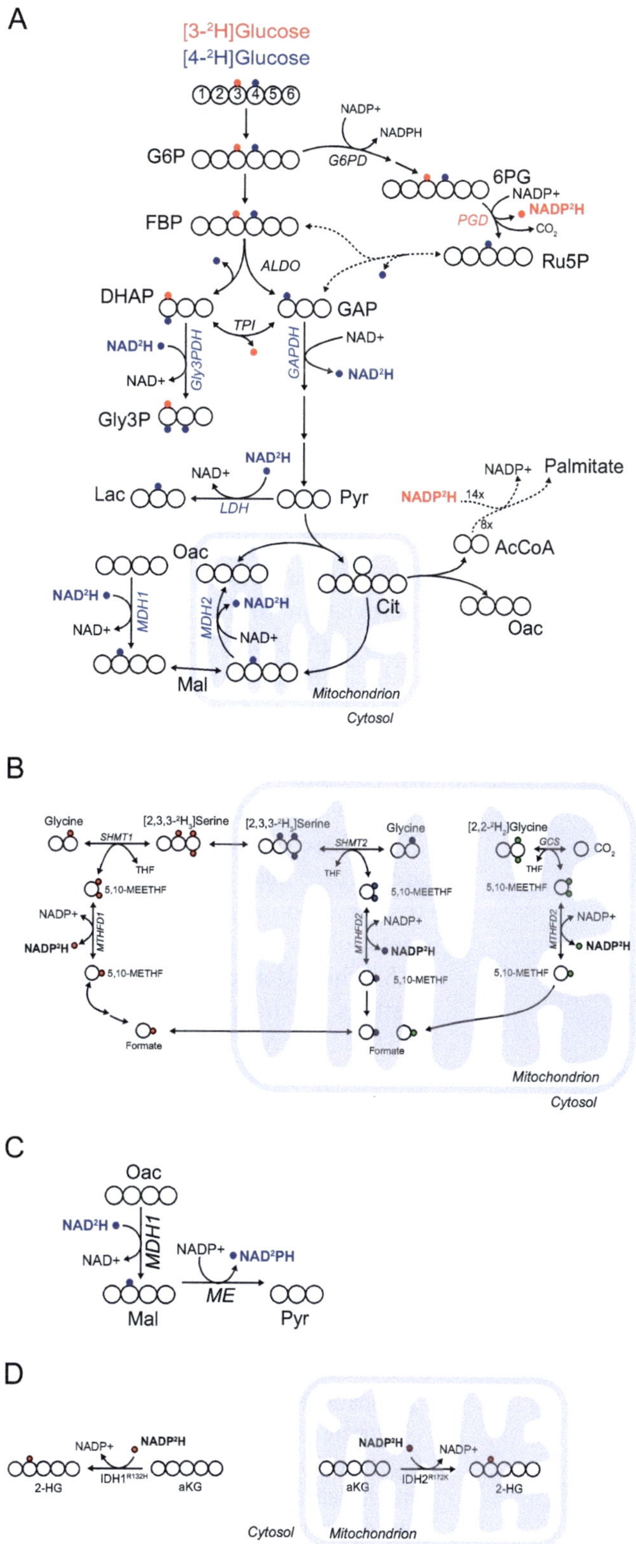

Fig. 1 Schematic for [3-^{2}H]glucose and [4-^{2}H]glucose metabolism (**a**), folate-mediated one-carbon metabolism with [2,3,3-^{2}H]serine and glycine cleavage system with [2,2-^{2}H]glycine (*see* **Note 1**) (**b**), malic enzyme-dependent NADPH

1.1.2 Oxidative Pentose Phosphate Pathway (PPP)-Dependent NADPH Metabolism

NADPH metabolism can be traced with [1-^{2}H]glucose or [3-^{2}H]glucose. Deuterium isotopes are transferred from glucose to NADPH by glucose-6-phosphate dehydrogenase (G6PD) and 6-phosphogluconate dehydrogenase (6PGD) (Fig. 1a). Although both [1-^{2}H]glucose and [3-^{2}H]glucose tracers label NADPH, [3-^{2}H]glucose is preferable for tracing NADPH produced by the oxidative pentose phosphate pathway because the isotope is lost to water in other pathways. In contrast, the deuterium from [1-^{2}H]glucose can label metabolites downstream of glycolysis, including lipogenic acetyl-CoA [16, 17]. Notably, labeling on NADPH is often underestimated due to H-D exchange reactions catalyzed by flavin enzymes [23]. As fatty acid and cholesterol synthesis require NADPH, label is also transferred from [3-^{2}H]glucose to newly synthesized lipids [16, 17].

1.1.3 Folate-Dependent NADPH Metabolism

One carbon metabolism is important for nucleotide synthesis, methylation, and mitochondrial function [24, 25]. Serine contributes significantly to the one carbon pool, and [2,3,3-^{2}H]serine and [3,3-^{2}H]serine will generate ^{2}H-labeled one carbon (1C) folates downstream of serine hydroxymethyltransferases (SHMTs). Oxidation of ^{2}H-labeled 1C-folates by methylenetetrahydrofolate dehydrogenases (MTHFD) can label NAD(P)H metabolism in the mitochondria and cytosol (Fig. 1b) [16, 26, 27]. These serine tracers will also label glycine and formate, which are further metabolized by the glycine cleavage system or recycled to one carbon units, respectively. As folate-mediated one-carbon metabolism reactions occur in both the cytosol and mitochondria [28], these deuterium serine tracers can be coupled to cell reporter systems to characterize the directionality of these reactions (e.g., SHMT1/2, MTHFD1/2) [16, 26].

Fig. 1 (continued) metabolism with [4-^{2}H]glucose (**c**), production of compartment-specific 2-HG and label incorporation from deuterated NADPH (**d**). *G6P* Glucose-6-phosphate, *FBP* Fructose-1,6-bisphosphate, *DHAP* Dihydroxyacetone phosphate, *GAP* Glyceraldehyde-3-phosphate, *Gly3P* Glycerol-3-phosphate, *Lac* Lactate, *Pyr* Pyruvate, *Oac* Oxaloacetate, Mal, Malate, *Cit* Citrate, *AcCoA* Acetyl-CoA, *G6PD* Glucose-6-phosphate dehydrogenase, *PGD* Phosphogluconate dehydrogenase, *ALDO* Fructose-bisphosphate aldolase, *TPI* Triose-phosphate isomerase, *Gly3PDH* Glycerol-3-phosphate dehydrogenase, *GAPDH* Glyceraldehyde-3-phosphate dehydrogenase, *LDH* Lactate dehydrogenase, *MDH* malate dehydrogenase, *SHMT* Serine hydroxymethyltransferase, *MTHFD* Methylenetetrahydrofolate Dehydrogenase, *GCS* Glycine cleavage system, *THF* tetrahydrofolate, *5,10-MEETHF* 5,10-methylenetetrahydrofolate, *5,10-METHF* 5,10-methenyltetrahydrofolate, *ME* Malic enzyme, *2-HG* (D)2-hydroxyglutarate, *aKG* alpha-ketoglutarate, *IDH* isocitrate dehydrogenase

1.1.4 Glycine Cleavage-Dependent NADPH Metabolism

Exogenous glycine or glycine produced from serine via SHMT2 can enter the cleavage system in the mitochondria and then be converted to 5,10-methenyltetrahydrofolate from 5,10-methylenetetrahydrofolate by NAD+-dependent MTHFD2 (Fig. 1b).

1.1.5 Malic Enzyme-Dependent NADPH Metabolism

The cytosolic NADP+-dependent isoform of malic enzyme (ME1) catalyzes the decarboxylation of malate to pyruvate [29, 30]. As such, this isozyme regenerates cytosolic NADPH, which contributes to lipogenesis and maintenance of reduced glutathione pools. While the hydride transfer mediated by malic enzyme can be observed by using [2,2,3,3-^{2}H]dimethyl succinate, [4-^{2}H]glucose is preferred because dimethyl succinate is an atypical metabolite and requires additional enzyme reactions and transport before it can be utilized by malic enzyme in the cytosol [16, 18]. As noted in Fig. 1a, [4-^{2}H]glucose labels cytosolic NADH and subsequently [2-^{2}H] malate (Fig. 1c). In turn, ME1-mediated generation of NADPH can be assessed by quantifying label on cytosolic 2-HG reporters or lipids.

1.1.6 Reductive Carboxylation

IDH1 and IDH2 are reversible, NADP+-dependent enzymes. Under conditions of mitochondrial dysfunction (e.g., hypoxia, complex I deficiency), the reductive reaction or exchange flux becomes significant relative to the oxidative decarboxylation reaction while the specific contributions of IDH1 and IDH2 to reductive carboxylation has remained unclear [31–33]. Since the oxidative pentose phosphate pathway generates ^{2}H-labeled NADPH specifically in the cytosol, [3-^{2}H] glucose allows for interrogation of IDH1-catalyzed reductive carboxylation. While the specific wiring of cytosolic and mitochondrial IDH metabolism is likely to be cell-type specific, we have used this approach to confirm that IDH1 contributes to reductive carboxylation flux [19]. Furthermore, anchorage-independent cultures use this pathway to transfer reducing equivalents into the mitochondrial matrix [34].

1.2 Compartment-Specific Reporter System

Since the time required to separate organelles (several minutes to hours) is much longer than the turnover rates of NAD(P)H (seconds to minutes), it is difficult if not impossible to reliably quantify isotope enrichment in compartment-specific metabolite pools. To address this issue, we developed a reporter system which exploits the neomorphic activity of oncogenic IDH1 and IDH2 enzymes [35, 36]. We generated cell lines that express IDH1-R132H (mtIDH1-C) or IDH2-R172K (mtIDH2-M) in a doxycycline-dependent manner, such that cells produce (D)2-hydroxyglutarate (2-HG) in the cytosol or mitochondria upon induction [16]. When combined with deuterium tracers that label NADPH pools, 2-HG labeling serves as an effective readout of compartment-specific NADPH labeling (Fig. 1d).

The impact of reporter-driven IDH1 or IDH2 expression was measured by quantifying NAD(P)+ and NADPH pools after doxycycline treatment. We observed minimal alterations in these metabolites [16]. Although the ectopic expression of IDH enzymes could also affect TCA metabolism, measurement of 2-HG production flux relative to other α-ketoglutarate dependent reaction fluxes is small, suggesting that the expression of mutant IDH has minimal impact on α-ketoglutarate pools [37].

In the cell lines tested (A549, H1299), the doxycycline-induced production of 2-HG by FLAG-tagged mutant IDH enzymes was verified by quantifying 2-HG after 24 h of doxycycline addition [16]. Importantly, 2-HG levels in the reporter system were far (10- to 100-fold) lower than that observed in tumor cell lines expressing endogenous (heterozygous) IDH1 mutations. Furthermore, doxycycline (at 0.1 μg/mL) and mutant IDH expression did not affect proliferation rates or pool sizes of central carbon metabolites, NAD+, NADH, NADP+, and NADPH in these cells [16]. However, the doxycycline concentration may need to be optimized for specific cell lines and applications to yield low but sufficient 2-HG production.

1.3 Experiment Setup

The experimental conditions, culture format, and extraction time points need to be planned in advance of the experiment. A typical experiment timeline can be found in Fig. 2. The type of deuterium tracer will be dictated by the hypothesis of the study.

For most cells, including reporter lines, culture in 6-well plates should be sufficient for analysis, but this may vary depending on cell type. In general, 0.5–1 million cells provide enough material for detection of relevant metabolites. If necessary, samples can be pooled to obtain adequate cells for analysis. Cell seeding density may also need to be adjusted, as cells should be semiconfluent at the time of extraction. In general, three replicate wells per condition is typical, but the number of replicates should be dictated by the experimental design. Additionally, plan for parallel plates for quantification of biomass via cell count or protein measurement.

The turnover for NADPH and NADH is rapid and this allows for the deuterium labeling from the tracer to reach isotopic steady-state within 30 min of incubation [16]. As such, the incubation duration of cells with the tracer does not typically need to be more than 24 h. Tracing with cells expressing mtIDH1-C and mtIDH2-M, however, requires the addition of doxycycline and expression of reporters for 2-HG production. To ensure that NADPH is labeled to steady state prior to initiating 2-HG production, cells are typically incubated with the deuterium tracer for about 24 h before addition of doxycycline. Performing the metabolite extraction 24–48 h post-doxycycline addition allows for detectable 2-HG to be produced for quantitation of labeling. To measure the contribution of NADPH for fatty acid or cholesterol synthesis, the

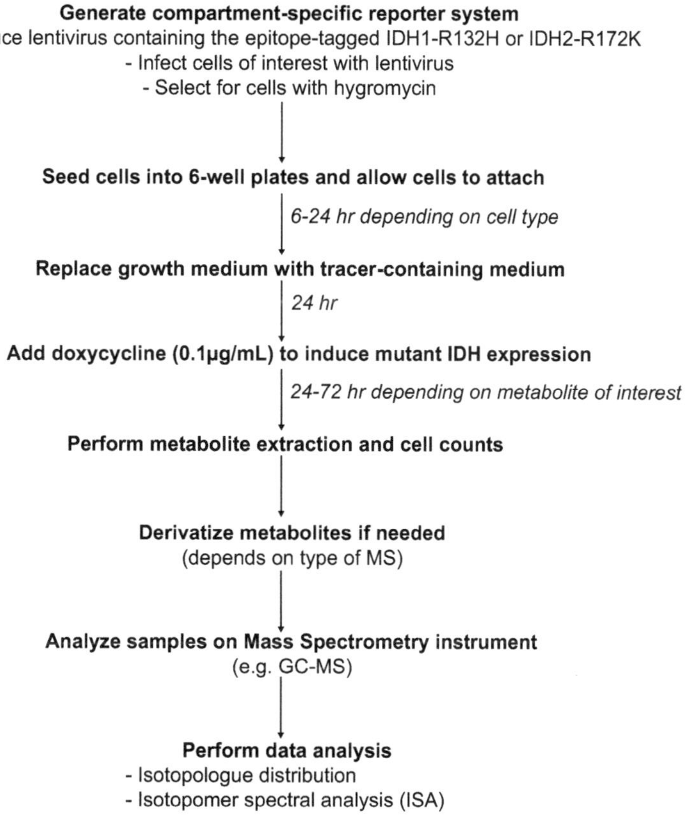

Fig. 2 Experiment outline for tracing studies using inducible expression of epitope-tagged R132H-IDH1 or R172K-IDH2 to study compartment-specific redox metabolism

incubation duration needs to be longer to allow for sufficient fatty acid or cholesterol turnover/synthesis. As is the case for any tracing or flux study, the appropriate length of incubation will depend on cell type and growth (metabolic) rate. For example, for H1299 and A549 cell lines, a 48–72 h incubation with the tracer post-doxycycline addition is sufficient.

2 Materials

2.1 Chemicals

1. Basal medium without deuterium tracer (*see* **Note 2**).
2. Chloroform Chromasolv® for HPLC (Sigma-Aldrich, 366927).
3. Deuterium tracers:
   (a) [3-^{2}H]glucose (Omicron Biochemicals, GLC-034).
   (b) [4-^{2}H]glucose (Omicron Biochemicals, GLC-035).

(c) [2,2-^{2}H]glycine (Cambridge Isotope Laboratories, DLM-1674-5).
(d) [2,3,3-^{2}H]serine (Cambridge Isotope Laboratories, DLM-582-0.5).
(e) [3,3-^{2}H]serine (Cambridge Isotope Laboratories, DLM-161-PK).

4. Doxycycline hydrochloride (Sigma-Aldrich, D3447).
5. Fetal bovine serum or serum of choice for HEK293T cells.
6. Hexane Chromasolv® for HPLC (Sigma-Aldrich, 34859).
7. Hygromycin B (Gibco, 10687010).
8. Internal standard (e.g., norvaline (Sigma-Aldrich, N7502), [U-$^2H_{31}$]palmitate (D31) (Cambridge Isotope Laboratories, DLM-215-1)).
9. Lentiviral packaging and envelope expressing plasmids: pMDLg/pRRE (Addgene plasmid #12251), pRSV-Rev (Addgene #12253), pMD2.G (Addgene, #12259).
10. Methanol Chromasolv® for HPLC (Sigma–Aldrich, 34860), equilibrated to −80 °C.
11. Methoxyamine hydrochloride (MOX) (Supelco, 33045-U).
12. Milli-Q-purified water or equivalently pure water.
13. *N*-Tertbutyldimethylsilyl-*N*-methyltrifluoroacetamide (MTBSTFA) with 1% *tert*-butyldimethylchlorosilane (Regis Technologies, 1-270143-200).
14. Opti-MEM (Gibco) or medium of choice for HEK293T cells.
15. Phosphate-buffered saline (PBS) (e.g., Corning, 21-031-CM).
16. Polybrene (Milipore, TR-1003) or similar transfection reagent (*optional*).
17. pSLIK-IDH1-R132H-FLAG (Addgene, plasmid #66803).
18. pSLIK-IDH2-R172K-FLAG (Addgene, plasmid #66807).
19. Pyridine (Sigma-Aldrich, 270407).
20. Medium buffering agent (e.g., sodium bicarbonate ($NaHCO_3$) and/or HEPES).
21. Saline solution (NaCl 0.9% (w/v)).
22. Sulfuric acid (H_2SO_4) (Sigma-Aldrich, 339741).
23. Tetracycline-free (or tetracycline-negative) FBS (*see* **Note 3**).

2.2 Consumables and Equipment

1. 0.22 μm polyethersulfone (PES) filter.
2. 0.45 μm polyethersulfone (PES) filter.
3. 1.5 mL microcentrifuge tubes (e.g., Eppendorf interlocking caps (Cat. No. 22363204)).

4. Amicon Ultra-15 Centrifugal Filter Unit (Millipore Sigma, UFC910024) (*optional*).
5. Analytical glass sample vials (e.g., Agilent 5182-7616) with glass inserts (Agilent 5183-2088) for fatty acid methyl esters (FAMEs).
6. Analytical platform (e.g., gas chromatograph coupled to mass spectrometry).
7. Analytical sample vials with inserts (e.g., Thermo Scientific, C4000-11) for media samples and metabolites.
8. Freeze dryer or rotary vacuum evaporator (e.g., CentriVap Concentrator (Labconco)), precooled to 4 °C.
9. Heating block for microcentrifuge tubes.
10. Humidified incubator for cell culture.
11. Microcentrifuge tube shaker, precooled to 4 °C.
12. Nitrogen/air flow evaporator for microcentrifuge tubes.
13. Refrigerated table centrifuge for microcentrifuge tubes, precooled to 4 °C.
14. Sterile filter unit to sterilize 500 mL medium (e.g., Nalgene Nunc, 156-4020).
15. Syringe filter (e.g., Pall Laboratories, 4612) with sterile syringe (e.g., 20 mL, AIR-TITE, 5200-X00V0) for small volumes of medium.
16. Tissue culture plates (e.g., 6-well plate).

3 Methods

3.1 Generation of Stably Expressing Inducible Forms of FLAG-Tagged Mutant IDH Cell Lines

1. Generate lentiviruses by transfecting HEK293T cells with the pSLIK-hygro-IDH1-R132H or pSLIK-hygro-IDH2-R172K plasmids along with the lentiviral packaging plasmids pMDLg/pRRE and pRSV-Rev and the envelope plasmid pMD2.G (*see* **Note 4**).
2. Collect supernatant containing lentiviral particles 48 h after transfection. Filter supernatant through 0.45 μm filter and add polybrene and/or concentrate if desired (*see* **Note 5**). Freeze virus at −80 °C for storage.
3. Infect subconfluent cell of choice with harvested virus at optimized multiplicity of infection (MOI) (*see* **Note 6**).
4. Allow infected cells to recover for 24 h before placing cells under selection with 350 μg/mL hygromycin for 10 days.
5. Freeze or bank cells if desired.

3.2 Media Preparation

For tracing experiments, use basal medium that does not contain the deuterium component such as basal media that does not contain glucose, glutamine, and/or amino acids (*see* **Note 2**). The tracer medium can be formulated with serum (*see* **Note 3**) and/or appropriate antibiotics, if applicable.

1. Preparation of basal medium (if necessary).
   (a) Dissolve medium powder in Milli-Q-purified water according to manufacturer instructions.
   (b) Add appropriate amount of buffering component (e.g., sodium bicarbonate, HEPES).
   (c) Adjust pH to optimal value for cell of interest (typically 7.3–7.4).
   (d) Filter-sterilize medium through 0.22 μm filter and store at 4 °C until use.
2. Preparation of deuterium tracer-containing medium.
   (a) Calculate volume of tracer medium required for experiment. Include a contingency factor since some medium may be lost during transfer and filtration.
   (b) Weigh out required amount of tracer for experiment and add to basal medium (*see* **Note 7**).
   (c) Add appropriate amount of serum, antibiotics and/or other medium supplements (*see* **Note 8**).
   (d) Adjust pH to optimal value for cell of interest (typically 7.3–7.4), if necessary.
   (e) Filter-sterilize medium through 0.22 μm filter and store at 4 °C until use.
3. Preparation of regular growth medium for quantification of biomass via cell count or protein measurement.
   (a) Repeat steps in Subheading 3.2, **step 2** but use nondeuterium tracer components.

3.3 Cell Culture Setup

1. Determine experimental conditions and plate layout. Plan for two sets of replicate wells: one set for metabolite extraction and another set for biomass quantification (*see* **Note 9**).
2. Seed cells in 6-well plates with appropriate volume of regular growth medium (*see* **Notes 10** and **11**).
3. Incubate cells in humidified cell culture incubator and allow cells to attach (typically 6–24 h depending on cell type) and reach target cell density or confluence.
4. When cells are ready, remove the regular growth medium by aspiration and wash cells once with PBS.
5. Add prewarmed tracer or growth medium.

6. Return cells to humidified cell culture incubator and incubate for 24 h (*see* **Note 12**).
7. Add doxycycline to medium (final concentration of 0.1 μg/mL) and return cells to incubator until metabolite extraction (*see* **Note 13**).

3.4 Quenching, Cell Lysis and Metabolite Extraction

This protocol employs a modified Bligh and Dyer method using methanol–water–chloroform for a 6-well plate. Volumes can be adjusted if using different culture formats (e.g., reduce volumes by half for 12-well plate). Rapid quenching is important to halt metabolism and preserve the native metabolites for analysis. As such, extract only one plate at a time to minimize waiting times between samples. A blank extraction from wells without cells can be performed in parallel to account for potential non-cell derived metabolites if using nonstandard tissue culture plates. Perform the following steps on ice. The following steps were adapted from Cordes and Metallo [38].

1. Aspirate medium from each well.
2. Add 1 mL saline solution to each well to rinse well (*see* **Note 14**). Aspirate solution from well.
3. Add 500 μL cold MeOH to each well and swirl plate around to ensure coverage of all cells.
4. Add 200 μL Milli-Q-purified water containing internal standard norvaline (e.g., 0.005 μg/μL) to each well. Swirl plate to mix.
5. Scrape cells off from each well.
6. Transfer cell extract from wells into precooled (−20 °C) 1.5 mL microcentrifuge tubes.
7. Add 500 μL chloroform with internal standard [U-$^{2}H_{31}$]palmitate (D31) solution (e.g., 0.002 μg/μL) to each vial (*see* **Note 15**).
8. Cap each tube and invert three times.
9. Vortex tubes for 5–10 min.
10. Centrifuge tubes at 21,000 × *g* at 4 °C for 10 min. Upon completion of centrifugation, 2 distinct phases will form:
    (a) Upper MeOH–H_2O layer containing polar metabolites ("aqueous phase").
    (b) Lower $CHCl_3$ layer containing nonpolar metabolites (e.g., fatty acids, triglycerides, cholesterol) ("organic phase").
    (c) The middle contains a thin layer of protein which can be utilized if desired (not analyzed here).

11. Transfer the upper aqueous layer into a new tube. Do not disturb the protein interphase. Evaporate the sample in a freeze dryer or rotary vacuum evaporator at 4 °C (*see* **Note 16**).
12. Transfer the bottom layer into a new tube but avoid collecting proteins or upper layer into the same pipette (*see* **Note 17**). Evaporate the sample under air flow at room temperature in a fume hood (*see* **Note 18**).
13. Cap and store dry samples at −80 °C until ready for further processing (*see* **Note 19**).
14. Perform biomass quantification (e.g., cell counts or protein measurement). This step can also be performed just prior to extraction.

3.5 Metabolite Derivatization for Gas Chromatography–Mass Spectrometry (GC-MS) Measurement

Samples should be derivatized prior to measurement on the GC-MS. A detailed protocol for derivatizing with *N*-tertbutyldimethylsilyl-*N*-methyltrifluoroacetamide (MTBSTFA) and/or methoxyamine hydrochloride (MOX) for polar metabolites can be found in Cordes and Metallo [38] (*see* **Note 20**). Similarly, a protocol describing the transesterification of fatty acids with sulfuric acid–methanol–hexane solution to prepare fatty acid methyl esters is described in Cordes and Metallo [38].

3.6 Metabolite Measurement on Mass Spectrometry (MS)-Based Platform

A variety of MS-based platforms exists to measure metabolite abundance and labeling. Refer to Cordes and Metallo [38] for details on GC-MS setup and setting parameters.

3.7 Metabolite Identification and Analysis

A variety of commercially available and academic software algorithms can be used to identify and quantify metabolite abundances and isotopolog distributions [39, 40].

1. Metabolite identification.

   Metabolites are identified by their characteristic fragmentation pattern and retention time (*see* **Note 21**). Coelution of closely related metabolites such as enantiomers is common on most MS-based platforms such as the GC-MS and may not be able to be distinguished readily (*see* **Note 22**). Libraries with the mass spectral and retention time are widely available (*see* **Note 23**).
2. Isotopolog distribution.

   Isotopolog (or mass isotopomer) distributions depict the fractional abundance of each isotopolog normalized to the sum of all possible isotopologs (total pool). The isotopolog distributions should be corrected for natural isotope abundances. Figure 3 shows some examples of the isotopolog distribution of select metabolites when traced with the specified deuterium tracers.

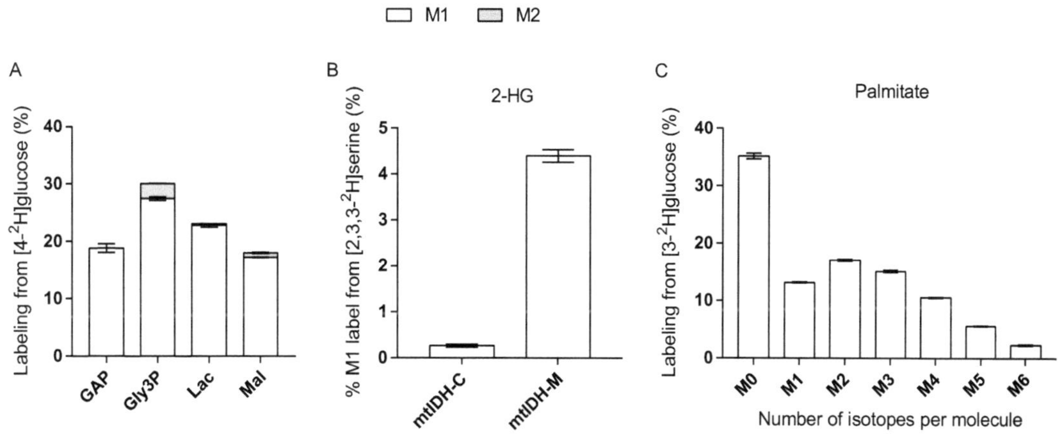

Fig. 3 Examples of isotopolog distributions for glyceraldehyde-3-phosphate (GAP), glycerol-3-phosphate (Gly3P), lactate (Lac), and malate (Mal) from [4-^{2}H]glucose (**a**), 2-HG from [2,3,3-^{2}H]serine (**b**), and palmitate from [3-^{2}H]glucose (**c**). Refer to Fig. 1 for atom transition maps. Data are plotted as mean ± standard deviation of three biological replicates.

3. Isotopomer spectral analysis (ISA).

   ISA is a method for evaluating biosynthesis using stable isotopes. The goal of ISA is to estimate the fractional contribution of the labeled metabolite to the precursor pool (known as D) and the fraction of newly synthesized product (known as g (t)). For example, when tracing with [3-^{2}H]glucose, D represents the fractional contribution of the deuterium to cytosolic (lipogenic) NADPH and g(t) represents the newly synthesized fatty acid (e.g., palmitate) (Fig. 4). We used ISA as well as TCA metabolite and 2HG reporter isotope enrichment to validate that kinetic isotope effects do not impact results when tracing through NADH (i.e., [4-^{2}H]glucose) or NADPH (i.e., [3-^{2}H] glucose), respectively (*see* **Note 24**) [16].

4 Notes

1. MTHFD2 and MTHFD2L have been shown to use either NAD+ or NADP+ under different physiological conditions [41]. When the ^{2}H serine tracer is combined with the reporter cell line, it is also possible to get some label from deuterium tracers that label the NADH pool because nicotinamide nucleotide transhydrogenase (NNT) catalyzes the transfer of deuterium from NADH to NADPH which may label 2-HG in the mitochondria.
2. Use your typical basal medium without the deuterium tracer. For example, if you usually culture cells in DMEM and will trace with [3-^{2}H]glucose, use glucose-free DMEM. Similarly, if

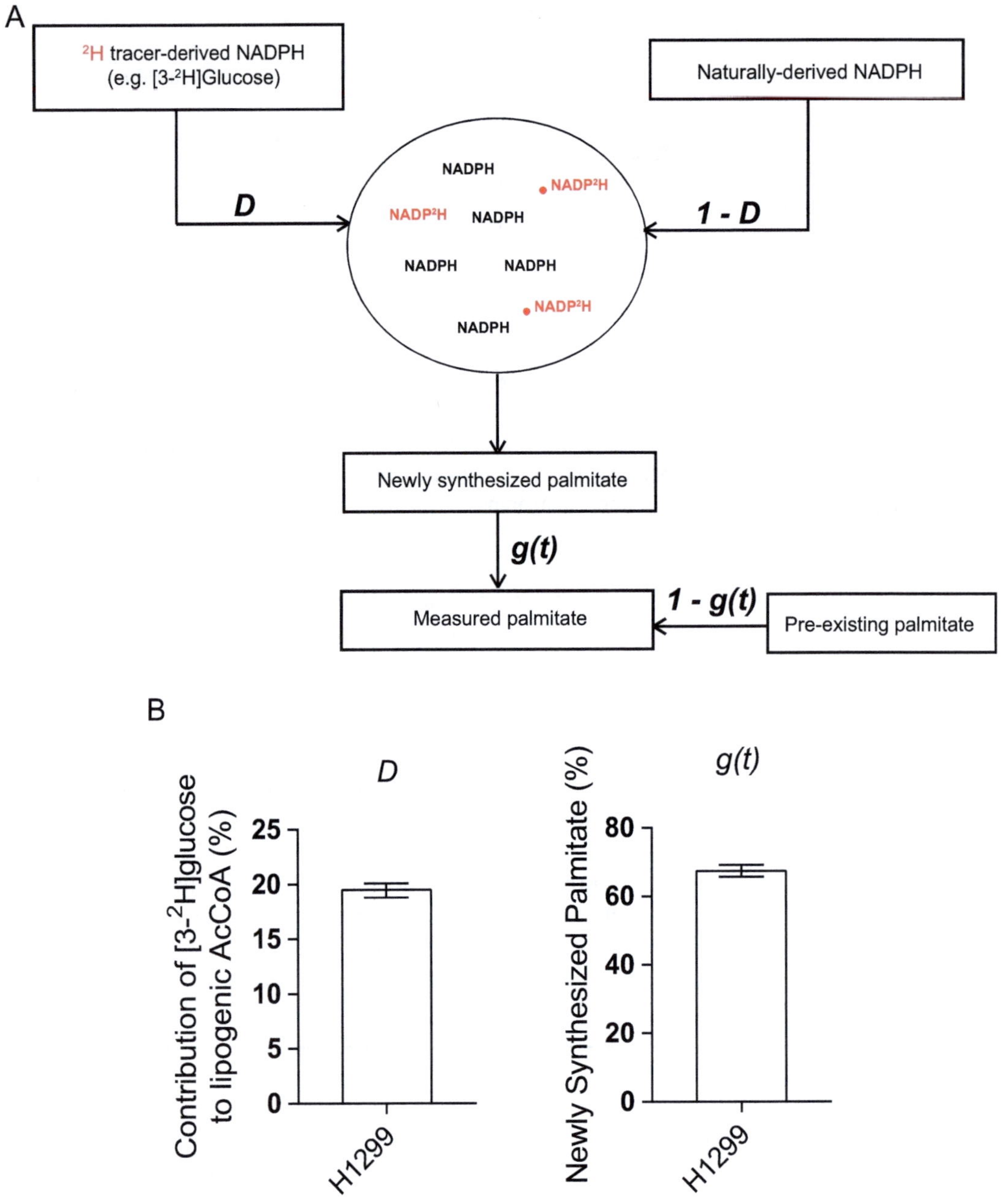

Fig. 4 Overview of isotopomer spectral analysis (ISA). ISA can be applied to estimate the relative enrichment of lipogenic (cytosolic) NADPH from the tracer (e.g. [3-^{2}H]glucose) (**a**). The measured pool of fatty acid (e.g., palmitate) is composed of both existing and newly synthesized fatty acids which may have ^{2}H incorporated. The fractional contribution of the labeled metabolite to the precursor pool is represented by parameter D and the fraction of newly synthesized product is represented by g(t). These parameters are estimated for a given tracer and isotopolog distribution, and the 95% confidence interval is determined via parameter continuation/ sensitivity analysis. Example of ISA applied to H1299 cells (**b**). ^{2}H from [3-^{2}H]glucose accounted for 19% of lipogenic AcCoA and 66% of the palmitate was newly synthesized. Data are plotted as mean $\pm$ 95% confidence interval. *AcCoA* acetyl-CoA

tracing with [3,3-^{2}H]serine, use amino acid- or serine-free DMEM. Performing the tracing experiment without removing the component will result in dilution of isotopic labeling of metabolite of interest.

3. Tetracycline-free (or tetracycline-negative) FBS should be used in reporter cell line experiments to limit production of "background" 2-HG prior to label incorporation. For nonreporter tracing experiments, dialyzed serum (e.g., dialyzed FBS) is preferred because regular serum contains various small molecules such as hormones, cytokines, and amino acids which may result in dilution of isotopic labeling of metabolites. As the dialysis process removes vitamins, cofactors [42], and signaling lipids [43] that may be important for some cells, it is important to validate the growth phenotype of these cells in dialyzed serum prior to the tracing experiment.
4. Prior to working with lentiviral vectors, ensure all safety specifications provided by your institution or governing body are met. Protocols for lentivirus production are widely available [13, 44, 45]. Briefly, we plate 2.5–4.5 million HEK293T cells with standard DMEM (high glucose) + 10% FBS in a 10 cm dish. After 24 h, the transfection cocktail containing Lipofectamine 3000 and plasmids (listed in Subheading 3.1, **step 1**) is prepared in Opti-MEM and added in a drop-wise manner to the cells.
5. The protocol for harvesting and concentrating the lentivirus you use will be dictated by the cell line as some cells are sensitive to the components that are typically added to the harvested virus solution (e.g., polybrene). In this case, skip the polybrene addition or substitute with other reagents (e.g., protamine sulfate). Concentrating the virus stock can be advantageous to minimize the volume of virus stock that needs to be added to the cells. Briefly, filter the supernatant containing the lentiviral particles through a 0.45 μm filter, transfer to Amicon Ultra-15 Centrifugal filters (or other similar filters), centrifuge at 3000 × *g* for 30 min at room temperature, aliquot the concentrated virus to sterile microcentrifuge tubes or vials, freeze virus stocks in −80 °C freezer.
6. The MOI is dependent on the type of cells. It is common to titrate the virus stock before use. The volume of virus stock added depends on the MOI and virus titer. If there is significant cell death postinfection, you may want to decrease the amount of virus stock added to the cells.
7. In most cases, target the same concentration of the component in the regular medium. For example, if tracing with DMEM and the media usually contains 0.4 mM serine, add 0.4 mM [2,3,3-^{2}H]serine. Since glucose is usually in excess in

commercially available media, you can reduce the amount of deuterium glucose tracer to 10 mM or 15 mM instead of the typical 25 mM if desired. The exact concentration of these components will depend on the cell type and experiment design. Also, note that the molecular weight of deuterium tracer is higher than the nondeuterium component.

8. Add the necessary components to the medium for cells to grow normally. For example, if cells are routinely grown in 10% serum, use the same concentration in the tracer-containing medium (*see* **Note 2**). Similarly, it is acceptable to add antibiotics such as penicillin–streptomycin or gentamicin to the tracer-containing medium. If working with cells that require cofactors or additional supplements, those components should be added to the tracer-containing medium as long as they do not impact the isotopic labeling of metabolites. For example, if tracing with [2,3,3-^{2}H]serine and your growth medium is routinely supplemented with 1× nonessential amino acid solution which contains serine, you may want to supplement the tracing medium with a custom nonessential amino acid stock that does not contain serine.

9. If running the experiment for the first time, you may want to run control cultures in parallel that lack any labeled substrates to gauge background (which is typically minimal). Overloaded metabolites can generate "false" signal upon integration of MS data, which can be significant in low labeling experiments. It is recommended to monitor MS signal or ideally run extracted sample with unlabeled substrates.

10. 6-Well plates are typically used in the reporter system to ensure adequate signal is obtained for all metabolites. In our experience, more material is required to reliably quantify labeling on intermediates such as glyceraldehyde phosphate (GAP) and glycerol-3-phosphate (Gly3P).

11. It is important to plate equal amounts of cells per well to minimize variability across replicates and experimental conditions. It is not advised to seed cells for varying experimental conditions differently (i.e., seeding double the amount of cells for a condition that is known to slow growth). If needed, additional wells may be plated and pooled at time of extraction. The actual cell number will be dictated by the cell type and duration of tracing. Performing the experiment in a 6-well plate is preferred to obtain sufficient material for metabolite detection on the GC-MS. In some cases, a 12-well plate experiment may also be acceptable.

12. This incubation step allows for ^{2}H label from the deuterium tracer to label NAD(P)H prior to doxycycline addition.

Doxycycline-induced cells will use the labeled NADPH which results in generation of labeled 2-HG.

13. Doxycycline is added to induce production of 2-HG. Performing the metabolite extraction 24–48 h post-doxycycline addition allows for detectable 2-HG to be produced for quantitation of labeling. The doxycycline concentration may need to be optimized for specific cell lines and applications.
14. Saline solution instead of PBS solution is used to prevent the contamination of phosphate during GC-MS measurement. The large phosphate amount in PBS can interfere with the metabolite derivatization and/or MS analysis.
15. Do not add chloroform directly into the culture dishes as the chloroform will dissolve the plastic culture plates. Using the chloroform containing the internal standard [U-$^{2}H_{31}$]palmitate (D31) is important when quantifying fatty acids. If the experiment analysis only requires polar metabolites or isotopic labeling of fatty acids, you may use chloroform without the internal standard.
16. If your lab does not have a freeze dryer or rotary vacuum evaporator, you may dry the upper aqueous phase under air or nitrogen flow in a fumehood. Be careful to start with low airflow to prevent loss of material since high airflow can force the sample out of the tube.
17. To avoid collecting proteins or upper layer into the same pipette, you can gently aspirate bubbles as you go through the interphase to avoid collecting the interphase or upper aqueous layer.
18. It is important to dry the sample completely because water can hinder the derivatization reaction. If sample tubes were kept in the cold (e.g., fridge or freezer) after the initial drying step, we recommended drying them again prior to derivatizing the samples.
19. Nondried aqueous phase extracts can be stored at -20 °C or preferably -80 °C overnight, though dry storage is preferable. Nondried organic phase extracts cannot be stored for extended periods due to the potential for extraction from the plastic.
20. While there are other derivatizing reagents and/or methods, we have not validated the ^{2}H label detection on metabolites with other derivatizing reagents.
21. ^{2}H labeling causes (sometimes significant) retention shifts in compounds, such that ^{2}H-labeled species elute earlier than unlabeled compounds. As an example, [U-$^{2}H_{31}$] palmitate elutes much earlier than unlabeled palmitate. For compounds labeled metabolically with NADP^{2}H described here, retention

time shifts are not significant enough to require changes to MS data integration and analysis.

22. (L)2-HG and (D)2-HG elute together on typical GC-MS and LC-MS/MS platforms. (L)2-HG is produced endogenously by MDH and LDH, so it is labeled by NAD^{2}H rather than NADP^{2}H [19]. As such, endogenously produced (L)2-HG can dilute signal in the reporter system and/or confound results. Baseline and doxycycline-induced 2-HG levels should be compared to gauge the impact from (L)2-HG dilution in specific cell lines/applications, as some cells produce appreciable (L)2-HG that could impact results.
23. Libraries with the mass spectral and retention time can be found online. For MTBSTFA derivatives measured on the GC-MS, refer to Table 1 in Cordes and Metallo [38] and Supplementary Table 1 in [16].
24. Given the ~2-fold increase in molecular weight compared to H, reactions involving ^{2}H can exhibit lower rate constants or "kinetic isotope effects." While this is true for in vitro enzyme assays, such kinetic effects will only be relevant when the reaction in question is rate limiting. In the context of cellular physiology (e.g. the numerous reactions involved in NAD(P)H metabolism or lipogenesis) this issue is more complex. To determine whether kinetic isotope effects impact downstream isotopic labeling from [3-^{2}H]glucose or [4-^{2}H] glucose we titrated enrichment and compared substrate enrichment to downstream labeling. We observed a linear relationship, indicating that ^{2}H-labeled substrates are not "competed away" from reactions by unlabeled species [16].

Acknowledgments

We thank members of the Metallo Lab for their feedback and discussion. C.M.M. is funded by NIH (R01 CA188652) and NSF CAREER (#145442) grants. E.W.L. is supported by a training grant from the NIH (T32 EB009380).

References

1. Pollak N, Dölle C, Ziegler M (2007) The power to reduce: pyridine nucleotides--small molecules with a multitude of functions. Biochem J 402:205–218. https://doi.org/10.1042/BJ20061638
2. Williamson DH, Lund P, Krebs HA (1967) The redox state of free nicotinamide-adenine dinucleotide in the cytoplasm and mitochondria of rat liver. Biochem J 103:514–527
3. Veech RL, Guynn R, Veloso D (1972) The time-course of the effects of ethanol on the redox and phosphorylation states of rat liver. Biochem J 127:387–397. https://doi.org/10.1042/BJ1270387
4. Lunt SY, Vander Heiden MG (2011) Aerobic glycolysis: meeting the metabolic requirements of cell proliferation. Annu Rev Cell Dev Biol

27:441–464. https://doi.org/10.1146/annurev-cellbio-092910-154237

5. Diehn M, Cho RW, Lobo NA, Kalisky T, Dorie MJ, Kulp AN, Qian D, Lam JS, Ailles LE, Wong M, Joshua B, Kaplan MJ, Wapnir I, Dirbas FM, Somlo G, Garberoglio C, Paz B, Shen J, Lau SK, Quake SR, Brown JM, Weissman IL, Clarke MF (2009) Association of reactive oxygen species levels and radioresistance in cancer stem cells. Nature 458:780–783. https://doi.org/10.1038/nature07733
6. Schafer ZT, Grassian AR, Song L, Jiang Z, Gerhart-Hines Z, Irie HY, Gao S, Puigserver P, Brugge JS (2009) Antioxidant and oncogene rescue of metabolic defects caused by loss of matrix attachment. Nature 461:109–113. https://doi.org/10.1038/nature08268
7. Jeon S-M, Chandel NS, Hay N (2012) AMPK regulates NADPH homeostasis to promote tumour cell survival during energy stress. Nature 485:661–665. https://doi.org/10.1038/nature11066
8. Jiang P, Du W, Mancuso A, Wellen KE, Yang X (2013) Reciprocal regulation of p53 and malic enzymes modulates metabolism and senescence. Nature 493:689–693. https://doi.org/10.1038/nature11776
9. Young JD (2013) Metabolic flux rewiring in mammalian cell cultures. Curr Opin Biotechnol 24:1108–1115. https://doi.org/10.1016/J.COPBIO.2013.04.016
10. Dong W, Keibler MA, Stephanopoulos G (2017) Review of metabolic pathways activated in cancer cells as determined through isotopic labeling and network analysis. Metab Eng 43:113–124. https://doi.org/10.1016/J.YMBEN.2017.02.002
11. Badur MG, Metallo CM (2018) Reverse engineering the cancer metabolic network using flux analysis to understand drivers of human disease. Metab Eng 45:95–108. https://doi.org/10.1016/J.YMBEN.2017.11.013
12. Metallo CM, Walther JL, Stephanopoulos G (2009) Evaluation of ^{13}C isotopic tracers for metabolic flux analysis in mammalian cells. J Biotechnol 144:167–174. https://doi.org/10.1016/J.JBIOTEC.2009.07.010
13. Zhao D, Badur MG, Luebeck J, Magaña JH, Birmingham A, Sasik R, Ahn CS, Ideker T, Metallo CM, Mali P (2018) Combinatorial CRISPR-Cas9 metabolic screens reveal critical redox control points dependent on the KEAP1-NRF2 regulatory Axis. Mol Cell 69:699–708.e7. https://doi.org/10.1016/j.molcel.2018.01.017
14. Maddocks ODK, Labuschagne CF, Adams PD, Vousden KH (2016) Serine metabolism supports the methionine cycle and DNA/RNA methylation through De novo ATP synthesis in cancer cells. Mol Cell 61:210–221. https://doi.org/10.1016/J.MOLCEL.2015.12.014
15. Ben-Sahra I, Hoxhaj G, Ricoult SJH, Asara JM, Manning BD (2016) mTORC1 induces purine synthesis through control of the mitochondrial tetrahydrofolate cycle. Science 351:728–733. https://doi.org/10.1126/science.aad0489
16. Lewis CA, Parker SJ, Fiske BP, McCloskey D, Gui DY, Green CR, Vokes NI, Feist AM, Vander Heiden MG, Metallo CM (2014) Tracing compartmentalized NADPH metabolism in the cytosol and mitochondria of mammalian cells. Mol Cell 55:253–263. https://doi.org/10.1016/j.molcel.2014.05.008
17. Fan J, Ye J, Kamphorst JJ, Shlomi T, Thompson CB, Rabinowitz JD (2014) Quantitative flux analysis reveals folate-dependent NADPH production. Nature 510:298–302. https://doi.org/10.1038/nature13236
18. Liu L, Shah S, Fan J, Park JO, Wellen KE, Rabinowitz JD (2016) Malic enzyme tracers reveal hypoxia-induced switch in adipocyte NADPH pathway usage. Nat Chem Biol 12:345–352. https://doi.org/10.1038/nchembio.2047
19. Badur MG, Muthusamy T, Parker SJ, Ma S, McBrayer SK, Cordes T, Magana JH, Guan K-L, Metallo CM (2018) Oncogenic R132 IDH1 mutations limit NADPH for De Novo lipogenesis through (D)2-hydroxyglutarate production in fibrosarcoma cells. Cell Rep 25:1018–1026. https://doi.org/10.1016/J.CELREP.2018.09.074
20. Chen L, Zhang Z, Hoshino A, Zheng HD, Morley M, Arany Z, Rabinowitz JD (2019) NADPH production by the oxidative pentose-phosphate pathway supports folate metabolism. Nat Metab 1:404–415. https://doi.org/10.1038/s42255-019-0043-x
21. Katz J, Rognstad R, Kemp RG (1965) Isotope discrimination effects in the metabolism of tritiated glucose. J Biol Chem 240: PC1484–PC1486
22. Rendina AR, Hermes JD, Cleland WW (1984) Use of multiple isotope effects to study the mechanism of 6-phosphogluconate dehydrogenase. Biochemistry 23(25):6257–6262
23. Zhang Z, Chen L, Liu L, Su X, Rabinowitz JD (2017) Chemical basis for deuterium labeling of fat and NADPH. J Am Chem Soc 139:14368–14371. https://doi.org/10.1021/jacs.7b08012

24. Ducker GS, Rabinowitz JD (2017) One-carbon metabolism in health and disease. Cell Metab 25(1):27–42. https://doi.org/10.1016/j.cmet.2016.08.009
25. Newman AC, Maddocks ODK (2017) One-carbon metabolism in cancer. Br J Cancer 116:1499. https://doi.org/10.1038/BJC.2017.118
26. Ducker GS, Chen L, Morscher RJ, Ghergurovich JM, Esposito M, Teng X, Kang Y, Rabinowitz JD (2016) Reversal of cytosolic one-carbon flux compensates for loss of the mitochondrial Folate pathway. Cell Metab 23:1140–1153. https://doi.org/10.1016/J.CMET.2016.04.016
27. Gao X, Lee K, Reid MA, Sanderson SM, Qiu C, Li S, Liu J, Locasale JW (2018) Serine availability influences mitochondrial dynamics and function through lipid metabolism. Cell Rep 22:3507–3520. https://doi.org/10.1016/J.CELREP.2018.03.017
28. Tibbetts AS, Appling DR (2010) Compartmentalization of mammalian Folate-mediated one-carbon metabolism. Annu Rev Nutr 30:57–81. https://doi.org/10.1146/annurev.nutr.012809.104810
29. Loeber G, Dworkin MB, Infante A, Ahorn H (1994) Characterization of cytosolic malic enzyme in human tumor cells. FEBS Lett 344:181–186. https://doi.org/10.1016/0014-5793(94)00386-6
30. Pongratz RL, Kibbey RG, Shulman GI, Cline GW (2007) Cytosolic and mitochondrial malic enzyme isoforms differentially control insulin secretion. J Biol Chem 282:200–207. https://doi.org/10.1074/jbc.M602954200
31. Wise DR, Ward PS, Shay JES, Cross JR, Gruber JJ, Sachdeva UM, Platt JM, DeMatteo RG, Simon MC, Thompson CB (2011) Hypoxia promotes isocitrate dehydrogenase-dependent carboxylation of α-ketoglutarate to citrate to support cell growth and viability. Proc Natl Acad Sci U S A 108:19611–19616. https://doi.org/10.1073/pnas.1117773108
32. Metallo CM, Gameiro PA, Bell EL, Mattaini KR, Yang J, Hiller K, Jewell CM, Johnson ZR, Irvine DJ, Guarente L, Kelleher JK, Vander Heiden MG, Iliopoulos O, Stephanopoulos G (2012) Reductive glutamine metabolism by IDH1 mediates lipogenesis under hypoxia. Nature 481:380–384. https://doi.org/10.1038/nature10602
33. Mullen AR, Wheaton WW, Jin ES, Chen P-H, Sullivan LB, Cheng T, Yang Y, Linehan WM, Chandel NS, DeBerardinis RJ (2012) Reductive carboxylation supports growth in tumour cells with defective mitochondria. Nature 481:385–388. https://doi.org/10.1038/nature10642
34. Jiang L, Shestov AA, Swain P, Yang C, Parker SJ, Wang QA, Terada LS, Adams ND, McCabe MT, Pietrak B, Schmidt S, Metallo CM, Dranka BP, Schwartz B, DeBerardinis RJ (2016) Reductive carboxylation supports redox homeostasis during anchorage-independent growth. Nature 532:255–258. https://doi.org/10.1038/nature17393
35. Dang L, White DW, Gross S, Bennett BD, Bittinger MA, Driggers EM, Fantin VR, Jang HG, Jin S, Keenan MC, Marks KM, Prins RM, Ward PS, Yen KE, Liau LM, Rabinowitz JD, Cantley LC, Thompson CB, Vander HMG, Su SM (2009) Cancer-associated IDH1 mutations produce 2-hydroxyglutarate. Nature 462:739. https://doi.org/10.1038/NATURE08617
36. Ward PS, Patel J, Wise DR, Abdel-Wahab O, Bennett BD, Coller HA, Cross JR, Fantin VR, Hedvat CV, Perl AE, Rabinowitz JD, Carroll M, Su SM, Sharp KA, Levine RL, Thompson CB (2010) The common feature of leukemia-associated IDH1 and IDH2 mutations is a Neomorphic enzyme activity converting α-ketoglutarate to 2-hydroxyglutarate. Cancer Cell 17:225–234. https://doi.org/10.1016/J.CCR.2010.01.020
37. Grassian AR, Parker SJ, Davidson SM, Divakaruni AS, Green CR, Zhang X, Slocum KL, Pu M, Lin F, Vickers C, Joud-Caldwell C, Chung F, Yin H, Handly ED, Straub C, Growney JD, Vander Heiden MG, Murphy AN, Pagliarini R, Metallo CM (2014) IDH1 mutations Alter citric acid cycle metabolism and increase dependence on oxidative mitochondrial metabolism. Cancer Res 74:3317–3331. https://doi.org/10.1158/0008-5472.CAN-14-0772-T
38. Cordes T, Metallo CM (2019) Quantifying intermediary metabolism and lipogenesis in cultured mammalian cells using stable isotope tracing and mass spectrometry. In: High-throughput metabolomics. Methods in molecular biology. Humana, New York, NY, pp 219–241
39. Yoo H, Antoniewicz MR, Stephanopoulos G, Kelleher JK (2008) Quantifying reductive carboxylation flux of glutamine to lipid in a brown adipocyte cell line. J Biol Chem 283:20621–20627. https://doi.org/10.1074/jbc.M706494200
40. Young JD (2014) INCA: a computational platform for isotopically non-stationary metabolic flux analysis. Bioinformatics 30:1333–1335. https://doi.org/10.1093/bioinformatics/btu015

41. Shin M, Momb J, Appling DR (2017) Human mitochondrial MTHFD2 is a dual redox cofactor-specific methylenetetrahydrofolate dehydrogenase/methenyltetrahydrofolate cyclohydrolase. Cancer Metab 5:11. https://doi.org/10.1186/s40170-017-0173-0
42. Kuri-Harcuch W, Wise LS, Green H (1978) Interruption of the adipose conversion of 3T3 cells by biotin deficiency: differentiation without triglyceride accumulation. Cell 14:53–59. https://doi.org/10.1016/0092-8674(78)90300-8
43. Aoki J, Taira A, Takanezawa Y, Kishi Y, Hama K, Kishimoto T, Mizuno K, Saku K, Taguchi R, Arai H (2002) Serum lysophosphatidic acid is produced through diverse phospholipase pathways. J Biol Chem 277:48737–48744. https://doi.org/10.1074/jbc.M206812200
44. Wang X, McManus M (2009) Lentivirus production. J Vis Exp 32:1499. https://doi.org/10.3791/1499
45. Benskey MJ, Manfredsson FP (2016) Lentivirus production and purification. Humana Press, New York, NY, pp 107–114

[illegible]

Interrogating Redox Metabolism in Cultured Cells with Deuterium Tracing 71

[illegible]

Chapter 5

Large-Scale Profiling of Cellular Metabolic Activities Using Deep ^{13}C Labeling Medium

Nina Grankvist, Jeramie D. Watrous, Mohit Jain, and Roland Nilsson

Abstract

The recently developed deep labeling method allows for large-scale profiling of metabolic activities in human cells or tissues using isotope tracing with a highly ^{13}C enriched culture medium in combination with liquid chromatography–high resolution mass spectrometry. This method generates mass spectrometry data sets where endogenous cellular products can be identified, and active pathways can be determined from observed ^{13}C mass isotopomers of the various metabolites measured. Here we describe in detail the experimental procedures for deep labeling experiments in cultured mammalian cells, including synthesis of the deep labeling medium, experimental considerations for cell culture, metabolite extractions and sample preparation, and liquid chromatography–mass spectrometry. We also outline a workflow for the downstream data analysis using publicly available software.

Key words Cell culture, Stable isotope tracing experiments, Metabolism, Metabolomics, LC-HRMS

1 Introduction

Cellular metabolism is central to a range of human disorders, including hundreds of rare inborn errors of metabolism, the "metabolic syndrome" cluster comprising diabetes, obesity and cardiovascular disease, and disorders of cell proliferation such as cancer [1] and immunological disease [2]. Yet the metabolic phenotypes of human cells are still largely unexplored: the set of metabolites synthesized by human cells is still only partially known [3], and the specific metabolic activities of most human cell types remain to be investigated [4–6]. This scarcity of information on human metabolism is at least in part due to lack of unbiased experimental methods for "profiling" metabolic activities in living cells and tissues.

To address this shortcoming, we recently introduced the deep labeling technique for profiling metabolic activities in cultured mammalian cells [7], an isotope tracing method where a highly ^{13}C-enriched, custom designed culture medium is used to label virtually every molecule synthesized by the cells. Here, we describe

Deepak Nagrath (ed.), *Metabolic Flux Analysis in Eukaryotic Cells: Methods and Protocols*, Methods in Molecular Biology, vol. 2088, https://doi.org/10.1007/978-1-0716-0159-4_5,

in depth how to perform deep labeling experiments, including synthesis of the required ^{13}C deep labeling medium, experimental considerations for cell culture and metabolite extraction, use of modern, liquid chromatography–high resolution mass spectrometry (LC-HRMS) to analyze samples, and data processing using computational methods for unbiased LC-HRMS peak detection. The most demanding experimental step is to synthesize the deep labeling medium; with this medium in hand, the isotope tracing experiments are straightforward, and can easily be tailored to various cell types and experimental conditions of interest, for example to explore metabolic effects of drugs, RNAi knockdown, or CRISPR gene editing. Deep labeling experiments with modern LC-HRMS systems yields rich data sets, typically covering hundreds or even thousands of endogenous molecules, whose isotopic state allows identifying a variety of active and inactive metabolic pathways.

2 Materials

Here we describe the materials required for the RPMI-1640 medium formulation [8]. For other medium formulations, the concentrations must naturally be altered, but the general procedure is the same. To provide the closest possible control culture, we describe synthesis of both a ^{12}C "unlabeled" RPMI-1640 medium where all components have natural isotope distribution, and one deep labeling medium where all major nutrients are fully ^{13}C. Both media have the same final concentrations of all components, specified in Table 1. All solutions should be prepared with deionized water, using a water purifier such as the Milli-Q® Advantage A10 System (Millipore, Bedford, MA).

2.1 Preparations of Stock Solutions

Except where noted, stock solutions will be made at 50 times the final medium concentration ("50× stocks"). This makes it easy to synthesize fresh medium whenever needed, since equal volumes of all stocks are pipetted into the final solution.

1. Prepare 50× stocks of all components listed in Tables 2 and 3 (glucose, amino acids, and various salts), with the exception of tyrosine (see below). We recommend preparing 40 mL stock solutions of the ^{12}C components, sufficient for 2 L of medium, and storing in 50 mL tubes. For the ^{13}C tracers, we recommend preparing 3 mL stocks. The excel tables provided can be used to adjust these volumes as needed.
2. Since tyrosine is difficult to dissolve in water (the solubility is 0.45 g/L at 25 °C), prepare a 25× stock of this amino acid. To fully dissolve tyrosine, add ~100 μL of 1 M HCl (without HCl, the solution is usually a bit cloudy).

Table 1
RPMI-1640 specification concentrations according to Moore et al. (1967) [8]

	Compound according to original formulation	Molecular weight (g/mol)	Medium conc. (mg/L)	Medium conc. (mM)
Nutrients	D-Glucose	180.156	2000	11.10
	Glycine	75.067	10	0.133
	L-Arginine	174.204	200	1.148
	L-Asparagine	132.119	50	0.378
	L-Aspartate	133.103	20	0.150
	L-Cysteine[a]	121.154	50.4	0.416
	L-Glutamate	147.130	20	0.136
	L-Glutamine	146.146	300	2.053
	L-Histidine	155.157	15	0.097
	L-Isoleucine	131.175	50	0.381
	L-Leucine	131.175	50	0.381
	L-Lysine–HCl	182.648	40	0.219
	L-Methionine	149.208	15	0.101
	L-Phenylalanine	165.192	15	0.091
	L-Proline	115.132	20	0.174
	L-Serine	105.093	30	0.285
	L-Threonine	119.120	20	0.168
	L-Tryptophan	204.229	5	0.024
	L-Tyrosine	181.191	20	0.110
	L-Valine	117.148	20	0.171
	Glutathione, reduced	307.321	1	0.003
	L-Hydroxyproline	131.131	20	0.153
	Myoinositol	180.156	35	0.1943
	Choline chloride	139.623	3	0.0215
Salts	$Ca(NO_3)_2$[a]$\cdot 4H_2O$	236.146	100	0.423
	Na_2HPO_4[a]$\cdot 7H_2O$	268.062	1512	5.640
	$MgSO_4$[a]$\cdot 7H_2O$	246.466	100	0.406
	KCl	74.5483	400	5.366
	$NaHCO_3$	84.0058	2000	23.81
	NaCl	58.4398	6000	102.7
Vitamins	Biotin	244.309	0.2	0.0008
	Folate	441.404	1	0.0023
	Niacinamide	122.127	1	0.0082
	2-Pantothenate[a] Ca	257.299	0.25	0.0010
	Pyridoxine[a]–HCl	206.646	1	0.0048
	Riboflavin	376.369	0.2	0.0005
	Thiamine[a]–HCl	338.263	1	0.0030
	Cyanocobalamin (B12)	1355.39	0.005	0.000004
	4-Aminobenzoic acid	137.138	1	0.0073

[a]Cysteine concentration was recalculated from the original description of the RPMI-1640 in terms of cystine (the cys–cys dimer), whose mass differs slightly due to the loss of two hydrogens in the S–S bond formation

Table 2
Reagents

	Compound	Compound MW (g/mol)[b]	Product/Formulation[a]	Vendor[c]	Cat. No.[a]	Product amount (g)	Product MW (g/mol)[b]
^{13}C tracers	U-^{13}C-D-glucose	186.050	U-^{13}C-D-glucose	CIL	CLM-1396-0	1	186.084
	U-^{13}C-glycine	77.032	U-^{13}C-glycine	CIL	CLM-1017-0	0.5	77.039
	U-^{13}C-L-arginine	180.098	U-^{13}C-L-arginine–HCl	CIL	CLM-2265-H	0.1	216.556
	U-^{13}C-L-asparagine	136.048	U-^{13}C-L-asparagine–H_2O	CIL	CLM-8699-H	0.05	154.063
	U-^{13}C-L-aspartate	137.032	U-^{13}C-L-aspartate	CIL	CLM-1801-H	0.25	137.051
	U-^{13}C-L-cysteine	124.101	U-^{13}C-L-cysteine	CIL	CLM-4320-H	0.1	124.030
	U-^{13}C-L-glutamate	152.042	U-^{13}C-L-glutamate	CIL	CLM-1800-H	0.25	152.070
	U-^{13}C-L-glutamine	151.058	U-^{13}C-L-glutamine	CIL	CLM-1822-H	0.1	151.086
	U-^{13}C-L-histidine	161.051	U-^{13}C-L-histidine–HCl–H_2O	CIL	CLM-2264-0	0.1	215.524
	U-^{13}C-L-isoleucine	137.069	U-^{13}C-L-isoleucine	CIL	CLM-2248-H	0.1	137.115
	U-^{13}C-L-leucine	137.069	U-^{13}C-L-leucine	CIL	CLM-2262-H	0.1	137.115
	U-^{13}C-L-lysine	152.084	U-^{13}C-L-lysine–2HCl	CIL	CLM-2247-H	0.1	225.000
	U-^{13}C-L-methionine	154.120	U-^{13}C-L-methionine	CIL	CLM-893-H	0.1	154.068
	U-^{13}C-L-phenylalanine	174.033	U-^{13}C-L-phenylalanine	CIL	CLM-2250-H	0.25	174.109
	U-^{13}C-L-proline	120.044	U-^{13}C-L-proline	CIL	CLM-2260-H	0.1	120.080
	U-^{13}C-L-serine	108.040	U-^{13}C-L-serine	CIL	CLM-1574-H	0.1	108.053
	U-^{13}C-L-threonine	123.049	U-^{13}C-L-threonine	CIL	CLM-2261-0	0.1	123.072
	U-^{13}C-L-tryptophan	215.035	U-^{13}C-L-tryptophan	CIL	CLM-4290-H	0.1	215.127
	U-^{13}C-L-tyrosine	190.032	U-^{13}C-L-tyrosine	CIL	CLM-2263-H	0.1	190.104
	U-^{13}C-L-valine	122.060	U-^{13}C-L-valine	CIL	CLM-2249-H	0.25	122.096
^{12}C nutrients	D-Glucose	180.156	D-Glucose	Sigma	D9434-250G	250	180.156
	Glycine	75.067	Glycine	Sigma	50,046-50G	50	75.067
	L-Arginine	174.204	L-Arginine	Sigma	A8094-25G	25	174.204
	L-Asparagine	132.119	L-Asparagine	Sigma	A4159-25G	25	132.119
	L-Aspartate	133.103	L-Aspartic acid	Sigma	A7219-100G	100	133.103
	L-Cysteine	121.154	L-Cysteine	Sigma	C7352-25G	25	121.154
	L-Glutamate	147.130	L-Glutamic acid	Sigma	G8415-100G	100	147.130
	L-Glutamine	146.146	L-Glutamine	Sigma	G8540-25G	25	146.146
	L-Histidine	155.157	L-Histidine	Sigma	H6034-25G	25	155.157

	L-Isoleucine	131.175	L-Isoleucine	Sigma	I7403-25G	25	131.175
	L-Leucine	131.175	L-Leucine	Sigma	L8912-25G	25	131.175
	L-Lysine	146.190	L-Lysine–HCl	Sigma	L8662-25G	25	182.648
	L-Methionine	149.208	L-Methionine	Sigma	M5308-25G	25	149.208
	L-Phenylalanine	165.192	L-Phenylalanine	Sigma	P5482-25G	25	165.192
	L-Proline	115.132	L-Proline	Sigma	P5607-25G	25	115.132
	L-Serine	105.093	L-Serine	Sigma	S4311-25G	25	105.093
	L-Threonine	119.120	L-Threonine	Sigma	T8441-25G	25	119.120
	L-Tryptophan	204.229	L-Tryptophan	Sigma	T8941-25G	25	204.229
	L-Tyrosine	181.191	L-Tyrosine	Sigma	T8566-25G	25	181.191
	L-Valine	117.148	L-Valine	Sigma	V0513-25G	25	117.148
	Trans-4-hydroxy-L-proline	131.131	L-Hydroxyproline (trans)	Sigma	H54409-25G	25	131.131
	Glutathione (reduced)	307.321	Glutathione	Sigma	G6013-5G	5	307.321
Salts	$Ca(NO_3)_2 + 4H_2O$	236.146	Calcium nitrate (tetrahydrate)	Sigma	C2786-500G	500	236.146
	Na_2HPO_4	141.957	Disodium phosphate (Na_2HPO_4)	Sigma	S0876-500G	250	141.957
	$MgSO_4$	120.361	Magnesium sulfate ($MgSO_4$)	Sigma	M2643-500G	500	120.361
	KCl	74.5483	Potassium chloride (KCl)	Sigma	P5405-250G	250	74.5483
	$NaHCO_3$	84.0058	Sodium bicarbonate	Sigma	S5761-500G	100	84.0058
	NaCl	58.4398	Sodium chloride	Sigma	S7653-1KG	1000	58.4398
Vitamins			Vitamin solution 100×	Sigma	R7256	100 mL	

[a]For tracers, we give the monoisotopic mass of the stated mass isotopomer, assuming 99% atom purity

[b]The product formulation differs somewhat due to counterions, hydration, and so on; hence, the product MW can differ from the MW of the nutrient

[c]*CIL* Cambridge Isotope Laboratories. There are also low concentrations of salts in the R7256 vitamin mix, but these can be neglected

Table 3
Stock solutions

	Component	Molecular weight (g/mol)	Medium conc. (mM)	Stock conc. factor	Stock concentration (mM)	Stock volume (mL)	Mass for × mL (stock volume) stock solution (mg)[a]
^{13}C Tracers	U-^{13}C-D-Glucose	186.084	11.101	50	555.07	3	309.9
	U-^{13}C-Glycine	77.039	0.133	50	6.66	3	1.54
	U-^{13}C-L-arginine–HCl	216.556	1.148	50	57.40	3	37.29
	U-^{13}C-L-asparagine–H_2O	154.063	0.378	50	18.92	3	8.75
	U-^{13}C-L-aspartate	137.051	0.150	50	7.51	3	3.09
	U-^{13}C-L-cysteine	124.030	0.416	50	20.81	3	7.74
	U-^{13}C-L-glutamate	152.070	0.136	50	6.80	3	3.10
	U-^{13}C-L-glutamine	151.086	2.053	50	102.64	3	46.52
	U-^{13}C-L-histidine–HCl–H_2O	215.524	0.097	50	4.83	3	3.13
	U-^{13}C-L-isoleucine	137.115	0.381	50	19.06	3	7.84
	U-^{13}C-L-leucine	137.115	0.381	50	19.06	3	7.84
	U-^{13}C-L-Lysine–2HCl	225.000	0.219	50	10.95	3	7.39
	U-^{13}C-L-methionine	154.068	0.101	50	5.03	3	2.32
	U-^{13}C-L-phenylalanine	174.109	0.091	50	4.54	3	2.37
	U-^{13}C-L-proline	120.080	0.174	50	8.69	3	3.13
	U-^{13}C-L-serine	108.053	0.285	50	14.27	3	4.63
	U-^{13}C-L-threonine	123.072	0.168	50	8.39	3	3.10
	U-^{13}C-L-tryptophan	215.127	0.024	50	1.22	5	1.32
	U-^{13}C-L-tyrosine	190.104	0.110	25	2.76	3	1.57
	U-^{13}C-L-valine	122.096	0.171	50	8.54	3	3.13
^{12}C nutrients	D-Glucose	180.156	11.101	50	555.07	40	4000
	Glycine	75.067	0.133	50	6.66	40	20
	L-Arginine	174.204	1.148	50	57.40	40	400
	L-Asparagine	132.119	0.378	50	18.92	40	100
	L-Aspartic acid	133.103	0.150	50	7.51	40	40
	L-Cysteine	121.154	0.416	50	20.81	40	101
	L-Glutamic acid	147.130	0.136	50	6.80	40	40
	L-Glutamine	146.146	2.053	50	102.64	40	600
	L-Histidine	155.157	0.097	50	4.83	40	30

	L-Isoleucine	131.175	0.381	50	19.06	40	100
	L-Leucine	131.175	0.381	50	19.06	40	100
	L-Lysine–HCl	182.648	0.219	50	10.95	40	80
	L-Methionine	149.208	0.101	50	5.03	40	30
	L-Phenylalanine	165.192	0.091	50	4.54	40	30
	L-Proline	115.132	0.174	50	8.69	40	40
	L-Serine	105.093	0.285	50	14.27	40	60
	L-Threonine	119.120	0.168	50	8.39	40	40
	L-Tryptophan	204.229	0.024	50	1.22	40	10
	L-Tyrosine	181.191	0.110	25	2.76	40	20
	L-Valine	117.148	0.171	50	8.54	40	40
	Trans-4-hydroxy-L-proline	131.131	0.153	50	7.63	40	40
	Glutathione (reduced)	307.321	0.003	50	0.16	40	2
Salts	Calcium nitrate (tetrahydrate)	236.146	0.423	50	21.17	40	200
	Disodium phosphate (Na_2HPO_4)	141.957	5.640	50	282.02	40	1601
	Magnesium sulfate ($MgSO_4$)	120.361	0.407	50	20.33	40	98
	Potassium chloride (KCl)	74.5483	5.366	50	268.28	40	800
	Sodium bicarbonate	84.0058	23.8	50	1190.39	40	4000
	Sodium chloride	58.4398	103.4	50	5172.00	40	12,090

[a]This is the mass to weight up for preparing the stock. Note that this mass is based on the product's MW including any hydrates and counterions

3. Sterile filter all stocks with Filtropur S 0.2 μm syringe filters into new labeled tubes.
4. Store all ^{12}C stocks at 4 °C, with the following exceptions: (1) sodium bicarbonate and disodium phosphate solutions should be stored at room temperature, as they crystallize easily when stored at 4 °C, and are then problematic to redissolve; (2) glutamine and asparagine should be aliquoted into suitable volumes and stored at −20 °C, since these amino acids are somewhat unstable in water solution for long term storage (aliquoting avoids repeated freeze–thaw cycles). The ^{13}C tracer stocks, which are very precious chemicals, should be aliquoted into smaller volumes (0.5–1 mL) and stored at −20 °C.

2.2 Dialyzed Serum

Always use heat-inactivated fetal bovine serum (FBS) to minimize enzyme activities present in serum (*see* **Note 1**).

1. Dialyze FBS in dialysis tubing with a 10 kDa molecular weight cutoff (SnakeSkin 10K MWCO, Thermo Fisher Scientific #88245) against deionized water during 1–2 h at room temperature, to remove proteins and other large macromolecules (*see* **Note 2**). During this time, replace the water at least twice. Then allow the dialysis to proceed over night at 4 °C.

2.3 Preparations of Unlabeled and Deep Labeled Custom Made Media

1. Medium should be prepared immediately before usage to minimize loss of nutrients due to serum activities. Before preparing medium, carefully consider the design of your cell culture experiments to determine the volume needed; *see* Subheading 3.
2. When preparing for example 100 mL of ^{12}C medium, start with adding 41 mL deionized water, 1 mL 100× vitamin stock solution, followed by 2 mL of all 50× stocks and 4 mL of the 25× tyrosine stock, for a total volume of 100 mL; *see* Table 4.
3. Add the desired amount of dialyzed FBS, antibiotics and optionally a pH indicator.
4. Adjust pH to 7.4 using 1 M HCl and 1 M NaOH as needed.
5. Repeat **steps 2–4** for the preparation of the deep labeling medium (Table 4).
6. Sterile filter the media with either Filtropur S 0.2 μm syringe filter, Filtropur V25 (250 mL) or Filtropur V50 (500 mL) depending on volume. Store media at 4 °C.

Table 4
Medium synthesis

	Components	Stock conc. Factor	Volume for 1 L medium (mL)
Vitamins	MQ water		410
	Vitamin solution 100×	100	10
Nutrients	D-Glucose	50	20
	Glycine	50	20
	L-Arginine–HCl	50	20
	L-Asparagine:H_2O	50	20
	L-Aspartic acid	50	20
	L-Cysteine	50	20
	L-Glutamic acid	50	20
	L-Glutamine	50	20
	L-Histidine–HCl–H_2O	50	20
	L-Isoleucine	50	20
	L-Leucine	50	20
	L-Lysine–2HCl	50	20
	L-Methionine	50	20
	L-Phenylalanine	50	20
	L-Proline	50	20
	L-Serine	50	20
	L-Threonine	50	20
	L-Tryptophan	50	20
	L-Tyrosine	25	40
	L-Valine	50	20
	Trans-4-hydroxy-L-proline	50	20
	Glutathione (reduced)	50	20
Salts	Calcium nitrate ($Ca(NO_3)_2$)	50	20
	Disodium phosphate (Na_2HPO_4)	50	20
	Magnesium sulfate ($MgSO_4$)	50	20
	Potassium chloride (KCl)	50	20
	Sodium bicarbonate ($NaHCO_3$)	50	20
	Sodium chloride (NaCl)	50	20
	Total volume		1000

3 Methods

3.1 Cell Culture Condition and Stable Isotope Tracing Experiment

The deep labeling method described previously [7] requires culturing cells in deep labeling medium for six population doublings to obtain a virtually complete renewal of biomass (*see* **Note 3**). This ensures isotopic steady-state in all components of the cell, and therefore allows estimating the fractional contribution of biosynthesis of various compounds.

1. Make a scheme for the cell culture experiment, taking into account the doubling time and recommended seeding density/split ratio of the cell line used to achieve six population

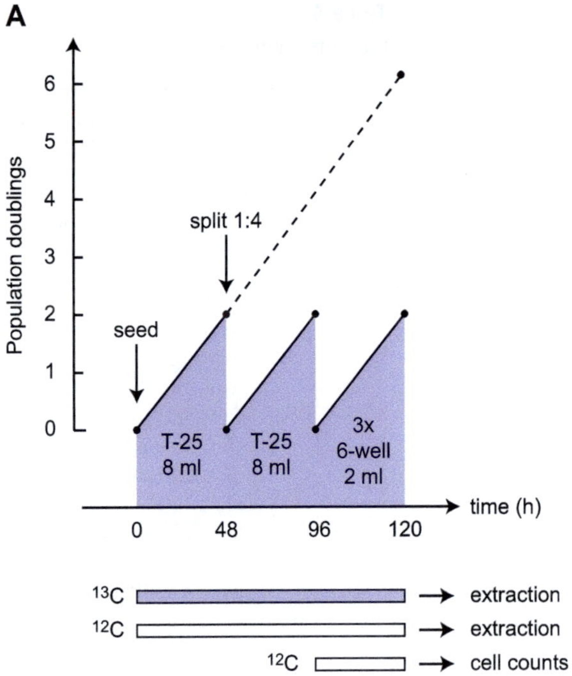

B

Required ^{13}*C medium volume*	
cultures	8 + 8 + 3*2 = 22 ml
6-well control	2 ml
resuspending pellets, 1 ml / split	2 ml
medium for LCMS	1 ml
	27 ml

Fig. 1 Design of deep labeling experiments. (**a**) Deep labeling requires a total of six population doublings (dashed line), usually divided into three or more subcultures (solid lines) to conserve ^{13}C medium. For most cell lines, T-25 cultures are suitable for the initial subcultures, followed by expansion to three 6-wells, as indicated. This culture scheme is carried out for both ^{13}C and ^{12}C media. One additional well is seeded and cultured in ^{12}C medium to be used for cell counting (see text). (**b**) Calculation of required ^{13}C medium volume for the experiment shown in (**a**), including control wells, extra medium for resuspending cell pellets when splitting cells, and 1 mL of medium for LCMS profiling

doublings without reaching confluence. An example for a cell line with a doubling time of 24 h is given in Fig. 1a. In this case, 6 days of culture is needed, and the culture is split twice. Slowly growing cell lines may require medium changes as well, just as during regular maintenance of the cell culture (medium should typically be changed every 48 h). The final subculture should be done in 6-well plates, seeded at a density to achieve ~85% confluence at 48 h; this usually yields extracts of 0.5–1 million cells per well for LCMS profiling. We recommend three independent cultures at the last step to obtain reliable results. This entire scheme is performed in parallel for ^{13}C deep labeling cultures and for ^{12}C control cultures.

2. Calculate the required amount of ^{12}C and ^{13}C medium required, as in Fig. 1b. Freeze an aliquot of 1 mL of each medium for LC-HRMS profiling.
3. Seed the initial cell cultures in appropriate cell culture plastic, in ^{13}C deep labeling medium and in ^{12}C control medium, respectively. Typically, T25 flasks with ~8 mL medium are suitable for this step. Subculture and/or change medium as needed.
4. In the final subculture, split the ^{13}C cultures to three 6-wells (each ~2 mL medium), and the ^{12}C cultures to four 6-wells, of which one is used for cell number estimation; *see* Subheading 3.2. At this time, also fill one 6-well with fresh medium only (no cells), and incubate for 48 h in the same conditions.

3.2 Cell Number Estimation

An estimate of cell number in the final cultures is important during preparation of samples for LC-HRMS, so that a suitable amount of cell extract is used for analysis.

1. Use one of the ^{12}C culture wells seeded as described in Subheading 3.1, **step 4** for the cell number estimation.
2. After 48 h (at the time of extraction of the other wells), remove the medium and wash the cells twice with PBS.
3. Add 0.5 mL Trypsin-EDTA (0.05%) and allow trypsinization for a few minutes (depending on cell lines characteristics).
4. To neutralize, add 0.5 mL ^{12}C medium (containing serum).
5. Measure or count the cell numbers in the cell suspension.

3.3 Metabolite Extraction

For best results, we recommend performing metabolite extraction from cells directly in the culture dish; this protocol is given below. We describe extraction of polar metabolites using cold methanol, a simple protocol that we have found to yield reproducible results (*see* also **Note 4**).

1. At the end of the last stage of the cell culture experiment, carefully remove 1 mL of medium from each well, taking care not to disturb cells (*see* **Note 5**). Remove the rest of the medium and rapidly wash the cells twice with 1 mL cool (+4 °C) PBS (*see* **Note 6**).
2. Remove the PBS and transfer the microplate to a bed of dry ice. Working quickly, add 1 mL (for 6-well plates) of −80 °C precooled HPLC grade methanol, to quench metabolic enzymes and extract polar metabolites. Scrape the cells (*see* **Note 7**) in the cold methanol with a cell scraper so that all cell material

detaches from the plastic, and collect the extract in 1.5 mL tubes.

3. Vortex the collected cell extract for 1 min.
4. Store in −80 °C until analysis (*see* **Note 8**).
5. To obtain a suitable blank sample, "extract" (rinse) an empty 6-well with cold methanol, and handle this solution in same way as cell extracts (**steps 2–4**). This ensures that compounds present in the microplate plastic, cell scrapers, tubes, and so on are included in the blank and discounted during data analysis (*see* Subheading 3.5, **step 4**).

3.4 Sample Preparation and Mass Spectrometry

To take full advantage of the deep labeling method, cell extracts should be analyzed with high resolution full-scan mass spectrometry to capture a wide variety of cellular products and metabolic intermediates. Below, we describe typical steps for analysis using liquid chromatography coupled to a high-resolution Orbitrap mass spectrometer, as previously described [7].

1. While the methanol used to quench metabolic reactions will have partially lysed the cells, complete cell lysis can be ensured by performing three freeze–thaw cycles where samples are alternated between 37 °C and −80 °C (methanol and dry ice) liquid baths in 60 s intervals.
2. To ensure complete metabolite extraction, vortex the samples for 30 s at high speed and sonicate for 2 min using a bath sonicator.
3. Dry down samples in vacuo using a vacuum concentrator, and resuspend in 25 μL of 80:20 methanol–water (HPLC grade).
4. Mix the resuspended samples by initial aspiration followed by 2 min of sonication and 30 s of hard vortex. Centrifuge the samples at 14,000 × *g* for 10 min at 4 °C and transfer supernatants to glass HPLC vials containing a 100 μL glass conical insert for injection into the LC-HRMS instrument. We recommend using a Thermo QExactive Orbitrap mass spectrometer coupled to a Thermo Vanquish UPLC system, or similar.
5. Perform chromatographic separation of polar metabolites using a Millipore (Sequant) Zic-pHILIC 2.1 × 150 mm 5 μm column maintained at 25 °C and flowing at 300 μL/min. Compounds are eluted via a 19 min linear gradient, starting from 90:10 acetonitrile–20 mM ammonium bicarbonate pH 9.6, to 45:55 acetonitrile–20 mM ammonium bicarbonate pH 9.6. We recommend an injection volume of 5 μL (one-fifth of the prepared sample volume), which corresponds to extract from ~80,000 cells. Note that higher injection volumes will result in poor peak shape on this column.

6. Operate the mass spectrometer at 35,000 resolution, 100 ms ion trap time, 1×10^6 AGC target, and a mass range from m/z 65–1000 for MS1 collection. For data-dependent (DDA) MS2 collection, use 17,500 resolution, 50 ms ion trap time, 2×10^5 AGC target, and a TopX of 8. Operate the heated electrospray ionization (HESI) source at 3.5 kV (−3.5 kV) spray voltage, sheath gas flow rate of 40 arbitrary units, auxiliary gas flow rate of 20 arbitrary units, sweep gas flow rate of 2 arbitrary units, capillary temperature of 275 °C, auxiliary gas heater temperature of 350 °C, and an S-lens RF of 45. Inject each sample in positive and negative ionization modes separately; do not use polarity switching between scans, as this tends to complicate downstream data analysis.

3.5 LC-HRMS Data Processing

The downstream data analysis of deep labeling experiments is difficult to describe for a general case, as software details change over time, and depend on the configuration of computer systems used. Here, we describe a series of steps used in our laboratory to process Thermo .raw instrument data files from deep labeling experiments and extract information on isotope-labeled peaks, using the Isotrack software to capture all LC-HRMS peaks above a predefined intensity threshold. An overview of the analysis process is shown in Fig. 2.

1. Perform an unbiased analysis of a set of replicated unlabeled cell extracts using the Isotrack software, available at https://github.com/Yaroslav-Lyutvinskiy/Isotrack (*see* **Note 9**). For the LC-HRMS method described (*see* Subheading 3.4), we recommend setting an LCMS peak apex intensity of at least 50,000, a minimum LC peak length of 6 s, and retaining only peaks that are detected in all replicates. This generates a list P_1 of all detectable LC-HRMS peaks satisfying these constraints, excluding natural isotopomers due to ^{13}C and ^{15}N (*see* **Note 10**).
2. Perform a manual inspection of at least 100 randomly chosen examples from P_1 to ensure that detected peaks are true chromatographic peaks, and that peaks are not misidentified across individual analyses due to retention time drift. Identified peaks can be visualized using a scientific computing platform such as R, python, or Mathematica, coupled with the mzAccess service [9] to directly access LC-HRMS data. We typically observe that >95% of all peaks detected by Isotrack that reproduce across three replicates are true peaks; if your error rate is substantially higher, repeat **step 1** with higher apex intensity threshold and/or a longer minimum trace length.
3. Use the m/z and retention time coordinates from the peak list P_1 to integrate base isotopomer peak areas from unlabeled

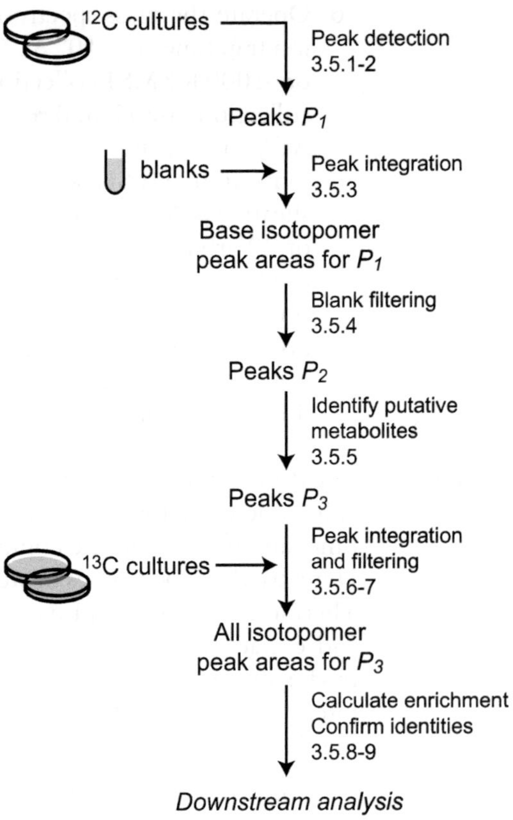

Fig. 2 Outline of deep labeling LCMS data analysis. The text sections corresponding to each step are indicated

samples and blanks. We recommend performing peak integration using a scientific computing platform such as R, python, or Mathematica, to allow customizing the workflow as needed and retaining scripts that can be used to reproduce the analysis.

4. Discard peaks from P_1 whose peak areas in unlabeled cell extracts are less than five times the peak area observed in blanks; call the resulting peak list P_2. This step removes compounds that derive from the environment, such as substances released from culture plastic or present in the atmosphere during analysis.

5. Identify putative metabolites from P_2 by matching estimated m/z of peaks against a database of known human compounds, such as the Human Metabolome Database (HMDB) [1]. Typically, matching peaks by m/z with a tolerance of ~5 ppm and considering common adducts (such as H^+, H^-) captures the majority of commonly observed small (roughly, 50—300 Da) metabolites, and most peaks in this range will be associated with a single sum formula, although many peaks will have more than one possible identity. Let P_3 denote the list of peaks that

successfully matched at least one metabolite, and associate each peak in P_3 with the number of carbons in the matched metabolite(s).

6. Perform peak integration as in step [3] above for each peak in P_3, and all its possible mass isotopomers $^{13}C_0$ through $^{13}C_n$ in all samples analyzed, based on the estimated number of carbons n obtained for each peak. This typically yields data for 10—20,000 mass isotopomers total on a full-scan high resolution instrument.
7. To avoid false isotopic peaks due to contamination by other ions of similar m/z, calculate for each peak the ratio of $^{13}C_k$ /$^{13}C_0$ peak areas, $k = 1, \ldots, n$, in the ^{12}C cultures, and discard mass isotopomers where this ratio exceeds the expected ratio.

$$\frac{p^k}{(1-p)^k}\binom{n}{k}$$

   Student's t-test may be used to determine if the observed ratio exceeds the expected by a reasonable statistical margin, given the observed variation in mass isotopomers (MI) fraction across replicates. However, since this variation is often very small, an additional arbitrary cutoff may be required to avoid removing isotopes with minor imperfections; we have used a value of 0.03 above the expected ratio.
8. To confirm effective ^{13}C labeling of cells, calculate for each peak the ^{13}C enrichment E, defined for a compound with n carbons and MI fractions $x_0, x_1, \ldots x_n$ as

$$E = \frac{1}{n}\sum_{k=0}^{n} kx_k$$

   You should observe $E > 95\%$ for glucose, amino acids and their immediate products. Central carbon metabolites such as intermediates of the TCA cycle commonly have E values around 85% (due to exchange with $^{12}CO_2$), while lipids and compounds containing choline or vitamin moieties (such as the pantothenate moiety of coenzyme A) exhibit lower E values.
9. Confirm identities for metabolites of interest by (1) matching against observed retention time of pure standards, (2) matching MS/MS data, if collected, or (3) confirming appearance expected mass isotopomers in deep labeling samples. For example, de novo synthesized purines (C_{10}) will exhibit a prominent $^{13}C_9$ mass isotopomer since all carbons except one derive from the deep labeling medium ^{13}C nutrients (the remaining carbon derives from CO_2).

3.6 Identifying Synthesized Metabolites and Active Pathways

Once metabolites identity has been established and their MI distribution is determined (as above), biosynthesized metabolites are easily identified as any metabolite with higher ^{13}C enrichment than the natural ^{13}C abundance ($E = 0.01$ approximately). With the described protocol, many biosynthesized compounds will typically be only partially ^{13}C labeled. This is due to incorporation of ^{12}C atoms from choline and the vitamins; from atmospheric $^{12}CO_2$ which is readily incorporated into central carbon metabolism by carboxylases/decarboxylases [7]; and from lipoprotein-derived lipids, which are quite abundant in serum and are not removed by dialysis.

Active metabolic enzymes can be identified by comparing observed MIs of enzyme substrates and products. An enzyme can be considered active if (1) the observed MIs of its product *p* are compatible with the corresponding MIs of its substrate(s); and (2) there is no other enzymatic reaction producing *p* from substrate(s) with compatible MIs. An example of this type of inference for NAD metabolism based on data from deep labeling experiments with HCT116 cells is given in Fig. 3. For active reactions that do not alter the carbon skeleton, we should observe the same MIs in substrates and products, as for example in the hydroxylation of $^{13}C_{11}$ tryptophan to $^{13}C_{11}$ 5-OH-tryptophan (Fig. 3a, b, g). For reactions or pathways that remove carbons from the substrate, the MIs will decrease accordingly, as in the conversion of $^{13}C_{11}$ tryptophan to $^{13}C_{10}$ kynurenine (Fig. 3c, g); while for reactions that condense smaller molecules into larger ones, the MIs will be added, as for example in the formation of $^{13}C_{14}$ NAD from $^{13}C_0$ nicotinamide, $^{13}C_9$ ATP and $^{13}C_5$ ribose (Fig. 3d–g). Generally, this type of inference works well for more "peripheral" metabolic systems, where there is usually one single enzyme producing each product *p*; however, for highly interconnected reaction networks in "central" metabolism, condition (2) often fails. To resolve such cases, additional isotope tracing experiments with specific tracers are needed. For "linear" pathways, consisting of many enzymes in sequence, each having a single substrate and product, presence of a labeled end product leads to the conclusion that all enzymes in the pathway must be active. In this way, a single identified metabolite can be informative for multiple reactions. For example, purine biosynthesis consists of several enzymes in such a linear arrangement, and all of these must be active in order to generate $^{13}C_9$ ATP (Fig. 3e). Conversely, inactive pathways are often identified by lack of ^{13}C atoms in products whose precursors are labeled. For example, the pathway of de novo synthesis of NAD from kynurenine is inactive in HCT116 cells, since activity of this pathway would generate $^{13}C_{20}$ NAD, which was not observed (Fig. 3f, g).

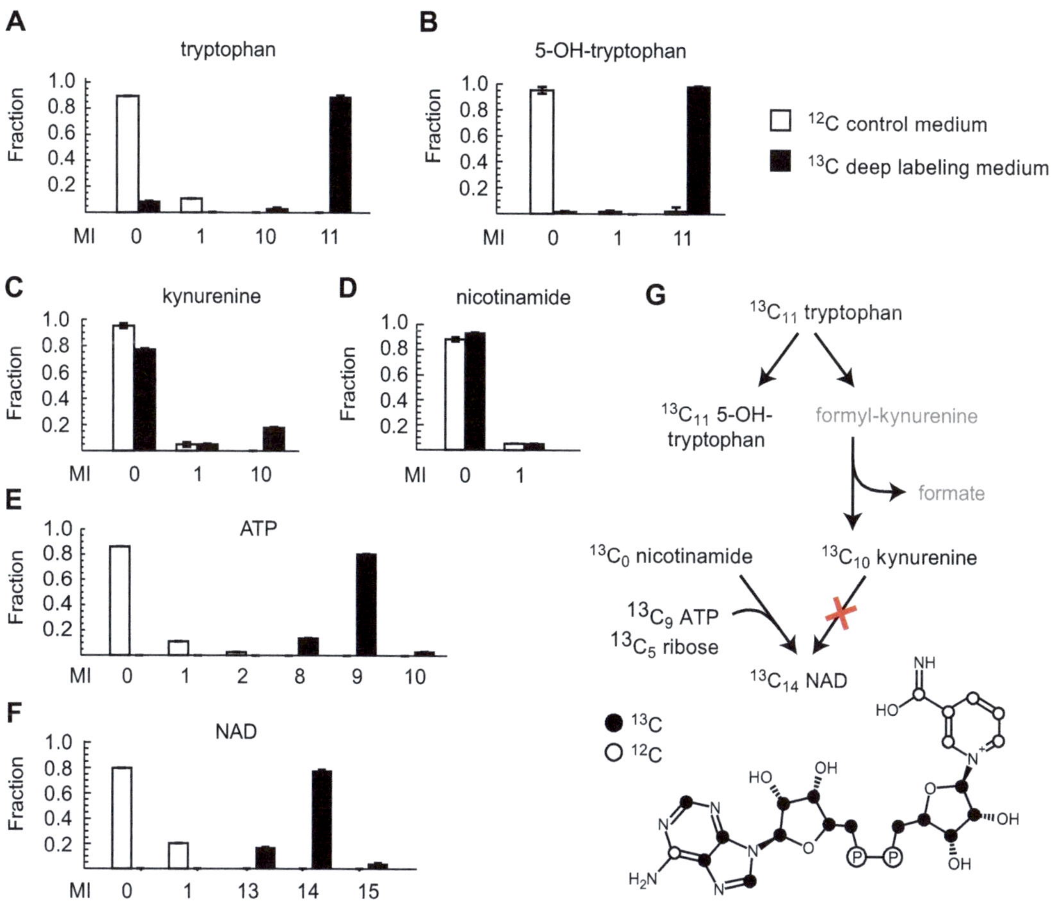

Fig. 3 Example of results from deep labeling experiments, with inference of active and inactive pathways. (**a**–**f**) Mass isotopomer (MI) distributions for indicated metabolites, calculated from ^{12}C control (open bars) and ^{13}C deep labeling experiments in HCT116 cells. Error bars denote standard deviation across biological triplicates. Undetectable mass isotopomers are not shown. (**g**) Above, schematic of tryptophan and NAD metabolism, with major isotopomers observed from (**a**–**f**) indicated. Metabolites listed in gray were not detectable on the LCMS method used. Below, structure of NAD showing the ^{13}C isotopomer consistent with the $^{13}C_{14}$ MI

4 Notes

1. Note that heat inactivation does not eliminate all enzyme activities from serum. Commercial FBS is typically heat-inactivated for 30 min at 56 °C, which does inactivate some enzymes, including asparaginase [10], but not all. Notably, both glutaminase [11] and arginase [12] are resistant to heat inactivation.
2. In some cases, a lower molecular weight cutoff can be needed to retain smaller proteins that are required by certain cell lines, such as peptide hormones and growth factors. Commercially

available dialyzed serum can also be used, but make sure to verify the molecular weight cutoff.

3. The fraction of newly synthesized biomass after N population doublings is $1 - 2^{-N}$, so after $N = 6$ doublings. 98% of the biomass must be been labeled from the medium.
4. The described protocol will extract polar metabolites only; this includes sugars, carboxylic acids, amino acids and various amine-containing compounds, nucleobases and nucleotides, and certain polar or amphipathic lipids such as phospholipids, and some fatty acids. However, nonpolar species such as triglycerides and sterols are not captured by this method. For a straightforward extraction method for nonpolar metabolites, *see* [13].
5. We do not describe further analysis of the medium samples in this chapter, but they are useful to preserve as controls and for troubleshooting. The samples of medium taken from cell cultures can be used for estimating what metabolites the cells have released into the medium. The fresh medium sample (directly from the flask) is a good control to verify ^{13}C enrichment of tracers in the medium, while the medium incubated 48 h without cells can be used to detect potential degradation of metabolites during this incubation time.
6. Washing with PBS is critical to remove residual medium metabolites that otherwise can contaminate the cell extracts. The most serious contaminant is glucose, which is on the order of 1–5 mM in spent RPMI-1640 medium, and similar in cytosol; since in a typical 6-well culture with $\sim 10^6$ cells, there is ~ 2 μL of cytosol total, compared to 2 mL medium. Hence, medium glucose outnumbers cytosolic glucose by at least 1000 to 1. After two rounds of washing with 1 mL PBS followed by extraction with 1 mL methanol, medium metabolites are expected to be reduced by at least a factor 10^6, removing glucose and other contaminants. However, one should be wary that prolonged washing also causes loss of membrane-permeable metabolites from cells.
7. Be careful when scraping the cells in methanol, since it is volatile and drips and splashes easily. To make this procedure easier, tilt the plate when scraping, allowing the suspension to accumulate at the bottom of the plate. Pipet up and down to ensure homogenous solution. During scraping, denatured protein may form fiber-like structures; this can safely be included in the extracts, as protein is removed later during LCMS sample preparation.
8. Methanol will not freeze when stored in −80 °C. If the samples will be shipped for analysis elsewhere, a tip is to add water to

bring the samples up to 50% water by volume the day of shipping to ensure sending frozen samples and reduce the risk of leakage.

9. For untargeted analysis of LCMS data, XCMS [14] and mzMine [15] are commonly used alternatives. However, we have noted that both of these software suites rely heavily on wavelet LCMS peak detection methods, which in our experience tend to give high false positive and false negative rates. The IsoTrack software avoids this problem to a large extent by taking advantage of the fact that chromatographic peaks on Orbitrap systems usually appear as isolated "traces" over time at a specific mass/charge ratio.
10. At the time of writing, the Isotrack software does not exclude ^{34}S mass isotopomers, and does not account for multiply charged species (such as a +2 charged $^{13}C_1$ mass isotopomer, which would differ in m/z from its base isotopomers by 0.5017). Therefore, candidate ions obtained from the described procedure may still contain a small fraction of such mass isotopomers.

Acknowledgments

This work was supported by grants from the Swedish Foundation for Strategic Research (FFL12-0220.006) and Karolinska Institutet to R.N., the National Institutes of Health (NIH) 1R01ES027595 and 1S10OD020025 to M.J., and NIH K01DK116917 to J.D.W.

References

1. Vander Heiden MG (2011) Targeting cancer metabolism: a therapeutic window opens. Nat Rev Drug Discov 10:671–684
2. Freitag J, Berod L, Kamradt T, Sparwasser T (2016) Immunometabolism and autoimmunity. Immunol Cell Biol 94:925–934
3. Viant MR, Kurland IJ, Jones MR, Dunn WB (2017) How close are we to complete annotation of metabolomes? Curr Opin Chem Biol 36:64–69
4. Jain M, Nilsson R, Sharma S, Madhusudhan N, Kitami T, Souza AL et al (2012) Metabolite profiling identifies a key role for glycine in rapid cancer cell proliferation. Science 336:1040–1044
5. Shlomi T, Cabili MN, Herrgard MJ, Palsson BØ, Ruppin E (2008) Network-based prediction of human tissue-specific metabolism. Nat Biotechnol 26:1003–1010
6. Agren R, Bordel S, Mardinoglu A, Pornputtapong N, Nookaew I, Nielsen J (2012) Reconstruction of genome-scale active metabolic networks for 69 human cell types and 16 cancer types using INIT. PLoS Comput Biol 8:e1002518
7. Grankvist N, Watrous JD, Lehmann KA, Lyutvinskiy Y, Jain M, Nilsson R et al (2018) Profiling the metabolism of human cells by deep 13C labeling. Cell Chem Biol 25:1419–1427
8. Moore GE, Gerner RE, Franklin H (1967) Culture of normal human leukocytes. JAMA 199:519–524
9. Lyutvinskiy Y, Watrous JD, Jain M, Nilsson R (2017) A web service framework for interactive

analysis of metabolomics data. Anal Chem 89:5713–5718

10. Wein J, Goetz IE (1973) Asparaginase and glutaminase activities in culture media containing dialyzed fetal calf serum. In Vitro 9:186–193
11. Katunuma N, Huzino A, Tomino I (1967) Organ specific control of glutamine metabolism. Adv Enzym Regul 5:55–58
12. Kihara H, de la Flor SD (1968) Arginase in fetal calf serum. Exp Biol Med 129:303–304
13. Jain M, Ngoy S, SA S, RA S, Rhee EP, Liao R et al (2014) A systematic survey of lipids across mouse tissues. Am J Physiol Endocrinol Metab 306:E854–E868
14. C a S, Want EJ, O'Maille G, Abagyan R, Siuzdak G (2006) XCMS: processing mass spectrometry data for metabolite profiling using nonlinear peak alignment, matching, and identification. Anal Chem 78:779–787
15. Pluskal T, Castillo S, Villar-briones A, Oresic M (2010) MZmine 2 : modular framework for processing , visualizing, and analyzing mass spectrometry-based molecular profile data. BMC Bioinformatics 11:395

Chapter 6

Analyzing the Metabolism of Metastases in Mice

Patricia Altea-Manzano, Dorien Broekaert, João A. G. Duarte, Juan Fernández-García, Mélanie Planque, and Sarah-Maria Fendt

Abstract

Metastasis formation is the leading cause of death in cancer patients. It has recently emerged that cancer cells adapt their metabolism to successfully transition through the metastatic cascade. Consequently, measuring and analyzing the in vivo metabolism of metastases has the potential to reveal novel treatment strategies to prevent metastasis formation. Here, we describe two different metastasis mouse models and how their metabolism can be analyzed with metabolomics and ^{13}C tracer analysis.

Key words Metastasis, Metabolism, In vivo metabolism, ^{13}C tracer analysis, Metabolomics, Mouse infusions

1 Introduction

Metabolic alterations in cancer enable malignant transformation, uncontrolled cell proliferation [1, 2] and promote tumor progression toward metastasis outgrowth [3–5]. The combination of cell intrinsic factors, such as genetic landscape and cell origin, and cell extrinsic factors, such as local nutrient concentrations and cell–cell interactions, regulate this reprogrammed tumor metabolism [4, 6–9]. Understanding the interplay between these cell intrinsic and extrinsic factors in shaping in vivo metastasis progression is highly relevant to develop new therapeutic strategies against metastasis formation. The analysis of metabolism during metastasis formation requires specific methods such as in vivo metabolomics and ^{13}C tracer analysis.

Infusing tumor-bearing mice and cancer patients with isotope-labeled nutrients (in vivo ^{13}C tracer analysis) is emerging as a powerful technique for analyzing tumor metabolism [10, 11]. Examination of label-enriched tumor metabolites after

---

Patricia Altea-Manzano, Dorien Broekaert, and João A. G. Duarte contributed equally to this work.

Deepak Nagrath (ed.), *Metabolic Flux Analysis in Eukaryotic Cells: Methods and Protocols*, Methods in Molecular Biology, vol. 2088, https://doi.org/10.1007/978-1-0716-0159-4_6,

intravenous infusion with nutrients labeled with stable isotopes, such as ^{13}C, ^{2}H or ^{15}N, can provide the first direct assessment of the activity of metabolic pathways in tissues [11, 12]. Using this cutting-edge technique, several studies have shown inter- and intra-tumor heterogeneity of nutrient preferences and pathway activity of intact tumors in mice and humans (Table 1). Based on such in vivo data, it emerges that cancer progression toward metastases formation is inherently linked to changes in the microenvironment [4, 9, 13–15]. However, the complete metabolic rewiring in secondary tumors still remains poorly studied, even though metabolic changes have an essential role in the metastatic outgrowth [4]. Therefore, it is highly relevant to measure and analyze the in vivo metabolism of cancer cells during the metastatic progression.

Here, we describe a protocol for determining the metabolism of mouse primary tumors (breast cancer) and metastases (lung) using stable isotope-labeled nutrients. In particular, we provide a protocol for the establishment of two different lung metastasis mouse models, catheterization, tracer administration by infusion, harvest of (cancer) tissue from mice for subsequent mass spectrometry (MS) analysis, and MS data interpretation.

2 Materials

Prepare and store all reagents at room temperature (unless indicated otherwise). Diligently follow all waste disposal regulations. Use a lab coat, gloves, and, if required, safety glasses at all times. Work, if necessary, under the chemical fume hood. Familiarize yourself with chemical/biological safety and ethical regulations before conducting the experiment.

2.1 Establishing a Lung Metastasis Mouse Model

2.1.1 Cell Preparation

1. Culture medium (*see* **Note 1**) containing 10% fetal bovine serum and 50 U/ml penicillin/streptomycin.
2. Cultured 4T1, B16F10, or EMT6.5 cells.
3. Trypsin–EDTA (0.25%).
4. Dulbecco's phosphate-buffered saline (DPBS).
5. 1.5 ml tube.
6. Ice.
7. Dry ice.
8. 37 °C water bath.
9. Cell culture flask.
10. 70% ethanol.
11. Falcon tube.

Table 1
Isotopic tracers used for measuring metabolic pathways in rodents and humans

Rodent	Human
$^{13}C_6$-glucose	
Healthy C57BL/6 mice [25] Normal tissues from mouse lung [26, 27]	Clear cell renal cell carcinoma [28]
Subcutaneous xenografts non-small cell lung carcinoma [27]	Lung cancer [27, 29–31]
Primary orthotopic human glioblastoma xenografts in mice [32, 33]	Glioblastoma [32, 34]
Lung cancer in genetically engineered mice [25, 26, 30, 35]	Adjacent lung tissue [27, 30, 31]
Pancreatic cancer in genetically engineered mice [25]	
Primary breast cancers and the corresponding lung metastases [13, 14]	
1,6-$^{13}C_2$-glucose	
Primary orthotopic human glioblastoma xenografts in mice [33]	
1,2-$^{13}C_2$-glucose	
Lung cancer in genetically engineered mice [26]	
3,4-$^{13}C_2$-glucose	
Genetically engineered mice and rats [36–38]	
$^{13}C_2$-acetate, 1-$^{13}C_1$-acetate	
Glioblastoma and brain metastases in mice [39]	Clear cell renal cell carcinoma [28]
	Fatty acid synthesis in healthy subjects [40]
$^{13}C_3$-lactate, 2-$^{13}C_1$-lactate	
Lung adenocarcinoma in Ncr nude mice [31] Subcutaneous xenografts non-small cell lung carcinoma and normal lung in mice [27]	Non-small cell lung cancer and adjacent lung tissue [27]
Lung cancer in genetically engineered mice [25]	
Pancreatic cancer in genetically engineered mice [25]	
Mammary carcinoma in rats [41]	
Genetically engineered mice (hepatocyte-specific) [42]	
Healthy C57BL/6 mice [25]	
$^{13}C_3$-pyruvate, 1-$^{13}C_1$-pyruvate	
Subcutaneous xenografts non-small cell lung carcinoma and normal lung in mice [27]	

(continued)

Table 1 (continued)

Rodent	Human
Lung cancer in genetically engineered mice [26]	
Genetically engineered mice (hepatocyte-specific) [42]	
$^{13}C_3$-alanine	
Subcutaneous xenografts non-small cell lung carcinoma and normal lung in mice [27]	
$^{13}C_5$-glutamine	
Primary orthotopic human glioblastoma xenografts [32, 33]	
Lung cancer in genetically engineered mice [25, 35]	
Healthy and pancreatic cancer in genetically engineered mice [25]	
3,4-$^{13}C_2$-acetoacetate	
Genetically engineered mice and rats [36–38]	
$^{13}C_4$-sodium β-hydroxybutyrate	
Genetically engineered mice and rats [36–38]	
$^{13}C_3$-propionate	
Genetically engineered mice and rats [36–38]	
^{13}C-labeled fatty acids	
Healthy and high fat-fed insulin-resistant Sprague–Dawley rats [43]	Healthy [44, 45] and type 2 diabetes subjects [46]
	Normoinsulinemic and hyperinsulinemic subjects [47]
2, 3-$^{2}H_2$-serine	
Colon cancer xenografts in CD1/nude mice [48]	
^{13}C-branched-chain amino acids	
Healthy C57BL/6 and insulin-resistant mice [49]	
^{13}C-bicarbonate/$^{13}CO_2$	
Lymphocytic leukemia in CDF1 mice [50]	
$^{2}H_2O$	
Healthy Sprague–Dawley rats and C57Bl/6 J male mice [51]	Healthy volunteers [51, 52]

2.1.2 Intravenous Injection of Cancer Cells

1. Balb/c mice 6 weeks old or C57BL/6 mice 8 weeks old (*see* **Note 2**).
2. Infrared lamp (*see* **Note 3**).
3. Insulin syringe with needle 29G, 0.5 ml, 12 mm length.
4. Single cell solution on ice prepared in Subheading 2.1.1.
5. Clean mouse cage.
6. Mouse tail vein restrainer.

2.1.3 Injection of Cancer Cells in the Mammary Fat Pad

1. Mouse shaver.
2. Hair removal cream.
3. Cotton swab.
4. Insulin syringe with needle 29G, 0.5 ml, 12 mm length.
5. Single cell solution on ice prepared in Subheading 3.1.1.

2.2 Mouse Surgery and ^{13}C Glucose Infusion

This technique can be performed as described by Broekaert and Fendt [10].

2.3 Collection of Frozen Tissue Samples

1. Liquid nitrogen.
2. Dry ice.
3. Ice.
4. Labeled polyzip bags.
5. Biosqueezer.
6. 0.9% blood bank saline.
7. 1 ml Single-use tuberculin syringe with ml graduation, Luer tip.
8. Microvette for capillary blood collection.

2.4 Metabolite Extraction

Work under the chemical fume hood when preparing extraction solutions.

1. Labeled 2 ml Eppendorf tubes (3 tubes per sample).
2. Grinding balls, stainless steel, 5 mm.
3. Grinding balls, stainless steel, 3 mm.
4. Analytical balance (range: 0.0001–50 g).
5. Forceps.
6. Liquid nitrogen.
7. Dry ice.
8. Ice.
9. MS grade chloroform.

10. Chromasolv grade methanol.
11. Cryomill.
12. Milli-Q water.
13. 400× norvaline–glutarate internal standard stock solution for analysis of polar metabolites: Dissolve 1 mg each of both norvaline and glutarate in 1 ml Milli-Q water (*see* **Note 4**). Aliquots can be stored for up to 1 year at −20 °C.
14. Extraction solution 1 (water + norvaline–glutarate + methanol): Prepare per sample 0.75 μl of 400× norvaline–glutarate stock (i.e., 0.75 μg of each norvaline and glutarate per sample) in 300 μl Milli-Q water. Add 500 μl methanol per sample and mix (final ratio water–methanol = 3:5). Store extraction solution 1 at −20 °C until extraction (*see* **Notes 4–6**).
15. 100× internal standard stock solution for analysis of fatty acids: Dissolve 1 mg of heptadecanoic acid (C_{17}) in 1 ml chloroform (*see* **Notes 4** and **5**). The C_{17} internal standard stock can be stored at −80 °C for 3–4 weeks.
16. Extraction solution 2: Prepare per sample 5 μl of 100× C_{17} stock (i.e., 5 μg C_{17} per sample) in 500 μl chloroform. Store this solution at −80 °C until extraction.
17. Chemical fume hood.
18. Refrigerated acid resistant vacuum centrifuge.

2.5 Derivatization of Polar Metabolites for Gas Chromatography-Mass Spectrometry (GC-MS) Analysis

Work under the chemical fume hood when preparing derivatization solutions

1. Methoxyamine derivatization solution: 20 mg/ml solution of *O*-Methoxyamine-HCl in anhydrous pyridine, counting for at least 20 μl of solution per sample. Weigh 20 mg methoxyamine in a safe lock Eppendorf tube. Wash a needle with pyridine using a syringe and discard the waste appropriately. Aliquot more than 1 ml of pyridine in a safe-lock Eppendorf tube using the washed needle. Add 1 ml of pyridine to the methoxyamine. Vortex until methoxyamine is entirely dissolved (*see* **Note** 7).
2. Glass GC-MS vials with insert and magnetic caps with septum.
3. Crimper.
4. *N*-(*tert*-butyldimethylsilyl)-*N*-methyl-trifluoroacetamide with 1% *tert*-butyldimethylchlorosilane (TBDMS) derivatization agent (*see* **Note 8**).
5. Heating block.
6. Safe-Lock Eppendorf tubes.
7. Analytical scale (range: 0.0001–50 g).

8. Syringe with needle.
9. Chemical fume hood.
10. Refrigerated acid resistant vacuum centrifuge.

2.6 Derivatization of Nonpolar Metabolites (Fatty Acids) for GC-MS Analysis

Work under the chemical fume hood when preparing derivatization solutions

1. 2% sulfuric acid in MS-grade methanol (*see* **Note 9**).
2. Hexane (≥97.0%, GC grade).
3. Saturated NaCl: Dissolve sodium chloride in autoclaved Milli-Q water until solution is saturated. From then on, work only with the saturated (top) fraction.
4. Glass GC-MS vials with insert and nonmagnetic caps with septum.
5. Crimper.
6. Heating block.
7. Safe-Lock Eppendorf tubes.
8. Chemical fume hood.
9. Refrigerated acid resistant vacuum centrifuge.

2.7 Preparation of Polar Metabolite Samples for LC-MS Analysis

1. 60% acetonitrile (LC-MS grade): prepare under the chemical fume hood.
2. Plastic vials with insert and plastic caps with septum.
3. Safe-Lock Eppendorf tubes.
4. Chemical fume hood.
5. Refrigerated acid resistant vacuum centrifuge.

2.8 Protein Quantification

1. BCA protein assay kit.
2. 0.2 M NaOH: Weigh 0.8 g NaOH and transfer to a glass recipient. Add autoclaved Milli-Q water to a volume of 100 ml and mix (*see* **Note 10**).
3. Autoclaved phosphate-buffered saline (PBS).
4. Heating block.
5. Multichannel pipette.
6. Pipetting reservoir.
7. 96-Well plate.
8. 37 °C incubator.

2.9 Collection of Lungs for Hematoxylin and Eosin (H&E) Staining

1. Formalin solution, neutral buffered, 10% (*see* **Note 11**).
2. 27 G needle.
3. 1 ml syringe without needle.

4. Mouse dissection tools: needle holder, scissors, standard forceps.
5. Closed container.
6. PBS.
7. Histology embedding cassettes.

3 Methods

Perform all procedures at room temperature unless otherwise specified.

3.1 Establishing a Lung Metastasis Mouse Model

3.1.1 Cell Preparation

1. Thaw one vial of 4T1, B16F10, or EMT6.5 cells by removing it rapidly from liquid nitrogen and immediately placing it into a 37 °C water bath (*see* **Note 12**).
2. Wash the vial with 70% ethanol before opening. Resuspend the contents of the vial in a falcon tube containing ten times the volume in the vial of the appropriate cell culture medium (RPMI for 4T1 and B16F10 cells, or MEM alpha for EMT6.5 cells). Spin down at 300 × *g* for 5 min. Aspirate the cell culture medium, and resuspend the cell pellet in the desired volume of the appropriate cell culture medium. Transfer the cell suspension to a cell culture flask.
3. When the cell culture flask reaches 80–90% confluency, "split" the cells in an appropriate ratio depending on growth rates. 4T1, B16F10 and EMT6.5 cells are fast growing cells and can be split using trypsin-EDTA (0.25%) in a 1/10 ratio to reach confluency again after 3 days. Expand the cells up to the desired amount (*see* **Note 13**). For intravenous injection, 100,000 cells per mouse will be needed. For injection in the mammary fat pad, 1 million cells per mouse will be needed (*see* **Note 14**).
4. When the desired amount of cells has been reached, dissociate the cells with trypsin-EDTA (0.25%) and wash three times with PBS to remove all Fetal Bovine Serum and penicillin/streptomycin. To wash the cells, spin down at 300 × *g* for 5 min. Count the cells and resuspend in the desired amount of PBS (*see* **Note 15**).
5. For intravenous injection, resuspend 100,000 cells in 100 μl. For injection in the mammary fat pad, resuspend 1 million cells in 50 μl (*see* **Note 16**). Transfer the cell suspension into an Eppendorf tube and put on ice.

3.1.2 Intravenous Injection of Cancer Cells (Lung Metastasis W/O Primary Tumor)

Before you start, take a clean cage and direct an infrared (IR) lamp toward the cage to be able to dilate the blood vessels in the tail of the mice (Fig. 1a) (*see* **Note 3**). Put a tail vein restrainer in front of you. Transfer the first 5 mice to the clean cage with direct IR light.

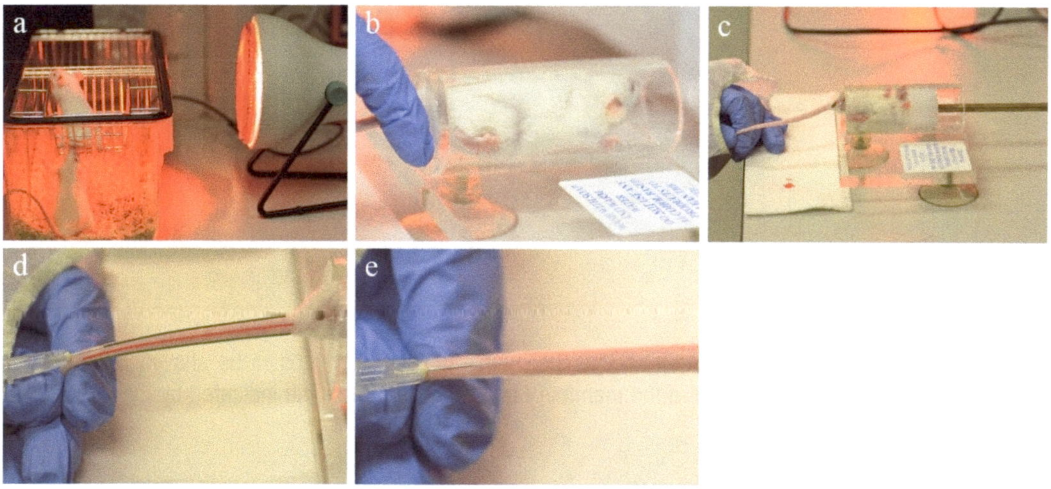

Fig. 1 Generation of lung metastasis mouse model without primary tumor by intravenous injection of cancer cells in the mouse tail vein. (**a**) Infrared lamp is directed toward the mice in the cage. (**b**) Position the mouse in a tail vein restrainer. (**c**) The tail is bent between the index finger and thumb. (**d**) Identify the lateral tail veins (indicated by the green lines) left and right from the ventral artery (indicated by the red line). (**e**) Insert the needle almost parallel to the tail

1. Resuspend the cells in the Eppendorf tube by pipetting up and down (*see* **Note 17**). Fill the insulin syringe (29G) with 100 μl cell suspension and remove all bubbles from the syringe (*see* **Note 18**).
2. Move the IR lamp further away from the clean cage, but keep the cage warm (*see* **Note 19**).
3. Take the first mouse and put it in a tail vein restrainer (Fig. 1b) (*see* **Note 20**). Hold the tail in your left hand (when right handed), identify the lateral tail vein and slightly bend the tail in between the index finger and thumb (Fig. 1c). To define the lateral tail vein you can place the mouse in front of you. The vein facing upward is the dorsal tail vein, the vein facing downward is the ventral artery (Fig. 1d red line) and the veins left and right from the dorsal tail vein are the lateral tail veins (Fig. 1d green lines). Insert the needle almost parallel to the tail, with the needle opening upward, and inject the cell suspension in one of the two lateral tail veins (Fig. 1d, e). The vein should change color when injecting correctly (*see* **Note 21**).
4. Put the mouse slowly back in its nonheated cage.

3.1.3 Injection of Cancer Cells in the Mammary Fat Pad (Spontaneous Metastasis from Primary Tumor)

1. Shave the mice 2 days before injection around the nipple (Fig. 2) (*see* **Note 22**). To remove every single hair, you can evenly spread a drop of hair removal cream on the nipple area using a cotton swab. Wait approximately 2 min so the skin can absorb the cream and remove the cream using wet tissues.

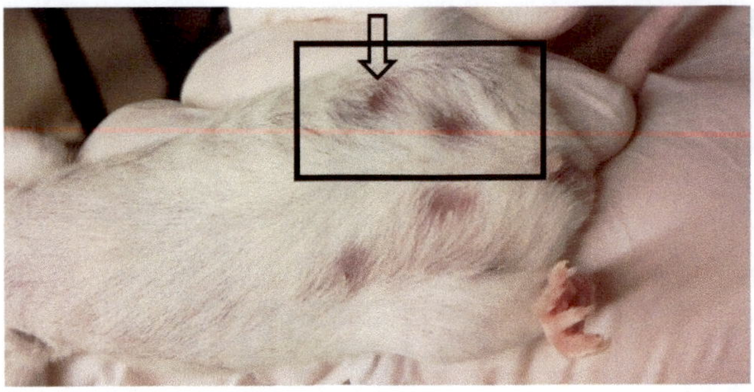

Fig. 2 Area around the mouse nipple that needs to be shaved. Cells will be injected in the mammary fat pad under the nipple indicated by the arrow

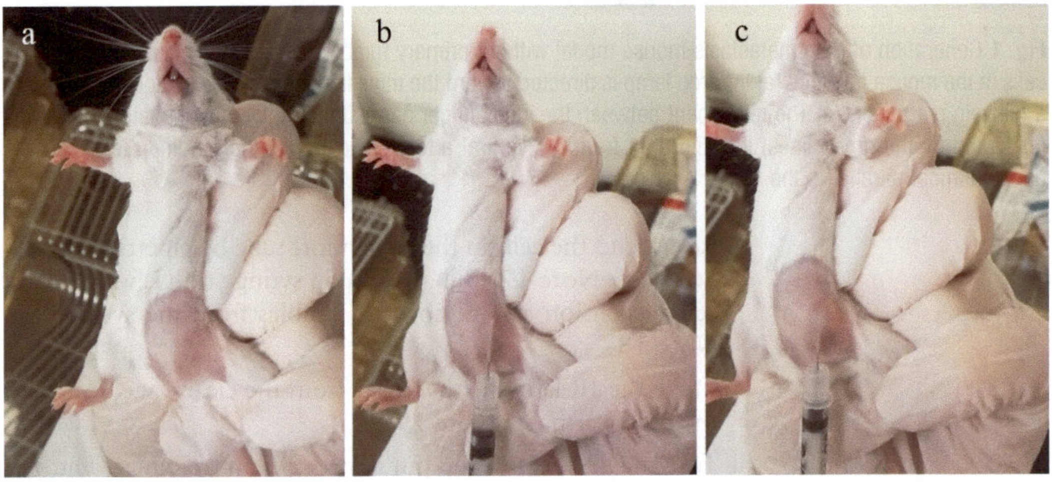

Fig. 3 Generation of spontaneous metastasis mouse model (with primary tumor) through injection of cancer cells in the mammary fat pad. (**a**) Pick up the mouse with your left hand and locate the nipple. (**b**) The needle was inserted about 10 mm distally from the nipple horizontally under the skin and moved to the nipple. (**c**) Inject when the needle tip is right under the nipple. A small bubble under the nipple is an indication of a good injection

2. Resuspend the cells in the Eppendorf tube by pipetting up and down (*see* **Note 17**). Fill the insulin syringe (29G) with 50 μl cell suspension and remove the bubbles.
3. Inject the cells subcutaneously under the nipple, by inserting the needle about 10 mm distally from the nipple horizontally under the skin. Move the needle horizontally and subcutaneously toward the nipple, and inject when the needle tip is right under the nipple, with the needle opening upward (*see* **Note 23**) (Fig. 3a, b). A small bubble under the nipple indicates a good injection (Fig. 3c) (*see* **Note 24**).

Table 2
Duration of described mouse models to generate lung metastases

Mouse strain	Cell line	Type of injection	Number of cells injected	Injection volume	Model duration
Balb/c	4T1	Mammary fat pad	1 million	50 μl	~3 weeks
Balb/c	4 T1	IV	100,000	100 μl	~2 weeks
Balb/c	EMT6.5	Mammary fat pad	1 million	50 μl	~3.5 weeks
Balb/c	EMT6.5	IV	100,000	100 μl	~2.5 weeks
C57BL/6 J	B16F10	IV	150,000	100 μl	~3 weeks

Mouse surgery and infusions can be performed as described before [10] (*see* **Note 26**)

3.2 Mouse Surgery and Infusion of Labeled Metabolites

The duration of the model and moment of infusion differs for each mouse model, based on the humane endpoints to sacrifice the mice defined by the ethics regulations where the experiment is conducted (*see* **Note 25**) (Table 2).

Mouse surgery needs to be performed approximately 1 week before infusion.

3.3 Collection of Snap-Frozen Tissue Samples

1. Sacrifice the mouse with an overdose of Dolethal (140 mg/kg, 2.8 μl per gram of animal weight of a 50 mg/ml solution) (*see* **Notes 27** and **40**).
2. Open the mouse and collect the blood by heart puncture. Put the blood on ice in a microvette for capillary blood collection, and spin down right after for 10 min at 10,000 × *g*. Transfer the plasma to an Eppendorf tube and store at −80 °C.
3. Collect the organs of interest as fast as possible. Wash the tissue in ice cold blood bank saline, remove the saline with a sterile compress, put the tissue in a labeled polyzip bag, squeeze it with the precooled biosqueezer, and put it into liquid nitrogen (*see* **Note 28**). Store the tissues at −80 °C.
4. When collecting tumor tissue, it is important to separate tumor and healthy tissue immediately during collection, as frozen tissue is not malleable enough to allow for a correct separation of tumor tissue. However, keep in mind to limit the collection time to prevent any potential tissue degradation (*see* **Note 29**).

3.4 Metabolite Extraction

Work under the chemical fume hood when handling the extraction solutions.

1. Weigh a piece of tissue (approximately 10 mg) while keeping it as cold as possible (*see* **Note 30**).

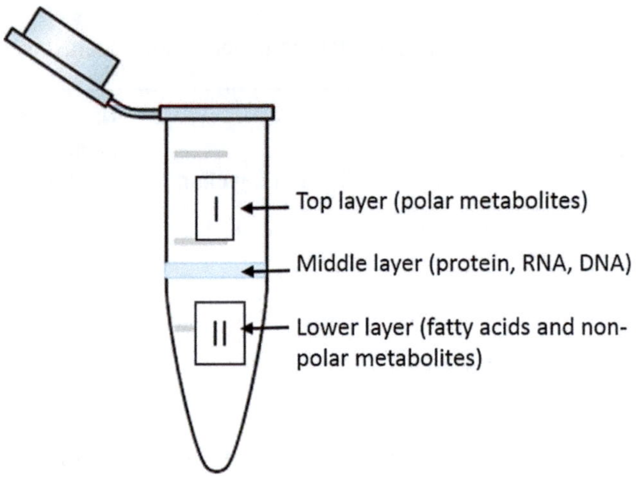

Fig. 4 Eppendorf tube with the three layers after metabolite extraction. The top layer consists of extraction solution 1 containing polar metabolites. The middle layer consists of protein, DNA and RNA. The lower layer consists of extraction solution 2 containing fatty acids and nonpolar metabolites

2. For grinding the tissue, add one 5 mm precooled grinding ball and one 3 mm precooled grinding ball to every tube with tissue. Cool down the Cryomill machine by opening the connection to the liquid nitrogen tank and grind the tissue for 30–40 s (frequency: 25 Hz). Place the sample tubes on dry ice immediately after grinding (*see* **Note 28**). Grounded tissue can be stored at −80 °C (*see* **Note 31**).
3. Transfer the tubes to a mixture of dry ice and ice. Add to each tube 800 μl of ice cold extraction solution 1, and 500 μl of ice-cold extraction solution 2. Vortex the samples for 10 min at 4 °C, and then centrifuge them for 10 min at 4 °C maximum speed. The contents of the tube will now have separated into 3 layers (Fig. 4).
4. Pipet the polar top phase (consisting of extraction solution 1) into a new Eppendorf tube, and place the tube on dry ice. This layer will contain polar metabolites such as amino acids and organic acids. The middle layer consists of protein, DNA, and RNA. With a new pipette, cross through the middle layer while releasing some air, so that proteins, DNA, and RNA do not enter the pipette tip. Then, collect the nonpolar bottom phase (consisting of extraction solution 2), transfer into a new Eppendorf tube (trying not to disturb the middle layer while removing the pipette tip), and place the tube on dry ice (*see* **Note 32**). This lower phase will contain fatty acids and other nonpolar metabolites (Fig. 4). Bring also the tube with the remaining protein/DNA/RNA into dry ice.

5. Dry down the collected samples in a vacuum centrifuge. The samples containing polar metabolites should be dried at 4 °C for 8 h, while the protein and fatty acid containing samples can be dried down at 20 °C for 1 h in an acid resistant vacuum centrifuge. When dry, store the samples at −80 °C.
6. Perform this extraction process one additional time in triplicate in identical Eppendorf tubes but without any sample. These samples are "mock" extractions and will serve to assess the presence of impurities at a later point.

3.5 Derivatization of Polar Metabolites for GC-MS Analysis

Work under the chemical fume hood when handling the derivatization solutions.

1. Add 20 μl of 20 mg/ml *O*-methoxyamine–HCl dissolved in pyridine to each sample, vortex briefly, and incubate for 90 min in a heating block at 37 °C. Subsequently, centrifuge the samples for 3–5 min at maximum speed, transfer 7.5 μl of the supernatant into a glass GC-MS vial with insert, and seal with a magnetic cap using a crimper (*see* **Note 33**).
2. Fill a glass vial with insert with *N*-(*tert*-butyldimethylsilyl)-*N*-methyl-trifluoroacetamide (TBDMS, 15 μl/sample + 25 μl extra per vial) and seal with a nonmagnetic cap (*see* **Note 34**). Fill a 2 ml glass vial without insert with pyridine, and seal with a nonmagnetic cap. Program the GC-MS autosampler to add 15 μl TBDMS to each sample, incubate for 1 h at 60 °C, and resuspend the sample prior to injection. Program also the autosampler to wash the injection needle with pyridine in between loading different samples (*see* **Notes 35** and **36**).
3. Measure the samples using a GC-MS instrument. In our case, as described before [13, 16], samples are measured on an Agilent 7890A GC system coupled to an Agilent 5975C Inert MS system. In brief, 1 μl of sample is injected into a DB35MS column in splitless mode, using an inlet temperature of 270 °C, and with helium at a flow rate of 1 ml/min as the carrier gas. Upon injection, the GC oven is first held at 100 °C for 1 min, then ramped up to 105 °C with a gradient of 2.5 °C/min, after that ramped up to 240 °C with a gradient of 3.5 °C/min, and finally ramped up to 320 °C with a gradient of 22 °C/min, followed by a 4 min post-run at 320 °C. The MS system is operated under electron impact ionization at 70 eV, and a mass range of 100–650 amu is scanned.

3.6 Derivatization of Nonpolar Metabolites (Fatty Acids) for GC-MS Analysis

Work under the chemical fume hood when handling the derivatization solutions.

1. Add 500 μl of 2% sulfuric acid in methanol to each sample and incubate for 3 h at 60 °C or overnight at 50 °C (*see* **Notes 28** and **33**).

2. Add 600 μl hexane and 100 μl saturated NaCl to each sample, vortex 10 min, and centrifuge 5 min at maximum speed. Transfer the upper hexane phase to a new Eppendorf tube, and dry the samples for 10 min at 20 °C in an acid resistant vacuum centrifuge (*see* **Note 37**).
3. Resuspend in a defined amount of hexane, depending on the amount and type of tissue (e.g., a sample containing 5 mg normal liver was resuspended in 440 μl hexane), vortex 10 min, transfer to a glass vial with insert, and seal with a nonmagnetic cap using a crimper.
4. Measure the samples using a GC-MS instrument. In our case, the approach is analogous to the one described above for polar metabolites, but using a higher helium carrier-gas flow rate (1.3 ml/min) and a different temperature gradient: upon injection of nonpolar metabolite samples, the GC oven is first held at 140 °C for 2 min, then ramped up to 185 °C with a gradient of 1 °C/min, and after that ramped up to 300 °C with a gradient of 20 °C/min, followed by a 2 min post-run at 300 °C. The MS instrument is operated in dual (positive + negative) polarity, with the following ion source parameters: gas (N_2) temperature = 270 °C; gas flow = 10 l/min; nebulizer pressure = 35 psi; sheath gas (N_2) temperature = 300 °C; sheath gas flow = 12 l/min; capillary voltage = 3500/−3000 V (positive/negative polarity); nozzle voltage = 500 V. Ion detection is performed in multiple reaction monitoring (MRM) mode, with a total cycle time of 1500 ms.

3.7 Protein Quantification

1. Resuspend the dried protein pellet in 200 μl 0.2 M NaOH and incubate for 20 min in a heating block 95 °C (*see* **Note 28**).
2. For the quantification continue as described in the manual of the BCA protein assay kit.
3. Use the protein amount to normalize the measured metabolite levels to the starting material used.

3.8 Collection of Lungs for H&E Staining

1. Sacrifice the mouse with an overdose of Dolethal (140 mg/kg, 2.8 μl per gram of animal weight of a 50 mg/ml solution) (*see* **Notes 27** and **38**).
2. Sanitize the mouse by spraying 70% ethanol on the fur along the abdominal region, so that any loose/dry hairs will not enter the region (Fig. 5a).
3. Carefully open the mouse thorax with a midline incision, cut away the diaphragm, and cut open the lateral chest walls without touching the lungs (Fig. 5b) (*see* **Note 38**).
4. Open the skin at the tracheal area and remove all the glands in that area until you can easily see and reach the trachea. Remove the connective tissue around the trachea using scissors. (Fig. 5c, d) (*see* **Note 39**).

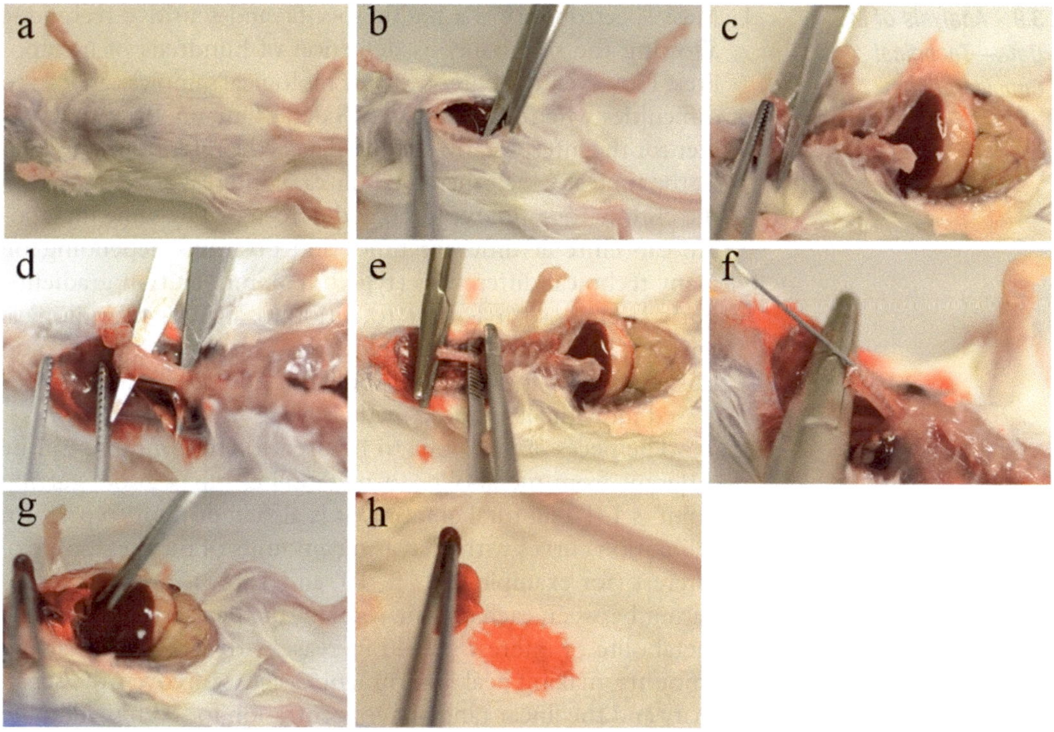

Fig. 5 Collection of lungs for H&E staining. (**a**) The mouse has been sanitized using 70% ethanol. (**b**) Open the mouse thorax with a midline incision. (**c**) Open the skin at the tracheal area and remove all the glands. (**d**) All glands and connective tissue have been removed around the trachea. (**e**) Clamp the cranial part of the trachea using a needle holder. (**f**) Insert the needle in the caudal part of the trachea parallel to the trachea. (**g**) Collect the lung. (**h**) Wash the lung in PBS, remove the blood and store in 10% neutral buffered formalin solution

5. Take a 1 ml syringe with 27G needle and fill with 1 ml 10% neutral buffered formalin solution. Clamp the cranial part of the trachea using a needle holder (Fig. 5e). Hold the trachea with the needle holder while you insert slowly the needle in the caudal part of the trachea parallel to the trachea with the needle opening toward you. Inject 1 ml 10% neutral buffered formalin solution (Fig. 5f) (*see* **Note 11**).
6. Collect the lung, wash for 1 min in PBS and store for 24 h at 4 °C in a closed container filled with 10% neutral buffered formalin solution. Ensure that the fixative is at least 10× the volume of tissue (Fig. 5g, h) (*see* **Note 40**).
7. After 24 h, wash the samples with 70% ethanol and proceed with embedding the lung in paraffin using an automated machine routinely used for tissue embedding in paraffin wax. When embedded, samples can be stored at room temperature [17, 18].

3.9 Analysis of MS Data—Technical Considerations

1. Mass spectrometry is a highly specific and sensitive technique allowing the simultaneous detection of hundreds of metabolites. However, the complexity of biological samples can lead to difficulties when interpreting MS data. One important parameter for the interpretation of MS data is the retention time, that is, the time at which the metabolites of interest are eluted from the chromatographic column and detected. The same metabolite can elute at different times and positions depending on many technical differences (type of column, elution gradients, buffers, etc.), so it is best to, prior to analysis of biological samples, establish an internal library of retention times. To do this, standards for each metabolite of interest, must be processed individually as described above.

   The necessary amount to be used for this purpose can be different from metabolite to metabolite, but a good starting point is 1–10 μg total amount. It is also important to perform calibration curves with increasing amounts of the metabolite of interest (for example 0, 1, 5, 10, and 15 μg). This way, comparing between each chromatogram, the peak arising from the metabolite of interest should increase with increasing starting amounts, making peak assignment easier. Moreover, the sensitivity and the linear range of concentration for each metabolite can be determined. The sensitivity (limit of detection) is defined as the lowest amount of metabolite that can be detected, and must be determined for each metabolite considering the slope of the calibration curve or a signal-to-noise ratio above 3 [19]. Once a retention time library has been established, it is possible to mix all metabolites of interest (provided that they do not coelute) in a single set of samples and perform standard curves for all metabolites in one go. Due to factors such as detector and derivatization efficiency, it is possible that the calibration curves from a specific run might not apply to a future run and, therefore, a new calibration curve should be performed each time.

2. The total amount of a metabolite present in the sample is directly proportional to its peak area. Thus, if information about the absolute amount of a metabolite (i.e., a metabolite concentration) in biological samples is necessary, it is important to create a standard curve. It is important to extract the samples used to generate the standard curve in the same way as the biological sample. Consequently, the peak area of the standard curve samples also need to be normalized to the internal standard (*see* Subheading 3.10.1). Due to the complexity of biological samples there is the possibility that metabolites will interact with each other, leading to shifts in retention times and/or peak intensity. This is known as the matrix effect.

   To correct for this effect, perform an additional 3 extractions for each of your biological samples using a similar amount of starting material to that of the biological samples, and pool them together (to prevent any possible metabolite differences between extraction). Afterward, split the samples into 3 aliquots and perform standard addition of $0\times$, $2\times$ and $5\times$ of the initial quantity of the metabolite of interest, as determined by the standard curves mentioned in Subheading 3.9, **step 1**.

If there are no matrix effects, this new calibration curve, performed in the presence of the biological matrix, should have the same slope as that of the pure metabolite, any deviation will be due to the matrix effect and the new standard curve should be used to quantify the metabolite.

Since the matrix effect is dependent on the metabolites present in the samples and these can vary wildly from tissue to tissue, it is recommended to perform matrix correction for each of the different biological origins of the samples. For example, a matrix effect present in a liver sample might not be present in a kidney sample; the same is true for liquid samples: a urine sample and plasma sample can possess different matrix effects that should be corrected separately.

3.10 Interpretation of MS-Derived Biological Data

3.10.1 Calculation of Total Metabolite Levels

1. For each sample and metabolite of interest, identify all possible mass isotopologs potentially contributing to that metabolite's total signal. We will designate these as $M + i$, where M is the mass of the unlabeled metabolite, and i ($i = 0, 1, \ldots, N$) denotes a specific isotopolog, based on the number of tracer atoms incorporated into the metabolite backbone (with N being the maximum number of tracer atoms that the metabolite can incorporate). For instance, if a ^{13}C tracer (such as $^{13}C_6$-glucose) is fed to the animal, there are 4 possible mass isotopologs potentially arising for pyruvate ($C_3H_4O_3$), namely, those due to the incorporation of 0, 1, 2, or 3 ^{13}C atoms into its carbon backbone (i.e., $M + 0$, $M + 1$, $M + 2$, and $M + 3$, respectively).
2. Determine the ion counts measured for each of those isotopologs (IC_{M+i}), by appropriately integrating the area under the corresponding spectral peaks using any commercially available peak integration software.
3. Add up the ion counts for all isotopologs corresponding to every given metabolite, to obtain total ion counts for each metabolite, $TC = \sum_{i=0}^{N} IC_{M+i}$. Furthermore, since some metabolites may be present as impurities arising from the extraction buffers or the tubes where the derivatization was performed, "mock" samples, as defined in Subheading 3.4, **step 6** can be used identify these impurities. Since these "mock" samples did not have any contact with any biological material, any metabolite present in these samples is considered an impurity. Consequently, metabolite abundances below or at the level of the mock should not be used for biological interpretation.

4. Standardize the total ion counts for each metabolite relative to those of the appropriate internal standard (e.g., norvaline, glutarate, or heptadecanoic acid for amino acids, organic acids and fatty acids, respectively) in the same sample. This removes the influence of any potential loss of material during the extraction process (*see* **Note 41**).
5. Using the previously mentioned calibration curves (standardized in an analogous manner), transform these standardized ion counts into absolute levels (concentrations, or amounts). This step can be skipped if all that is sought from the experiment is a comparison between the relative levels of the same metabolites among different samples (e.g., when comparing pyruvate levels between primary tumor and metastasis tissues).
6. Finally, normalize the above levels (absolute or relative) by the starting amount of biological sample (i.e., fluid volume, tissue weight, or cell number). These values can then be compared between different groups.

3.10.2 Analysis of Fractional Nutrient Contributions and Pathway Activities

1. For each sample and metabolite of interest, determine a (measured) fractional mass isotopolog distribution, or mass distribution vector (MDV), $\Upsilon = \{\Upsilon_{M+0}, \Upsilon_{M+1}, \ldots, \Upsilon_{M+N}\}$, based on the isotopologue counts IC_{M+i} and total ion counts TC defined in Subheading 3.10.1, according to

$$\Upsilon_{M+i} = \frac{IC_{M+i}}{TC} (i = 0, 1, \ldots, N) \quad (1)$$

2. For appropriate interpretation, the measured MDV must first be corrected to account for the natural abundance of heavier isotopes. This is because, for every metabolite, and independently of the presence of tracer atoms, there is a probability for each of the atoms in it to naturally appear in the form of a heavier isotope. Thus, these natural abundances must be corrected to avoid overestimating the label incorporation into the metabolites of interest. The theory underlying natural abundance correction and its practical implementation have been widely covered in the literature [20, 21]. Readers are thus encouraged to use one of the several programmatic tools already developed and tested to perform these corrections (see for instance https://pypi.org/project/IsoCor). For the purpose of this manuscript, it is sufficient to state that the latter all involve solving a linear system of equations of the form

$$\Upsilon = \overline{\overline{L}}\tilde{n}X \quad (2)$$

Where Υ is the measured MDV, $X = \{X_{M+0}, X_{M+1}, \ldots, X_{M+N}\}$ is the actual (natural abundance-corrected) MDV, and $\overline{\overline{L}}$ is a correction matrix, whose components depend both on the full atomic composition of the metabolite of interest upon measurement (including all atoms, not only those susceptible of tracer incorporation) and on the nature and number of tracer atoms that can be incorporated by the latter. Consequently, it is important to keep in mind that, if there is a chemical derivatization process (such as the aforementioned TBDMS derivatization), the atomic compositions of the whole derivatives (and not just those of the bare metabolites) must be taken into consideration when correcting for natural abundance. Conversely, the number of potentially labeled atoms considered for the correction will stem from the bare metabolite backbone, since the chemical derivatization process will not add labeled atoms to the metabolite. For instance, in the case of the TBDMS derivative of pyruvate, the formula to be used for natural abundance correction is $C_6H_{12}O_3NSi$ (corresponding to the full derivative ion), and not $C_3H_4O_3$ (corresponding to pyruvate only). However, the correction must consider that only up to 3 carbon atoms (and not 6) may be labeled. The same principle applies whenever any atoms are added to/removed from the metabolite as a result of an ionization process (e.g., H^+ removal upon negative electrospray ionization in LC-MS measurements).

3. After correcting for natural abundance, it is possible to calculate the fractional contribution (FC) of the tracer to each metabolite of interest, by using the equation

$$\mathrm{FC} = \frac{\sum_{i=0}^{N} i\tilde{n}X_{M+i}}{N} \tag{3}$$

where N is the number of atoms in the metabolite susceptible of tracer incorporation, and the X_{M+i} are the corrected fractional abundances of each of its different isotopologues. In isotope-labeling experiments with uniformly labeled tracers, FC represents the fraction of a given metabolite pool that originates from those labeled tracers, and thus can be used to infer the relative contributions of labeled and unlabeled nutrients toward the production of a given metabolite [12]. For a faithful comparison between FCs for different samples, the FC of each metabolite of interest must be standardized toward the actual FC of the original tracer in the same sample. For example, if $^{13}C_6$-glucose is infused, the FC of plasma glucose in each sample should be used to standardize the FC values for all other metabolites in that sample. This standardization corrects for any possible dilutions of the initial tracer in the bloodstream, which will significantly alter the distribution of label across the metabolic network. Following the example of $^{13}C_6$-glucose and pyruvate, let the FC values determined for each metabolite be 0.90 and 0.72, respectively. This would indicate that 0.72/0.9 = 80% of the measured pyruvate originated from glucose.

As an exception to this, it is possible to directly compare the FC of the same metabolite in two different tissues of the same mouse (for example, hepatic and renal pyruvate FC) as the enrichment of the tracer in the blood stream of the mouse is the same measure for both tissues.

4. In addition to obtaining information about differential nutrient contributions (by comparing the FCs for given metabolites between samples), it is also possible to infer conclusions about differential pathway activities by comparing the underlying MDVs. Indeed, since each isotopolog can only arise from a subset of specific metabolic pathways, comparing the relative abundances of specific isotopologs of a given metabolite can provide information on the relative activities of the metabolic pathways converging into that metabolite. For instance, when analyzing citrate labeling from a uniformly labeled $^{13}C_6$-glucose tracer, multiple isotopologs can be detected, including $M+2$ (if doubly labeled acetyl-CoA is incorporated into citrate via citrate synthase), $M+3$ (if uniformly labeled pyruvate is incorporated into oxaloacetate via pyruvate carboxylase, and from there further into citrate), or $M+5$ (if both pathways are active). Thus, comparing the fractional abundances of the citrate $M+2$, $M+3$, and $M+5$ isotopologues will provide information about the relative activities of those pathways.

   It is important to keep in mind that, unlike FC, individual isotopologs should never be corrected for the enrichment of the initial tracer in the blood stream.

4 Notes

1. Culture media needs to be adapted to the individual cell lines (e.g. RPMI 1640 Medium for 4T1 and B16F10 cells and MEM Alpha Medium for EMT6.5 cells).
2. Mouse strains need to be syngeneic with the injected cell line (e.g., murine 4T1 or EMT6.5 cell lines can be injected in Balb/c mice without any rejection response). The B16F10 cell line can be injected in C57BL/6 mice without any rejection response.
3. Alternative: heating chamber. You can heat the tail by putting it in warm water or in a heating chamber.
4. Use glass bottles and glass pipettes or “inert plastics” if possible. Normal plastic will get dissolved.
5. Work under the chemical fume hood.
6. When performing LC-MS measurements without infusion of labeled metabolites, add ^{13}C yeast internal standard solution to extraction solution 1, instead of norvaline–glutarate for normalization [10].

7. Work under the chemical fume hood. Seal both the methoxyamine and pyridine bottles with Parafilm and store in a desiccator, since humidification impairs the derivatization efficiency.
8. Store bottle at −20 °C, warm up to room temperature and dry immediately before use, since humidification impairs the derivatization efficiency.
9. Sulfuric acid needs to be of 99.999% purity, since water will inhibit the derivatization reaction. Add sulfuric acid dropwise to the methanol. Work under the chemical fume hood, wear safety glasses, use clean glassware and clean glass pipettes. Buffer can be stored up to 14 days at room temperature. If you would like to reuse the glass pipette for H_2SO_4, rinse the glass pipette three times with water and three times with isopropanol.
10. Work with safety glasses.
11. Work in a chemical fume hood or beneath a strong exhaust to not breathe the formalin damps.
12. Transport the vials in dry ice. Care should be taken on thawing, as liquid nitrogen may cause vials to explode. Liquid nitrogen is hazardous. Use gloves and protective face equipment during handling. DMSO can penetrate skin and carry dissolved products across the skin barrier. Handle DMSO with caution.
13. Keep the cells at least 1 week and maximum 4 weeks in culture before using them for the mouse model.
14. Culture more cells than needed in case the cells did not grow as fast as expected. It is better to have more flasks that are less confluent than flasks that are more than 90% confluent. Never use cells that were grown too confluent.
15. If the effect of culture medium is not important in your tumor model, the survival of cells is better when resuspended and injected in culture medium without Fetal Bovine Serum and penicillin/streptomycin instead of PBS.
16. Always take into account the dead volume of your syringes. Make the cell suspension for 5 extra mice.
17. The best results are obtained by resuspending the cells every time before filling a syringe.
18. Injecting bubbles can cause air embolisms. To easily remove all bubbles, take more than 100 μl, tap the syringe to relocate the bubbles to the needle tip and push out the air and extra cell suspension.
19. The mice should not suffer because of the heat. Make sure the cage is not too hot by holding your hand in the cage for a couple of seconds.

20. Put the mouse very tight with its nose positioned in a way it can breathe.
21. Start injecting distally so you can move proximally when correct insertion in the vein fails. When you feel pressure and the injection site thickens you are not injecting in the vein. Do not waste any cell suspension and try again more proximally. When the injection goes smoothly, you are in the vein. Inject slowly but smoothly. Inject the cell suspension every time in the same way to increase the reproducibility of your results.
22. Wounds that can have been caused by shaving will be healed at the time of injection.
23. Inject every mouse in the same way to have reproducible results.
24. The organ where the metastases will manifest first is cell type specific. For example, 4T1 and EMT6.5 metastatic cancer cells seed first into the lungs.
25. Humane endpoints need to be in accordance with the local regulations. Here, we determined them as follows: The size of the primary tumor cannot exceed 1.8 cm^3, loss of ability to ambulate, labored respiration, surgical infection or weight loss over 10% of initial body weight.
26. Upon detection of one of these symptoms the animal was euthanized: loss of ability to ambulate, labored respiration, surgical infection or weight loss over 10% of initial body weight.
27. Each method of anesthesia or euthanasia induces specific changes in the metabolome, which need to be considered when interpreting results. The procedure used for anesthesia, euthanasia, and tissue collection should be fully documented in all publications [22–24].
28. Work with safety glasses.
29. When collecting tumors or metastases, limit the amount of normal adjacent tissue that is collected along. It can be useful to split the tumor section into two parts lengthwise, and use one of the halves for histological determination of actual tumor area, and the other for metabolic extractions. Performing this extra analysis allows researchers to confirm the purity of the collected tumor tissue. However, this determination cannot be used to correct the metabolite abundances detected. As an example, if a specific collected tumor piece contains 90% tumor and 10% nontransformed tissue, this does not mean that, for each metabolite, 90% of its total amount arose from the tumor tissue.
30. Label the Eppendorf tube on the cap and on the side to avoid losing the label. Keep the tissues in plastic bags in a box with liquid nitrogen. Precool the Eppendorf tube in liquid nitrogen,

place the empty Eppendorf tube on the balance and tare the balance. Quickly break off a piece of tissue with a liquid nitrogen cooled tweezers. Place the tissue into the Eppendorf tube as fast as possible, write down the weight, and transfer the closed Eppendorf tube with the tissue into a tube holder floating in liquid nitrogen.

31. Wear face protection when using the Cryomill, as tubes could break if liquid nitrogen enters the tube. In a case where liquid nitrogen has entered a tube, open the tube as fast as possible and let the liquid nitrogen evaporate. Regrind the tissue if it is not completely ground after one round in the Cryomill. Beads can stay in the tube for the whole extraction process.
32. Work under a fume hood. Release air when going through the middle layer (protein phase) by depressing the pipette by 80% and press the rest out as bubbles while crossing the middle layer. Do not take the complete liquid from each phase to not disturb the middle (protein containing) layer, as the loss of some volume from each phase can later be corrected by the internal standards.
33. Work under the chemical fume hood. Samples must be completely dry before addition of the derivatization solution. If the samples were stored at −80 °C, redry them for 10 min in an acid resistant vacuum centrifuge at 20 °C.
34. While adding 20 μl *O*-methoxyamine–HCl dissolved in pyridine solution to your sample, wash the walls of the Eppendorf tubes by pipetting up and down.
35. Work under the chemical fume hood. The 25 μl extra volume of TBDMS is needed to correct for evaporation of TBDMS.
36. The derivatization of polar metabolites for GC-MS analysis can also be performed by hand. Add 20 μl 20 mg/ml methoxyamine dissolved in pyridine to each sample, vortex briefly and incubate for 90 min in a heating block at 37 °C. Subsequently, add 40 μl TBDMS, vortex briefly and incubate for 60 min in a heating block at 60 °C. Centrifuge the samples for 2 min at maximum speed, transfer the supernatant to a glass GC-MS vial with insert and seal with a nonmagnetic cap using a crimper.
37. Work under the chemical fume hood. Make sure to only collect the upper phase. Any addition of the lower phase will destroy the GC-MS column. Bubbles detected when flicking the vials are indicative of liquid from the lower phase in your sample. When using hexane, prepare an aliquot in a cleaned glass bottle, the aliquot should not be older than 2 weeks.
38. Wait until the mouse does not respond anymore when you pinch the toe.
39. Be careful to not rupture the trachea.

40. For the collection of lungs for H&E staining you will have to fix the complete lung. If you want to collect lung tissue for histology and metabolite analysis you will need 2 mice.
41. If no infusion or injection with stable isotopic tracers are performed, also stable and heavy labeled metabolites can be used as internal standard to calculate metabolite concentrations.

References

1. Sullivan LB, Gui DY, Vander Heiden MG (2016) Altered metabolite levels in cancer: implications for tumour biology and cancer therapy. Nat Rev Cancer 16:680–693
2. Lorendeau D, Christen S, Rinaldi G, Fendt S-M (2015) Metabolic control of signalling pathways and metabolic auto-regulation. Biol Cell 107:251–272
3. Lunt SY, Fendt S-M (2018) Metabolism – a cornerstone of cancer initiation, progression, immune evasion and treatment response. Curr Opin Syst Biol 8:67–72
4. Elia I, Doglioni G, Fendt S-M (2018) Metabolic hallmarks of metastasis formation. Trends Cell Biol 28:673–684
5. Doglioni G, Parik S, Fendt S-M (2019) Interactions in the (pre)metastatic niche support metastasis formation. Front Oncol 9:219
6. Vander Heiden MG, DeBerardinis RJ (2017) Understanding the intersections between metabolism and cancer biology. Cell 168:657–669
7. Rinaldi G, Rossi M, Fendt S-M (2018) Metabolic interactions in cancer: cellular metabolism at the interface between the microenvironment, the cancer cell phenotype and the epigenetic landscape. Wiley Interdiscip Rev Syst Biol Med 10:e1397
8. Elia I, Schmieder R, Christen S, Fendt S-M (2015) Organ-specific cancer metabolism and its potential for therapy. Handb Exp Pharmacol 233:321–353
9. Elia I, Fendt S-M (2016) In vivo cancer metabolism is defined by the nutrient microenvironment. Transl Cancer Res 5:S1284–S1287
10. Broekaert D, Fendt SM (2019) Measuring in vivo tissue metabolism using 13C glucose infusions in mice. Methods Mol Biol 1862:67–82
11. Faubert B, DeBerardinis RJ (2017) Analyzing tumor metabolism in vivo. Annu Rev Cancer Biol 1:99–117
12. Buescher JM, Antoniewicz MR, Boros LG, Burgess SC, Brunengraber H, Clish CB, DeBerardinis RJ, Feron O, Frezza C, Ghesquiere B et al (2015) A roadmap for interpreting 13 C metabolite labeling patterns from cells. Curr Opin Biotechnol 34:189–201
13. Christen S, Lorendeau D, Schmieder R, Broekaert D, Metzger K, Veys K, Elia I, Buescher JM, Orth MF, Davidson SM et al (2016) Breast cancer-derived lung metastases show increased pyruvate carboxylase-dependent Anaplerosis. Cell Rep 17:837–848
14. Elia I, Broekaert D, Christen S, Boon R, Radaelli E, Orth MF, Verfaillie C, Grünewald TGP, Fendt S-M (2017) Proline metabolism supports metastasis formation and could be inhibited to selectively target metastasizing cancer cells. Nat Commun 8:15267
15. Elia I, Rossi M, Stegen S, Broekaert D, Doglioni G, van Gorsel M, Boon R, Escalona-Noguero C, Torrekens S, Verfaillie C et al (2019) Breast cancer cells rely on environmental pyruvate to shape the metastatic niche. Nature 568:117–121
16. Lorendeau D, Rinaldi G, Boon R, Spincemaille P, Metzger K, Jäger C, Christen S, Dong X, Kuenen S, Voordeckers K et al (2017) Dual loss of succinate dehydrogenase (SDH) and complex I activity is necessary to recapitulate the metabolic phenotype of SDH mutant tumors. Metab Eng 43:187–197
17. Feldman AT, Wolfe D (2014) Tissue processing and Hematoxylin and eosin staining. Methods Mol Biol 1180:31–43
18. Hewitson TD, Wigg B, Becker GJ (2010) Tissue preparation for Histochemistry: fixation, embedding, and antigen retrieval for light microscopy. Methods Mol Biol 611:3–18
19. Shrivastava A, Gupta V (2011) Methods for the determination of limit of detection and limit of quantitation of the analytical methods. Chronicles Young Sci 2:21

20. Fernandez CA, Des RC, Previs SF, David F, Brunengraber H (1996) Correction of ^{13}C mass Isotopomer distributions for natural stable isotope abundance. J Mass Spectrom 31:255–262
21. Millard P, Letisse F, Sokol S, Portais J-C (2012) IsoCor: correcting MS data in isotope labeling experiments. Bioinformatics 28:1294–1296
22. Pierozan P, Jernerén F, Ransome Y, Karlsson O (2017) The choice of euthanasia method affects metabolic serum biomarkers. Basic Clin Pharmacol Toxicol 121:113–118
23. Overmyer KA, Thonusin C, Qi NR, Burant CF, Evans CR (2015) Impact of anesthesia and euthanasia on metabolomics of mammalian tissues: studies in a C57BL/6J mouse model. PLoS One 10:e0117232
24. Brooks SP, Lampi BJ, Bihun CG (1999) The influence of euthanasia methods on rat liver metabolism. Contemp Top Lab Anim Sci 38 (6):19–24
25. Hui S, Ghergurovich JM, Morscher RJ, Jang C, Teng X, Lu W, Esparza LA, Reya T, Le Zhan L, Yanxiang Guo J et al (2017) Glucose feeds the TCA cycle via circulating lactate. Nature 551:115–118
26. Davidson SM, Papagiannakopoulos T, Olenchock BA, Heyman JE, Keibler MA, Luengo A, Bauer MR, Jha AK, O'Brien JP, Pierce KA et al (2016) Environment impacts the metabolic dependencies of Ras-driven non-small cell lung cancer. Cell Metab 23:517–528
27. Faubert B, Li KY, Cai L, Hensley CT, Kim J, Zacharias LG, Yang C, Do QN, Doucette S, Burguete D et al (2017) Lactate metabolism in human lung tumors. Cell 171:358–371.e9
28. Courtney KD, Bezwada D, Mashimo T, Pichumani K, Vemireddy V, Funk AM, Wimberly J, McNeil SS, Kapur P, Lotan Y et al (2018) Isotope tracing of human clear cell renal cell carcinomas demonstrates suppressed glucose oxidation in vivo. Cell Metab 28:793–800
29. Fan TWM, Lane AN, Higashi RM, Farag MA, Gao H, Bousamra M, Miller DM (2009) Altered regulation of metabolic pathways in human lung cancer discerned by 13Cstable isotope-resolved metabolomics (SIRM). Mol Cancer 8:1–19
30. Sellers K, Fox MP, Bousamra M, Slone SP, Higashi RM, Miller DM, Wang Y, Yan J, Yuneva MO, Deshpande R et al (2015) Pyruvate carboxylase is critical for non–small-cell lung cancer proliferation. J Clin Invest 125:687–698
31. Hensley CT, Faubert B, Yuan Q, Lev-Cohain N, Jin E, Kim J, Jiang L, Ko B, Skelton R, Loudat L et al (2016) Metabolic heterogeneity in human lung tumors. Cell 164:681–694
32. Tardito S, Oudin A, Ahmed SU, Fack F, Keunen O, Zheng L, Miletic H, Sakariassen PØ, Weinstock A, Wagner A et al (2015) Glutamine synthetase activity fuels nucleotide biosynthesis and supports growth of glutamine-restricted glioblastoma. Nat Cell Biol 17:1556–1568
33. Marin-Valencia I, Yang C, Mashimo T, Cho S, Baek H, Yang X-L, Rajagopalan KN, Maddie M, Vemireddy V, Zhao Z et al (2012) Analysis of tumor metabolism reveals mitochondrial glucose oxidation in genetically diverse human Glioblastomas in the mouse brain in vivo. Cell Metab 15:827–837
34. Maher EA, Marin-Valencia I, Bachoo RM, Mashimo T, Raisanen J, Hatanpaa KJ, Jindal A, Jeffrey FM, Choi C, Madden C et al (2012) Metabolism of [U-13C]glucose in human brain tumors in vivo. NMR Biomed 25:1234–1244
35. Yuneva MO, Fan TWM, Allen TD, Higashi RM, Ferraris DV, Tsukamoto T, Matés JM, Alonso FJ, Wang C, Seo Y et al (2012) The metabolic profile of tumors depends on both the responsible genetic lesion and tissue type. Cell Metab 15:157–170
36. Kucejova B, Duarte J, Satapati S, Fu X, Ilkayeva O, Newgard CB, Brugarolas J, Burgess SC (2016) Hepatic mTORC1 opposes impaired insulin action to control mitochondrial metabolism in obesity. Cell Rep 16:508–519
37. Satapati S, Kucejova B, Duarte JAG, Fletcher JA, Reynolds L, Sunny NE, He T, Nair LA, Livingston KA, Livingston K et al (2015) Mitochondrial metabolism mediates oxidative stress and inflammation in fatty liver. J Clin Invest 125:4447–4462
38. Satapati S, Sunny NE, Kucejova B, Fu X, He TT, Méndez-Lucas A, Shelton JM, Perales JC, Browning JD, Burgess SC (2012) Elevated TCA cycle function in the pathology of diet-induced hepatic insulin resistance and fatty liver. J Lipid Res 53:1080–1092
39. Mashimo T, Pichumani K, Vemireddy V, Hatanpaa KJ, Singh DK, Sirasanagandla S, Nannepaga S, Piccirillo SG, Kovacs Z, Foong C et al (2014) Acetate is a bioenergetic substrate for human glioblastoma and brain metastases. Cell 159:1603–1614

40. Kim C-W, Addy C, Kusunoki J, Anderson NN, Deja S, Fu X, Burgess SC, Li C, Ruddy M, Chakravarthy M et al (2017) Acetyl CoA carboxylase inhibition reduces hepatic Steatosis but elevates plasma triglycerides in mice and humans: a bedside to bench investigation. Cell Metab 26:394–406.e6
41. Kennedy KM, Scarbrough PM, Ribeiro A, Richardson R, Yuan H, Sonveaux P, Landon CD, Chi J-T, Pizzo S, Schroeder T et al (2013) Catabolism of exogenous lactate reveals it as a legitimate metabolic substrate in breast cancer. PLoS One 8:e75154
42. Rauckhorst AJ, Gray LR, Sheldon RD, Fu X, Pewa AD, Feddersen CR, Dupuy AJ, Gibson-Corley KN, Cox JE, Burgess SC et al (2017) The mitochondrial pyruvate carrier mediates high fat diet-induced increases in hepatic TCA cycle capacity. Mol Metab 6:1468–1479
43. Vatner DF, Majumdar SK, Kumashiro N, Petersen MC, Rahimi Y, Gattu AK, Bears M, Camporez J-PG, Cline GW, Jurczak MJ et al (2015) Insulin-independent regulation of hepatic triglyceride synthesis by fatty acids. Proc Natl Acad Sci U S A 112:1143–1148
44. DeLany JP, Windhauser MM, Champagne CM, Bray GA (2000) Differential oxidation of individual dietary fatty acids in humans. Am J Clin Nutr 72:905–911
45. Sidossis LS, Coggan AR, Gastaldelli A, Wolfe RR (1995) Pathway of free fatty acid oxidation in human subjects. Implications for tracer studies. J Clin Invest 95:278–284
46. Blaak EE, Wagenmakers AJM (2002) The fate of [U-(13)C]palmitate extracted by skeletal muscle in subjects with type 2 diabetes and control subjects. Diabetes 51:784–789
47. Gallego S, Hermansson M, Liebisch G, Hodson L, Ejsing C, Gallego SF, Hermansson M, Liebisch G, Hodson L, Ejsing CS (2018) Total fatty acid analysis of human blood samples in one minute by high-resolution mass spectrometry. Biomol Ther 9:7
48. Ducker GS, Chen L, Morscher RJ, Ghergurovich JM, Esposito M, Teng X, Kang Y, Rabinowitz JD (2016) Reversal of cytosolic one-carbon flux compensates for loss of the mitochondrial Folate pathway. Cell Metab 23:1140–1153
49. Neinast MD, Jang C, Hui S, Murashige DS, Chu Q, Morscher RJ, Li X, Zhan L, White E, Anthony TG et al (2018) Quantitative analysis of the whole-body metabolic fate of branched-chain amino acids. Cell Metab 29(2):417–429
50. Strong JM, Anderson LW, Monks A, Chisena CA, Cysyk RL (1983) A 13C tracer method for quantitating de novo pyrimidine biosynthesis in vitro and in vivo. Anal Biochem 132:243–253
51. Busch R, Kim Y-K, Neese RA, Schade-Serin V, Collins M, Awada M, Gardner JL, Beysen C, Marino ME, Misell LM et al (2006) Measurement of protein turnover rates by heavy water labeling of nonessential amino acids. Biochim Biophys Acta 1760:730–744
52. Pinnick KE, Gunn PJ, Hodson L (2019) Measuring human lipid metabolism using deuterium labeling: in vivo and in vitro protocols. Methods Mol Biol 1862:83–96

Chapter 7

Robust Analytical Methods for the Accurate Quantification of the Total Biomass Composition of Mammalian Cells

Diana Széliová, Harald Schoeny, Špela Knez, Christina Troyer, Cristina Coman, Evelyn Rampler, Gunda Koellensperger, Robert Ahrends, Stephen Hann, Nicole Borth, Jürgen Zanghellini, and David E. Ruckerbauer

Abstract

Biomass composition is an important input for genome-scale metabolic models and has a big impact on their predictive capabilities. However, researchers often rely on generic data for biomass composition, e.g. collected from similar organisms. This leads to inaccurate predictions, because biomass composition varies between different cell lines, conditions, and growth phases. In this chapter we present protocols for the determination of the biomass composition of Chinese Hamster Ovary (CHO) cells. These methods can easily be adapted to other types of mammalian cells. The protocols include the quantification of cell dry mass and of the main biomass components, namely protein, lipid, DNA, RNA, and carbohydrates. Cell dry mass is determined gravimetrically by weighing a defined number of cells. Amino acid composition and protein content are measured by gas chromatography mass spectrometry. Lipids are quantified by shotgun mass spectrometry, which provides quantities for the different lipid classes and also the distribution of fatty acids. RNA is purified and then quantified spectrophotometrically. The methods for DNA and carbohydrates are simple fluorometric and colorimetric assays adapted to a 96-well plate format. To ensure quantitative results, internal standards or spike-in controls are used in all methods, e.g. to account for possible matrix effects or loss of material. Finally, the last section provides a guide on how to convert the measured data into biomass equations, which can then be integrated into a metabolic model.

Key words Biomass composition, DNA, RNA, Amino acids, Lipids, Carbohydrates, Chinese Hamster Ovary cells

1 Introduction

Chinese Hamster Ovary (CHO) cells are one of the most widely used organisms for the production of biopharmaceuticals, especially monoclonal antibodies and glycoproteins such as

---

Electronic supplementary material: The online version of this chapter (https://doi.org/10.1007/978-1-0716-0159-4_7) contains supplementary material, which is available to authorized users.

Deepak Nagrath (ed.), *Metabolic Flux Analysis in Eukaryotic Cells: Methods and Protocols*, Methods in Molecular Biology, vol. 2088, https://doi.org/10.1007/978-1-0716-0159-4_7, © Springer Science+Business Media, LLC, part of Springer Nature 2020

erythropoietin [1]. The advantages of CHO cells include the ability to perform human-like post-translational modifications, their safety and ease of cultivation in suspension culture. However, the generation of new producer cell lines is still a time-consuming and laborious process, which relies on screening a high number of clones until a suitable producer is found. It is not yet clear what attributes are important to ensure high productivity and product quality and which limitations hinder further improvements [2].

Constraint-based modeling is a toolset that promises not only to help understand CHO cells better, but also to help improve their performance [3]. Specifically, it can be used to identify metabolic bottlenecks, to optimize medium and feed composition, and to predict targets for metabolic engineering. The latter comprise gene deletions, knock-downs, and overexpressions predicted to increase productivity and final titers [4, 5]. However, the application of constraint-based modeling to CHO cells, as well as to mammalian cells in general, is still limited.

The recent publication of a community-built genome-scale metabolic model (GSMM) of CHO cells [6] brings us closer to a successful application of constraint-based modeling to CHO cells. A GSMM is a computable reconstruction of the biochemical capabilities of an organism [7]. However, the accuracy of the predictions made using GSMMs depends not only on the correct reconstruction of metabolic pathways, but also on an accurate representation of the biomass composition in the model. Biomass composition can vary between different organisms, cell lines, conditions, and growth phases and these variations have a significant impact on model predictions [8]. The main biomass components that need to be represented in the model are proteins, lipids, DNA, RNA, and carbohydrates. This includes the total amounts of the components per cell and, for macromolecules, also their composition.

However, the experimental data for CHO cell biomass composition are still limited. The CHO GSMM contains two biomass reactions [6], which are partly based on literature data for CHO cells, but also other cell lines, such as hybridoma cells. Predictions are markedly different using one or the other biomass reaction [9]. Currently, there are only few data sets available for CHO cell biomass composition [10–12], the majority of them being incomplete. Amino acid composition was measured for five different CHO cell lines [10] and revealed no significant differences among them. However, the total protein content, which is also an important parameter for modeling, was estimated rather than measured. Pan et al. [11] found large changes in biomass composition for one CHO cell line (amino acid and lipid composition, total protein, lipid and carbohydrate content) during fed-batch culture, but DNA and RNA content were not measured. Lipid composition was measured by Zhang et al. [12] for CHO, SP2/0, and HEK cell

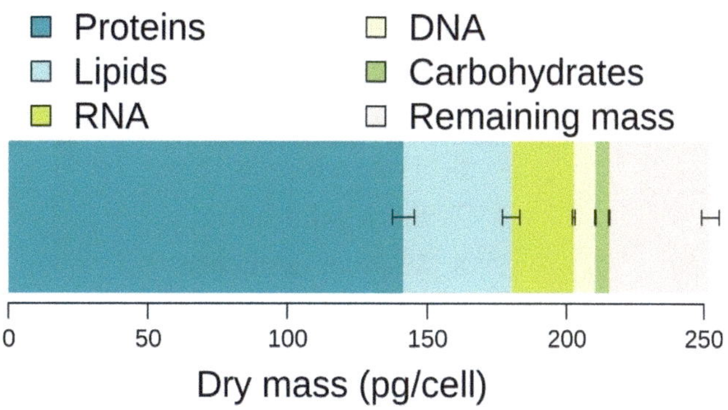

Fig. 1 Biomass composition of CHO-K1 in mid-exponential phase grown in HyClone ActiPro medium (GE Healthcare) supplemented with 8 mM glutamine and 0.2% Anti-Clumping Agent (GIBCO). The error bars are standard deviations from three biological replicates

lines in exponential and stationary phase, revealing big differences between the different mammalian cell lines.

The limited amount of available data and the importance of the biomass composition for model predictions motivated the development of the methods presented in this chapter. It contains detailed protocols for the quantification of dry mass, proteins, lipids, DNA, RNA, and carbohydrates as well as for the determination of protein amino acid and lipid species composition. Although these methods were developed for CHO cells, they can be easily adapted to other mammalian cell lines. Additionally we provide, as an example, a dataset of CHO-K1 biomass composition in mid-exponential phase of a batch culture, which can be downloaded from GitHub (https://github.com/diana-sz/MiMB_biomass.git). The measured macromolecular composition is shown in Fig. 1.

2 Materials

2.1 Sampling for Biomass Components

1. Phosphate-buffered saline (PBS).
2. Pluronic-F68 Non-ionic Surfactant.
3. Pipettes.
4. ViCell XR or other cell counting device.
5. Centrifuge for 200×*g*, the size depends on the volume of sampled cell culture.
6. Freezer for storage at −80°C.

2.2 Quantification of Cell Dry Mass

1. PBS or 0.9% (weight per volume (w/v)) sodium chloride (NaCl).
2. Pluronic-F68 Non-ionic Surfactant.

3. 50 mL tubes.
4. Pipettes.
5. Glass beakers (50 mL).
6. Tweezers.
7. Silica gel beads.
8. Vacuum desiccator.
9. ViCell XR or other cell counting device.
10. Centrifuge for 50 mL tubes at 170×*g*.
11. Oven for incubation at 90°C.
12. Precision balance (readability 0.0001×*g*, standard deviation ≤±0.0001).

2.3 Quantification of Proteinogenic Amino Acids

To avoid interferences, all chemicals should be of high purity (e.g., LC-MS grade). The presence of blanks should be tested by performing procedural blanks in parallel to sample preparation for each sample batch.

2.3.1 Chemicals

All solvents and reagents are diluted with LC-MS grade water.

1. Formic acid.
2. Hydrogen peroxide solution ≥ 30%.
3. Phenol.
4. 0.1 M hydrochloric acid (HCl).
5. 6 M HCl.
6. 0.1 M methanesulfonic acid (MSA).
7. 4 M MSA with 0.2% (w/v) tryptamine (3-(2-aminoethyl) indole).
8. 7 or 14 M sodium hydroxide (NaOH).
9. N-*tert*-butyldimethylsilyl-N-methyltrifluoroacetamide (MTBSTFA) with 1% *tert*-butyldimethylchlorosilane (TBDMSCl).
10. Pyridine anhydrous in airtight bottles.
11. Pyridine for syringe washing, not necessarily water free.
12. Water LC-MS grade.
13. Internal standard (ISTD): isotopically labeled "Cell Free" Amino Acid Mix (solid mix containing 20 amino acid (AA)s), (U-^{13}C, 97–99% + U-^{15}N, 97–99%), Cambridge Isotope Laboratories.
14. Solid standards for each amino acid which is quantified.
15. Certified amino acid standard for quality control.

Table 1
Preparation of calibration standards for protein amino acid analysis via GC-EI-MS/MS

Amino acid	*c* Stock solution (mmol L^{-1})	*c* Standard mix (μmol L^{-1})	Concentration of the calibration solutions (μmol L^{-1})
Ala	100.0	6000	15/45/150/450/1500/4500
Arg	100.0	2000	5/15/50/150/500/1500
Asp	50.0	6000	15/45/150/450/1500/4500
CysA	100.0	2000	5/15/50/150/500/1500
Glu	50.0	8000	20/60/200/600/2000/6000
Gly	100.0	6000	15/45/150/450/1500/4500
His	100.0	2000	5/15/50/150/500/1500
Ile	100.0	4000	10/30/100/300/1000/3000
Leu	100.0	6000	15/45/150/450/1500/4500
Lys	100.0	5000	12.5/37.5/125/375/1250/3750
MetS	100.0	2000	5/15/50/150/500/1500
Phe	100.0	4000	10/30/100/300/1000/3000
Pro	100.0	4000	10/30/100/300/1000/3000
Ser	100.0	4000	10/30/100/300/1000/3000
Thr	100.0	4000	10/30/100/300/1000/3000
Tyr	30.0	2000	5/15/50/150/500/1500
Val	100.0	4000	10/30/100/300/1000/3000

2.3.2 Preparation of Stock and Standard Solutions

Single standard stock solutions are prepared for the amino acids listed in Table 1 by weighing each solid standard substance in separate 1.5 mL brown glass chromatography vials.

For Gln, Asn, Met, and Cys no standards need to be prepared due to the decomposition, respectively, oxidation of these amino acids. Instead, Gln is quantified as Glu (sum parameter), Asn as Asp (sum parameter), Met as MetS, and Cys as CysA. Arginine is decomposed to Orn during derivatization. Hence Arg is detected and quantified as Orn, although the standard is prepared with Arg (*see* **Note 1**).

The solid standard substances are dissolved in 0.1 M HCl resulting in the standard stock concentrations listed in Table 1. In the next step, a standard mix is prepared at concentrations ranging from 2 to 8 mM (*see* Table 1 and **Note 2**).

The 1:4 dilution of the ISTD is used for standard preparation. Six standard solutions are prepared at the concentrations stated in Table 1, each standard containing 100 μL of the 1:4 diluted ISTD per 1 mL of standard solution. Following this procedure, the standards contain an equivalent of 25 μL of ISTD per mL standard, which corresponds to 25 μL of ISTD per mL hydrolysate. Consequently, no correction factor has to be applied for internal standardization. 40 μL aliquots of these standards are then dried in a vacuum centrifuge or at 40°C under a constant flow of nitrogen prior to derivatization and analysis. Dried standards can be stored at −80°C and need to be redried before derivatization.

2.3.3 Preparation of Internal Standard Solutions for HCl and MSA Hydrolysis

The isotopically labeled internal standard is an algal "cell free" ^{13}C, ^{15}N labeled amino acid mix, which contains all amino acids listed in Table 1 with the exception of methionine sulfone and cysteic acid. It additionally contains methionine, cystine, glutamine, and asparagine. Consequently, the internal standard has to be oxidized (for HCl analysis) and hydrolyzed (for HCl and MSA analysis) prior to its use in order to convert Gln and Asn to Glu and Asp, respectively, as well as Cys to CysA and Met to MetS. If this preparation step was omitted, the internal standard would change its composition during sample preparation and would hence fail as internal reference. For the preparation of the ISTD solutions for HCl and MSA analysis, the cell free amino acid standard therefore needs to be pretreated according to the (oxidation)/hydrolysis protocols for the samples.

2.3.4 Pretreatment of the Algal Amino Acid Mix for Internal Standard Preparation

1. Prepare two microcentrifuge screw cap tubes, each containing approximately 50 mg of solid internal standard.
2. Oxidize and hydrolyze the internal standards according to the procedures for HCl (Subheading 3.3.1, **steps 3** to **12** and **14** to **19**) and MSA (Subheading 3.3.2, **steps 4** to **10**). Duration of hydrolysis can be shortened to 20 h.
3. Reconstitute in 1.5 mL 0.1 M HCl or MSA, respectively, by vortexing for 15 min.
4. Dilute the internal standard 1:4 with LC-MS grade water before spiking calibration solutions, reference standards or samples.
5. The internal standard can be stored at −80°C.

2.3.5 Disposables

1. 2 mL screw cap micro tube (polypropylene (PP), e.g. Sarstedt—these tubes are tested for blanks and tightness at elevated temperatures).
2. 0.2 mL microinserts, 31 × 6 mm, clear glass, flat bottom.
3. Chromatography vials for screw caps (1.5 mL, 11 mm diameter, brown glass).
4. 11 mm screw caps with inert septum material (e.g.: polytetrafluoroethylene (PTFE)/silicone/PTFE).

5. Gas chromatography (GC) crimp vials for reagents and solvents (1.5 mL, 11 mm diameter, clear glass, wide opening).
6. For automated derivatization: 11 mm magnetic crimp caps with septum material, which is inert against pyridine and silylation agents (e.g.: PTFE/silicone/PTFE).
7. 11 mm nonmagnetic crimp caps with septum material, which is inert against pyridine and silylation agents (e.g.: PTFE/silicone/PTFE).
8. Single-use spatulas for preparation of stock solutions.

2.3.6 Laboratory Equipment, Instrumentation, and Software

1. Pipettes.
2. Vortex mixer/shaker.
3. Evaporator system (thermoblock for 40-85°C with N_2 dispenser).
4. Fume hood.
5. Centrifuge for 2 mL microcentrifuge tubes with cooling function for the rotors (4°C) for 10,000×*g*.
6. Acid resistant vacuum centrifuge.
7. Oven (110–115°C).
8. Freezer (−80°C).
9. GC-triple quadrupole mass spectrometer (MS/MS) (7890B GC with 7010B triple quadrupole GC/MS, Agilent Technologies, or equivalent system).
10. Programmed Temperature Vaporizer (CIS6 or PTV, Gerstel, or equivalent cooled injection system).
11. GC-column, 5% diphenyl 95% dimethyl polysiloxane as stationary phase (30 m, 0.25 mm inner diameter, 0.25 mm film thickness).
12. Non-polar GC-guard column, 3 m × 0.25 mm inner diameter.
13. T-Piece Purged Ultimate Union Assembly (inert) with corresponding metal ferrules (Agilent Technologies, or other column connectors).
14. Micro Balance (minimum readability 10 μg, maximum standard deviation 10 μg).
15. MPS2 Multi-Purpose Sampler (Gerstel) or equivalent autosampler for automated derivatization, equipped with heated shaker, syringes for sample, reagent and solvent handling, cooled trays, reservoirs for reagents, solvents and washing solutions.
16. 10 μL glass syringe for GC autosampler (injection).

17. 50 or 100 μL glass syringe for reagent handling for manual derivatization or for automated derivatization on the sample preparation robot.
18. MassHunter (Agilent Technologies) or equivalent software for quantitative selected reaction monitoring (SRM) evaluation.
19. Maestro (Gerstel) or equivalent software for automated derivatization.

2.4 Lipid Quantification

To avoid background contamination all solvents have to be LC-MS grade, the glassware should be heated up to 450°C in a muffle furnace for 3 h and all plasticware should be of high quality.

2.4.1 Chemicals

1. PBS.
2. Methyl *tert*-butyl ether (MTBE).
3. Methanol (MeOH).
4. 2-propanol.
5. 2-propanol/methanol/chloroform (4:2:1, v/v/v) with 7.5 mM ammonium formate (AF) (*see* **Notes 3** and **4**).
6. SPLASH® LIPIDOMIX® Mass Spec Standard (Avanti Polar Lipids) (*see* **Note 5**).
7. Cardiolipin Mix I (Avanti Polar Lipids).
8. 0.1% (w/v) AF in ultrapure water.

2.4.2 Disposables

1. 2 and 5 mL microcentrifuge tubes, PP (e.g., Eppendorf, Safe-Lock Tubes).
2. Screw neck vials, brown, flat bottom, small opening, 1.5 mL.
3. Lids, PP with PTFE septum, not sliced.
4. Inserts for small opening, clear, conical, 0.15 mL.
5. 96 well plate (e.g., Eppendorf twin.tec® PCR Plate 96, skirted, 150 μL, PCR clean, colorless).
6. Adhesive aluminum foils.
7. Chips with spraying nozzles of 5 μm (Advion, HD_A_384).
8. Rack of 384 spray pipette tips (Advion).

2.4.3 Laboratory Equipment, Instrumentation, and Software

1. Pipettes.
2. Vortex mixer.
3. Waterbath sonicator cooled to 4°C.
4. Muffle furnace for 450°C.
5. Thermomixer (*see* **Note 6**).
6. Evaporator system N_2.

7. Centrifuge for 2 mL and 5 mL microcentrifuge tubes (3000×*g*) (*see* **Note 7**).
8. Freezer (−80°C).
9. Triversa NanoMate® (Advion) for direct sample infusion.
10. Q Exactive HF (Thermo Fisher Scientific) or alternative high resolution mass spectrometer.
11. ProteoWizard (v3 or higher).
12. ChipSoft (Advion, v8 or higher).
13. Nanoware (Advion, installation for polarity switching).
14. Xcalibur (Thermo Fisher Scientific, v4 or higher).
15. LipidXplorer (v1.2.7 or higher).

2.5 RNA Quantification

1. RNase-free water.
2. TRIzol or TRI reagent.
3. Chloroform.
4. 70% (volume per volume (v/v)) ethanol (EtOH) prepared with nuclease-free water.
5. Purified RNA (700–1000 ng/μL, *see* **Note 8**).
6. Isopropanol.
7. Nuclease-free tubes 1.5 mL (e.g., Sarstedt 72.692.005).
8. RNaseZap wipes (Sigma-Aldrich R2020) or similar product for removal of RNases.
9. Pipettes.
10. Fume hood.
11. Vortex.
12. A centrifuge for 1.5 mL tubes and 12,000×*g* with a cooling function (4°C).
13. NanoDrop or similar device for spectrophotometric measurements.
14. Thermoblock for incubation at 55°C.

2.6 DNA Quantification

1. PBS.
2. Quant-iT™ PicoGreen™ dsDNA Assay Kit (ThermoFisher Scientific P7589)—contains 20× TE buffer, PicoGreen™ dsDNA reagent and λ DNA standard.
3. Proteinase K (QIAGEN 19131).
4. Lysis buffer AL (QIAGEN 19075).
5. RNase A 100 mg/mL (QIAGEN 19101).

Table 2
Preparation of DNA standards

Concentration (ng/mL)	1× TE (μL)	2 μg/mL λ DNA (μL)	Lysis buffer (μL)
1000	162	180	18
800	198	144	18
600	234	108	18
400	270	72	18
300	288	54	18
200	306	36	18
100	324	18	18
50	333	9	18
0 (blank)	342	0	18

6. 1× TE: dilute 20× stock solution from the PicoGreen kit with nuclease-free water (e.g., 38 mL water + 2 mL 20× TE, *see* **Note 9**).
7. 2 μg/mL λ DNA: Dilute 100 μg/mL stock solution 50× with 1× TE (e.g., 637 μL 1× TE + 13 μL 100 μg/mL λ DNA (*see* **Note 10**)).
8. Lysis buffer for preparation of DNA standards (to have the same matrix as in the samples): mix 100 μL PBS, 10 μL proteinase K, 2 μL RNase A, 100 μL AL, and 848 μL 1× TE.
9. DNA standards: prepare according to Table 2.
10. 1× PicoGreen dye: dilute the stock solution 200× with 1× TE. Wrap the tube in aluminum foil to protect it from light. This reagent always has to be prepared fresh before performing the assay (*see* **Notes 11** and **12**).
11. 1.5 mL tubes.
12. Black 96-well plate with a flat bottom.
13. Aluminum foil.
14. Multichannel pipette reservoir.
15. Pipettes.
16. Vortex.
17. Microplate shaker.
18. Thermoblock for 56°C.
19. Centrifuge for 1.5 mL tubes for short spin.
20. Fluorescence microplate reader.

Table 3
Preparation of glucose standards

Concentration (μg/mL)	200 μg/mL glucose (μL)	PBS (μL)
0 (blank)	0	1000
10	50	950
20	100	900
40	200	800
70	350	650
100	500	500
130	650	350
160	800	200

2.7 Carbohydrate Quantification

Use plastics resistant to sulfuric acid (e.g., PP). Test them beforehand and make sure that the sulfuric acid does not dissolve them or cause discoloration, which might affect the results of the assay.

1. PBS.
2. Concentrated sulfuric acid.
3. 200 μg/mL glucose solution in PBS. Store at 4°C.
4. Glucose standards: prepare according to Table 3 (*see* **Note 13**). The standards can be stored at 4°C.
5. 2 mg/mL anthrone in concentrated sulfuric acid. This reagent always has to be made fresh before performing the assay and kept on ice/in a fridge (*see* **Note 14**).
6. 15 or 50 mL PP tubes.
7. 1.5 mL tubes.
8. 0.2 mL PCR strips.
9. 96-well microplate with a flat bottom (e.g., Nunc™ MicroWell™ 96-Well Microplates, ThermoFisher Scientific 269620).
10. Multichannel pipette reservoir.
11. Pipettes.
12. Vortex.
13. Centrifuge for short spin of PCR tubes.
14. Thermoblock or PCR thermal cycler for 92°C.
15. Absorbance microplate reader.

3 Methods

For all quantification methods, at least three biological replicates should be used. Run test samples before you apply a method to important samples in order to get familiar with the procedure and to ensure you have everything prepared. We recommend doing a run with 10 technical replicates, i.e. splitting a batch of cells into 10 aliquots and applying the whole procedure to these samples. This way you can estimate whether the handling of the samples and the procedure in general are repeatable. Note that all protocols require precise sampling (*see* Subheading 3.1), a step which is prone to "operator-effects". Lastly, randomize samples as much as possible, e.g. avoid batch-wise analysis of samples.

3.1 Sampling for Biomass Components

Taking an accurate number of cells is a crucial step for all biomass quantification methods. For most protocols, the cells need to be centrifuged, washed with PBS, and centrifuged again. It is advisable to remeasure the cell concentration after the washing step to account for possible cell loss during centrifugation. However, it was observed that cells become more shear sensitive after washing them with PBS, because protective compounds from the culture medium are removed. If an automated cell counter such as ViCell XR (Beckman Coulter) is used, the increased shear sensitivity of the cells can lead to underestimation of the cell number concentration or viability because of cell death during the measurement. To protect the cells from shear stress, samples should be supplemented with a shear protectant, such as 2% Pluronic F-68 [13]. This might not be applicable to other counting methods and can be tested by comparing the cell counts in samples with or without the shear protectant.

This section describes a general protocol for sampling, which should be adapted based on the required amount of cells.

1. Measure cell concentration and calculate the required number of cells. Always take more cells than needed, because an aliquot of the cell suspension will be used for cell counting and some cells might be lost during the centrifugation/resuspension steps.
2. Centrifuge samples for 8 min at 200×*g*, room temperature, discard supernatant.
3. Resuspend pellet in PBS by gently pipetting up and down. The amount of PBS should be such that the resulting cell concentration is within the range of the cell counting method.
4. Take a sample for cell counting, add Pluronic F-68 to a final concentration of 2% (v/v) and measure the cell concentration (*see* **Notes 15** and **16**).

5. Aliquot the cell suspension for the required assays (*see* **Note 17**).
6. Centrifuge again for 8 min, 200×*g*, room temperature. Discard supernatant and store the pellets at −80°C or proceed directly with analysis.

3.2 Quantification of Cell Dry Mass

Cell dry mass is determined by drying and weighing a defined number of cells. Cell number determination is a crucial step in the protocols presented here. Before drying, cells are washed twice and the cell count is determined again to account for possible loss of cells during the washing and centrifugation steps. Typical dry mass values for CHO cells range from 200 to 300 pg per cell.

3.2.1 Quantification

1. Place empty pre-labeled beakers and silica gel beads (e.g., on a Petri dish) into the incubator at 90°C overnight. You need one beaker per sample plus at least three additional ones as controls.
2. Transfer the hot beakers and silica beads into a vacuum desiccator and evacuate. Leave for at least 20 min.

Important: *Do not touch the beakers with bare hands or gloves (only with tweezers) and do not label them after the incubation at 90° C, because this can change their mass. Wear safety googles while evacuating and repressurizing the desiccator due to the risk of implosion. Do not move the evacuated desiccator and wear goggles when working near it.*

3. Centrifuge 0.5–1 × 10^8 cells for 10 min at 170×*g*, room temperature. Discard supernatant.
4. Wash the pellet with 15 mL PBS or 0.9% NaCl, centrifuge again at the same conditions, discard supernatant.
5. Resuspend pellet in 15 mL PBS or 0.9% NaCl.
6. Remeasure the cell concentration on ViCell XR in the presence of 2% (v/v) Pluronic F-68 to protect the cells against shear stress during measurement (e.g., mix 480 μL cell suspension with 120 μL 10% Pluronic F-68). *See* Subheading 3.1 for details on ViCell XR measurements.
7. Take out the beakers from the desiccator and weigh them immediately.
8. Transfer the entire cell suspension into the beakers and weigh immediately. Put the same volume of PBS or 0.9% NaCl without cells into three additional beakers each, as controls.
9. Place beakers at 90°C overnight (*see* **Note 18**).

10. On the next day, put the silica beads and the beakers in a vacuum desiccator. Evacuate and leave for 20 min. Weigh them immediately after taking them out (*see* **Note 19**).
11. Repeat **steps 9** and **10** until the weight is constant.

3.2.2 Data Evaluation

1. Calculate the average cell concentration after washing (if Pluronic F-68 was used for the cell counting, multiply by the dilution factor).
2. Calculate the total number of cells per beaker from the cell concentration and the volume of cell suspension that was transferred to the beaker.
3. Calculate the precise salt concentration from the control beakers.
4. Use the value for the salt concentration to calculate the mass of salt per each beaker. Subtract the mass of the salt from the mass of the dry cell suspension.
5. Divide the mass of cells by the number of cells per beaker.

3.3 Quantification of Proteinogenic Amino Acids

For the quantification of proteinogenic AAs, two independent thermal hydrolysis methods are employed using (1) HCl combined with prior oxidation with performic acid and (2) MSA [14–17]. For a schematic workflow, *see* Fig. 2. HCl analysis with prior oxidation is used for the quantification of the sulfur containing amino acids (cysteine/cystine and methionine), for the quantification of stable amino acids (e.g. proline and alanine) as well as for amino acids which need harsh conditions for hydrolysis (valine, leucine and isoleucine in certain structural arrangements) [16]. In contrast to this, MSA hydrolysis is used for less stable amino acids (e.g. serine, threonine, tyrosine). After hydrolysis the amino acids are quantified in either the HCl or MSA hydrolysate using GC—electron ionization (EI)—triple quadrupole mass spectrometry (MS/MS) as their *tert*-butyldimethylsilyl derivatives [18]. Quantification is based on internal standardization with isotopically labeled internal amino acid standards and external calibration with authentic AA standards of certified purity and stablility. Internal standards are added to the samples before hydrolysis (which is after oxidation for HCl hydrolysis) to compensate for errors during sample preparation.

During acidic hydrolysis, asparagine and glutamine are converted into aspartate and glutamate, respectively, resulting in a sum parameter for these two amino acids [14, 16]. Arginine is degraded to ornithine during derivatization and is hence quantified using arginine and isotopically labeled arginine as internal standard for calibration, but ornithine detection in GC-MS/MS. Serine, threonine and tyrosine are partially degraded during oxidation and hydrolysis with 6 M HCl. Since this degradation cannot be fully compensated by the use of the isotopically labeled internal

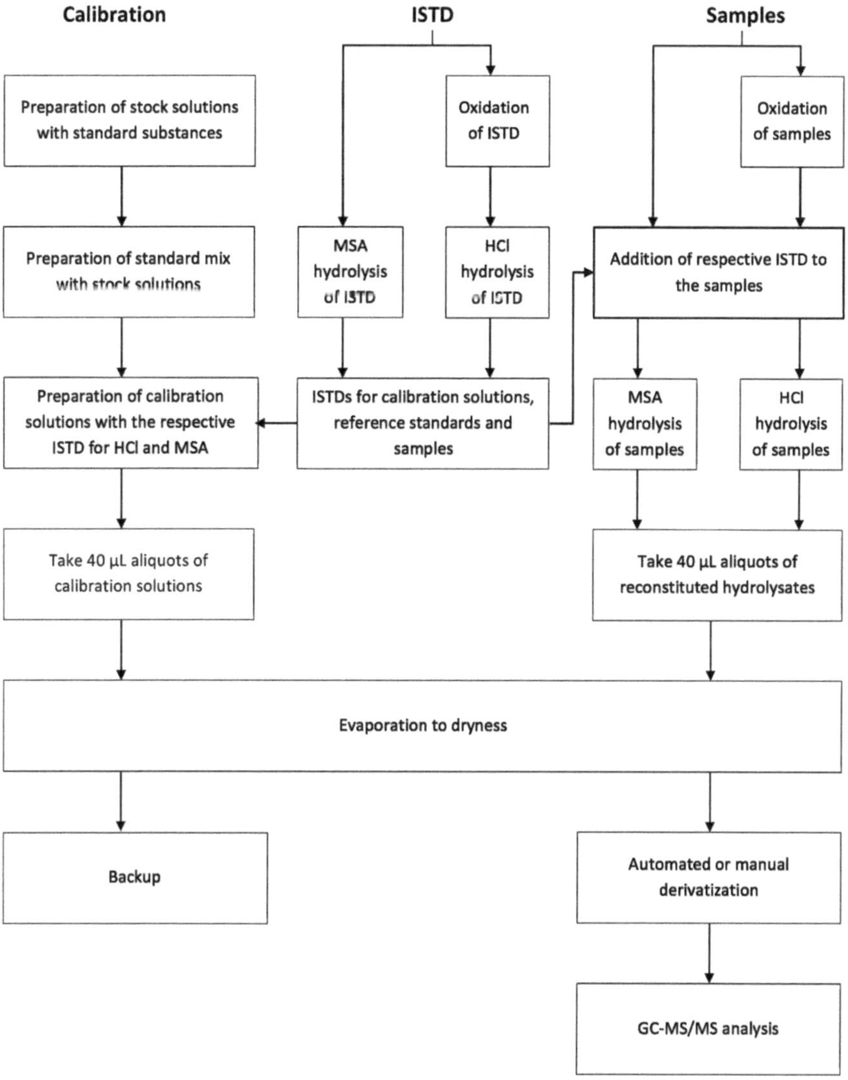

Fig. 2 Workflow for the quantification of protein amino acids

standards, these three amino acids are quantified in the MSA hydrolysates. For stabilization cysteine/cystine and methionine are oxidized with performic acid prior to HCl hydrolysis, forming cysteic acid and methionine sulfone, respectively [19]. It is known that hydrolysis of valine is incomplete when paired with isoleucine or valine, hydrolysis of leucine is incomplete when multiple molecules of leucine occur in sequence and hydrolysis of isoleucine is incomplete when paired with isoleucine or valine [16]. The concentrations of these amino acids are evaluated using the HCl hydrolysates, since the efficiency of HCl hydrolysis is higher than the efficiency of MSA hydrolysis. Tryptophan is degraded during both acidic hydrolysis methods (fully degraded in HCl hydrolysis,

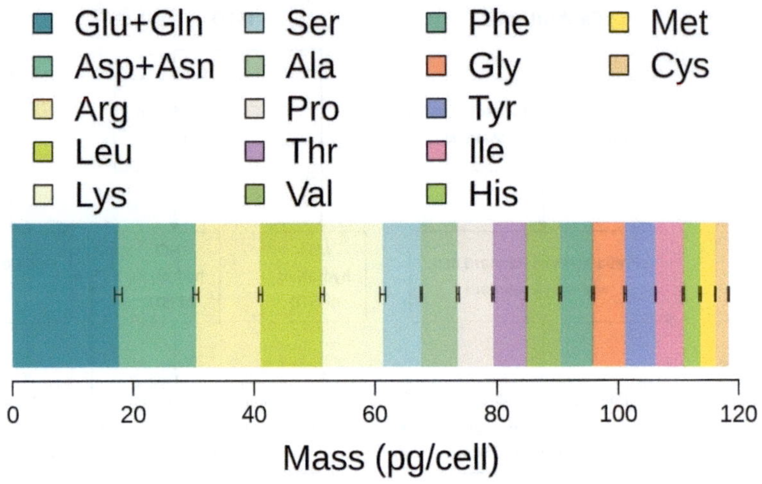

Fig. 3 Amino acid composition of CHO-K1 in mid-exponential phase grown in HyClone ActiPro medium (GE Healthcare) supplemented with 8 mM glutamine and 0.2% Anti-Clumping Agent (GIBCO). The error bars represent standard deviations from three biological replicates

partially degraded in MSA hydrolysis) and can hence not be evaluated (*see* **Note 20**) [14]. Consequently, 17 amino acids can be quantified using the presented method (*see* **Note 21**). The amino acid composition measured for CHO-K1 is shown in Fig. 3.

3.3.1 Oxidation and Hydrolysis with HCl

1. Take sample aliquots corresponding to 1 mg cell dry mass (CDM) as described in Subheading 3.1. Use 2 mL screw cap tubes for sampling (*see* Subheading 2.3.5, *see* **Note 22**). The method is valid for CDM ranging from 0.2 to 3 mg per sample.
2. Store cell pellets at −80°C until analysis.
3. Mix 9 mL formic acid with 1 mL hydrogen peroxide solution and add 50 μL phenol.
4. Wait 45 min until the solution has turned brown.
5. Add 500 μL of the reagent mix prepared in **steps 3** and **4** to each sample in 2 mL PP screw cap tubes, close tightly.
6. Vortex for 10 min.
7. Heat the samples to 50°C for 7 min.
8. Vortex for 1 min.
9. Heat the samples to 50°C for 8 min.
10. Quickly cool down the samples, e.g. in a −80°C freezer for 1 min.

11. Spin down the samples by centrifugation for approximately 2 min at 10,000×*g*.
12. Evaporate the reagent at 40°C under a constant flow of nitrogen until samples are completely dry (~1 h). Dry samples can be stored at −80°C.
13. Add 100 μL of ISTD solution, which has been diluted 1:4 with LC-MS grade water. Add ISTD to the calibration standards.
14. Add 1000 μL of 6 M HCl and vortex for 10 min.
15. Flush the tube for 20 s with nitrogen and close it rapidly.
16. Heat the samples at 110°C for 24 h.
17. Vortex the hydrolyzed sample for 1 min.
18. Centrifuge for 2–5 min at 10,000×*g*.
19. Dry the hydrolysates at 85°C under a constant flow of nitrogen until they are completely dry.
20. Add 1000 μL of 0.1 M HCl and vortex or shake for 15 min.
21. Centrifuge for 15 min at 10,000×*g*, centrifuge cooled to 4°C.
22. Transfer 40 μL aliquots of the sample to 1.5 mL chromatography vials equipped with straight inserts. Aliquots need to be taken immediately after centrifugation from the clear middle layer of the hydrolysates (take care not to take the aliquot from the lipid containing top layer or the bottom layer containing cell debris, *see* **Note 23**).
23. Dry the samples in the vacuum centrifuge or under a constant flow of nitrogen at 40°C until they are completely dry.
24. Analyze samples immediately or store at −80°C. In case of storage samples need to be redried before analysis (*see* **Note 24**).

3.3.2 Hydrolysis with MSA

1. Take sample aliquots corresponding to 1 mg CDM as described in Subheading 3.1. Use 2 mL screw cap tubes for sampling (*see* Subheading 2.3.5, *see* **Note 22**). The method is valid for CDM ranging from 0.2 to 3 mg per sample.
2. Store cell pellets at −80°C until analysis.
3. Add 100 μL of ISTD solution, which has been diluted 1:4 with LC-MS grade water, to each screw cap tube containing approximately 1 mg of dry sample.
4. Add 1000 μL of 4 M MSA containing 0.2% tryptamine and vortex for 10 min. Add ISTD to the calibration standards.
5. Flush the tube for 20 s with nitrogen and close it rapidly.
6. Heat to 115°C for 22 h.
7. Vortex the hydrolyzed sample.
8. Neutralize the sample with 0.5 mL 7 M NaOH (or 0.25 mL 14 M NaOH).

9. Centrifuge for 2–5 min at 10,000×*g*.
10. Dry the hydrolysate at 85°C under a constant flow of nitrogen until samples are completely dry.
11. Add 1000 μL of 0.1 M MSA and vortex for 15 min.
12. Centrifuge for 15 min at 10,000×*g*, centrifuge cooled to 4°C.
13. Transfer 40 μL aliquots of the sample to 1.5 mL chromatography vials equipped with straight inserts. Aliquots need to be taken immediately after centrifugation from the clear middle layer of the hydrolysates (take care not to take the aliquot from the lipid containing top layer or the bottom layer containing cell debris, *see* **Note 23**).
14. Dry the samples in the vacuum centrifuge or under a constant flow of nitrogen at 40°C until they are completely dry.
15. Analyze samples immediately or store samples at −80°C. In case of storage samples need to be redried before analysis (*see* **Note 24**).

3.3.3 Preparation of Quality Control (QC) Standards and quality control (QC) Samples

Certified standards should be diluted to a concentration approximately in the middle of the calibrated range. The dilutions must contain the same concentration of ISTD as the standards and samples. 40 μL aliquots are used for analysis.

QC samples are advisable if samples are measured in several batches to ensure comparability. For this purpose a number of 40 μL aliquots are taken from a CHO hydrolysate or pooled hydrolysates. QC samples are dried and stored at −80°C.

3.3.4 Setup of Measurement Sequence

At the beginning of each sequence, 1 μL of the pure derivatization reagent (MTBSTFA + 1% TBDMSCl) is repeatedly injected in order to condition the liner and the column. These conditioning injections are then followed by several pyridine injections. This procedure ensures that the column and the injector have been heated and deactivated prior to analysis. To complete the conditioning procedure at least three samples and a blank are injected.

Sequences need to be set up with calibration standards being injected at the beginning and at the end of the sequence. After a maximum of 10 sample injections, QC measurements (certified amino acid standard and/or QC sample) and a blank measurement (to ensure the absence of contaminants and carry-over) need to be implemented (*see* Subheading 3.3.3 Preparation of QC standards and QC samples). All samples need to be randomized.

3.3.5 Manual Derivatization

1. Redry the hydrolysates immediately before derivatization.
2. Wash 50 or 100 μL glass syringe with pyridine (*see* **Note 25**).
3. Quickly add 40 μL pyridine to each sample vial.
4. Quickly add 40 μL MTBSTFA + 1% TBDMSCl.

5. Crimp vials rapidly and check that the crimp septa are sufficiently tight (crimp caps cannot be turned any more).
6. Vortex for 1 min or shake for 5 min.
7. Centrifuge for 2 min at 10,000×*g* to spin down reagent and sample.
8. Heat for 1 h at 85°C in a thermoblock under the fume hood.
9. Cool samples to 4°C until GC-MS/MS analysis.

3.3.6 Automated Derivatization

Precise control of derivatization conditions (time, temperature, agitation) can be achieved by using automated derivatization. Dedicated sample preparation robots equipped with cooled trays, a heated shaker or vortex, syringes for injection and reagent handling as well as solvent and reagent reservoirs enable just-in-time injection without waiting time on the autosampler. The software of these systems calculates the onset of the automated derivatization such that the samples are analyzed promptly after derivatization.

Setup of an automated derivatization just-in-time online GC-MS/MS method on a sample preparation robot (e.g. Gerstel Multi Purpose Sampler):

1. Redry the hydrolysates immediately before derivatization.
2. Crimp dry samples with magnetic caps for automated derivatization. Check that the crimp septa are sufficiently tight (crimp caps cannot be turned anymore).
3. Place vials in a cooled tray of the sample preparation robot.
4. Set up the robot with pyridine (*see* **Note 25**) and MTBSTFA + 1% TBDMSCl in reagent/solvent reservoirs.
5. Set up the robot with pyridine in a solvent reservoir for syringe washing.
6. Program a sample preparation method based on the following steps:
   a. Add 40 μL of pyridine to sample, standard, QC sample/standard or blank vial.
   b. Add 40 μL of MTBSTFA + 1% TBDMSCl to the vial.
   c. Transport the vial to the heated shaker (85°C).
   d. Shake the vial with 250 rpm at 85°C for 1 h.
   e. Transport the vial back to the cooled tray (4°C).
   f. Wait for 5 min.
   g. Inject.
7. Start the sample preparation and GC-MS/MS sequence.

3.3.7 Transport and Storage of Samples

Derivatized samples should be analyzed immediately after derivatization due to limited stability of the derivatives. Derivatized samples should be stored at 4°C until analysis, hence cooling of the autosampler tray is beneficial. If cooled at 4°C, samples can be analyzed up to 48 h after derivatization. However, best results in terms of sensitivity and repeatability are achieved with just-in-time online derivatization.

3.3.8 GC-MS/MS Analysis

Tune and calibrate the MS/MS system according to the vendor description. After either manual or automated just-in-time online derivatization the amino acids are separated via GC on a non-polar capillary column. Detection is carried out by triple quadrupole mass spectrometry in SRM mode after electron ionization. GC-MS/MS parameters are listed in Table 4 and SRM settings are listed in Table 5.

3.3.9 Data Evaluation

Following visual control of the chromatograms in Masshunter Qual, the chromatograms are evaluated using Masshunter Quant (Agilent Technologies) or an equivalent software for data analysis. Quantification is based on internal standardization and external calibration with 6-point calibration curves. The concentration of the standards listed in Table 1 are chosen such that the expected amino acid concentrations in the CHO hydrolysates are within the calibration range for a CDM of 0.2–3 mg.

3.4 Lipid Quantification

For lipid quantification of CHO samples, lipids are extracted by MTBE using the SIMPLEX protocol [20] followed by direct infusion into a high-resolution mass spectrometer [21]. This shotgun lipidomics strategy enables high-throughput analysis of the most abundant lipid species. As all lipids are simultaneously infused, similar chemical behavior within one lipid class (same class defining head group but different fatty acyl chain composition) can be assumed and class-specific quantification via non-endogenous lipid standards can be performed [22]. If lipids such as cholesterol, cholesterol esters or cardiolipins are not detectable or quantifiable, Reversed Phase Liquid Chromatography (RP-LC) is an alternative method to achieve better sensitivity (not covered in this chapter).

Figure 4 shows the lipid composition measured for CHO-K1.

Next to the samples, solvent blanks (pure solvent prior analysis) and method blanks with internal standards are required (*see* **Note 26**). It is also helpful to use a quality control over the measurement by using a pooled sample or a standard reference material. Each biological replicate should be measured at least twice.

3.4.1 Lipid Extraction

1. Collect 10^7 cells, wash once with PBS, pellet them (*see* Subheading 3.1 for details about the sampling procedure) and freeze at −80°C in a 5 mL microcentrifuge tube until measurement (*see* **Note 27**).

Table 4
Parameters for GC-MS/MS analysis of protein amino acids

GC parameters			
Oven			
Equilibration time	0.5 min		
Oven program			
Ramp	**Rate** (°C/min)	**Temp.** (°C)	**Hold Time** (min)
–	–	160	1.00
1	20.00	250	0.00
2	5.00	270	0.00
3	30.00	310	3.50
Total run time	14.33 min		
Inlet			
Type	PTV		
Injection volume	1 µL		
Mode	Split		
Liner	Baffled siltek (Gerstel)		
Split ratio	1:30		
Equilibration time	0.5 min		
Temperature program			
Ramp	**Rate** (°C/s)	**Temp.** (°C)	**Hold Time** (min)
–	–	70	0.10
1	12.00	260	1.00
2	12.00	300	5.00
Analytical column			
Type	Phenomenex ZB-5MS		
	(30 m × 0.25 mm i.d. × 0.25 µm film thickness)		
Mode	Constant flow		
Carrier gas	Helium		
Flow	1.2 mL/min		
Precolumn			
Type	Non-polar guard column (3 m × 0.25 mm i.d.)		
Mode	Constant flow		

(continued)

Table 4
(continued)

GC parameters	
Carrier gas	Helium
Flow	1.1 mL/min
Column connector	Purged ultimate union (Agilent Technologies)

MS parameters	
Transfer line	280°C
Ionization mode	EI
Acquisition mode	SRM
Solvent delay	3.3 min
Source temperature	230°C
Quadrupole temperatures	150°C

Table 5
SRM settings for quantifier and qualifier transitions in GC-EI-MS/MS analysis

Compound	Type of transition	tr (min)	Precursor Ion (m/z)	Product Ion (m/z)	Collision Energy (V)	Resolution MS1, MS2
Alanine	Quant	4.02	260.2	232.2	5	Wide
	Quant ISTD	4.02	264.2	235.2	5	Wide
	Qual	4.02	232.2	147.1	15	Wide
	Qual ISTD	4.02	235.2	147.1	15	Wide
Glycine	Quant	4.13	246.1	218.2	5	Wide
	Quant ISTD	4.13	249.1	220.2	5	Wide
	Qual	4.13	218.2	147.1	5	Wide
	Qual ISTD	4.13	220.2	147.1	5	Wide
Valine	Quant	4.60	260.2	147.1	20	Wide
	Quant ISTD	4.60	265.2	147.1	20	Wide
	Qual	4.60	288.2	260.2	15	Wide
	Qual ISTD	4.60	294.2	265.2	15	Wide
Leucine	Quant	4.78	302.2	274.2	10	Wide
	Quant ISTD	4.78	309.2	280.3	10	Wide
	Qual	4.78	274.3	147.1	15	Wide
	Qual ISTD	4.78	280.3	147.1	15	Wide
Isoleucine	Quant	4.78	302.2	274.2	10	Wide
	Quant ISTD	4.78	309.2	280.3	10	Wide
	Qual	4.78	274.3	147.1	15	Wide
	Qual ISTD	4.78	280.3	147.1	15	Wide

(continued)

Table 5
(continued)

Compound	Type of transition	tr (min)	Precursor Ion (m/z)	Product Ion (m/z)	Collision Energy (V)	Resolution MS1, MS2
Proline	Quant	5.20	258.3	147.1	15	Wide
	Quant ISTD	5.20	263.3	147.1	15	Wide
	Qual	5.20	286.2	258.2	10	Wide
	Qual ISTD	5.20	292.2	263.2	10	Wide
Serine	Quant	6.25	390.2	362.3	15	Wide
	Quant ISTD	6.25	394.2	365.3	15	Wide
	Qual	6.25	362.4	230.2	10	Wide
	Qual ISTD	6.25	365.2	233.2	10	Wide
Threonine	Quant	6.42	404.4	376.4	15	Wide
	Quant ISTD	6.42	409.2	380.4	15	Wide
	Qual	6.42	376.4	244.2	10	Wide
	Qual ISTD	6.42	380.2	248.2	10	Wide
Phenylalanine	Quant	6.88	336.3	308.2	10	Wide
	Quant ISTD	6.88	346.2	317.2	10	Wide
	Qual	6.88	308.3	147.1	20	Wide
	Qual ISTD	6.88	317.2	147.1	20	Wide
Aspartate	Quant	7.20	418.4	376.3	5	Wide
	Quant ISTD	7.20	423.3	379.3	5	Wide
	Qual	7.20	418.4	390.3	10	Wide
	Qual ISTD	7.20	423.3	394.3	10	Wide
Glutamate	Quant	7.95	432.4	272.2	15	Wide
	Quant ISTD	7.95	438.4	277.2	15	Wide
	Qual	7.95	432.4	147.1	30	Wide
	Qual ISTD	7.95	438.4	147.1	30	Wide
Methionine	Quant	8.10	352.2	244.2	15	Wide
sulfone	Quant ISTD	8.10	358.2	248.2	15	Wide
	Qual	8.10	352.2	324.2	5	Wide
	Qual ISTD	8.10	358.2	329.2	5	Wide
Ornithine	Quant	6.31	286.1	154.1	10	Wide
	Quant ISTD	6.31	292.2	160.1	10	Wide
	Qual	6.31	286.1	258.2	10	Wide
	Qual ISTD	6.31	292.2	263.2	10	Wide
Cysteic acid	Quant	8.75	454.3	426.3	10	Wide
	Quant ISTD	8.75	458.3	429.3	10	Wide
	Qual	8.75	426.3	147.1	20	Wide
	Qual ISTD	8.75	429.2	147.1	20	Wide
Lysine	Quant	8.76	329.3	198.0	10	Wide
	Quant ISTD	8.76	336.3	204.0	10	Wide
	Qual	8.76	431.3	300.3	10	Wide
	Qual ISTD	8.76	439.3	307.3	10	Wide

(continued)

Table 5 (continued)

Compound	Type of transition	tr (min)	Precursor Ion (m/z)	Product Ion (m/z)	Collision Energy (V)	Resolution MS1, MS2
Histidine	Quant	10.37	440.4	280.2	30	Widest
	Quant ISTD	10.37	449.4	288.2	30	Widest
	Qual	10.37	440.4	147.1	40	Widest
	Qual ISTD	10.37	449.4	147.1	40	Widest
Tyrosine	Quant	10.68	466.3	147.1	30	Widest
	Quant ISTD	10.68	476.4	147.1	30	Widest
	Qual	10.68	438.4	147.1	30	Widest
	Qual ISTD	10.68	447.3	147.1	30	Widest

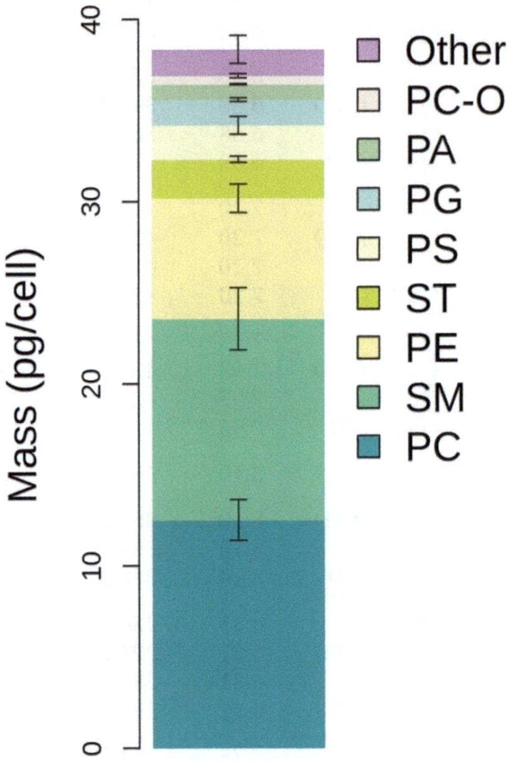

Fig. 4 Lipid composition of CHO-K1 in mid-exponential phase grown in HyClone ActiPro medium (GE Healthcare) supplemented with 8 mM glutamine and 0.2% Anti-Clumping Agent (GIBCO). PC = phosphatidylcholine, SM = sphingomyelin, PE = phosphatidylethanolamine, ST = cholesterol, PS = phosphatidylserine, PG = phosphatidylglycerol, PC-O = phosphatidylcholine (ether-linked), PA = phosphatidic acid, Other = phosphatidylinositol, diacylglycerol, cardiolipin, lysophosphatidylcholine, lysophosphatidylethanolamine, triglyceride. The error bars represent standard deviations from three biological replicates. Please note that cholesterol and cardiolipin were measured by RP-LC

2. Thaw the cells and resuspend the cell pellet in 1 mL MeOH (*see* **Note 28**).
3. Lyse the cells in a waterbath sonicator at 4°C (example settings for Bioruptor NGS (Diagenode): 30 s ON, 30 s OFF, 30 cycles, high intensity).
4. Place the samples on ice and add 10 μL SPLASH® LIPIDOMIX® Mass Spec Standard and, if it is of interest, 5 μL Cardiolipin Mix I (*see* **Note 29**).
5. Add 3375 μL ice-cold MTBE (*see* **Note 30**).
6. Incubate the samples while shaking (950 rpm) at 4°C for 1 h (*see* **Note 6**).
7. Add 846 μL of 0.1% (w/v) AF and mix to induce phase separation.
8. Centrifuge for 5 min at 3000×*g*.
9. Transfer 1 mL of the upper phase into a 2 mL Eppendorf tube (*see* **Notes 30** and **31**).
10. Dry the lipid extracts under nitrogen flow.
11. The remaining lower aqueous phase contains metabolites and proteins and can be used to e.g. quantify the protein content of the sample (see SIMPLEX protocol [20]). Results from protein quantification can be used to normalize lipid quantities. Here, we use the cell number for normalization, so the lower phase is simply discarded.
12. Resolubilize the lipid extracts in 500 μL 2-propanol/methanol/chloroform (4:2:1, v/v/v) with 7.5 mM AF (volume can be changed if concentration is too high or too low), vortex briefly and centrifuge for 5 min at 10,000×*g*. Transfer the supernatant into the muffled glass vials and either measure or store the samples at −80°C until further use. Measurements should be performed as soon as possible after the extraction (ideally on the same or the next day).

3.4.2 Measurement

1. Calibrate the mass spectrometry (MS) instrument according to the vendor description.
2. Attach the Triversa NanoMate to the MS. It is constructed as an ion source and can be installed via a dedicated rack. No spray shield is needed. The NanoMate has to be further connected to the input peripheral control of the MS, a computer, nitrogen supply and power supply (*see* **Note 32**).
3. Configure Xcalibur in the Instrument Configuration.
4. Place the alignment chip into the NanoMate and align the chip position (Interface Settings/Calibrate Device).
5. Replace the alignment chip with the nozzle chip.

6. Under Interface settings set the plate cooler temperature to 10°C.
7. Use the parameter settings in the ChipSoft software as indicated in Table 6. Adapt if necessary.
8. Perform Spray optimization (Method Manager/Spray Optimization) with a solvent blank (*see* **Note 4**). The spray current should be stable and above 50 nA. If necessary the tip position can be changed. The gas pressure and voltage in the settings are

Table 6
ChipSoftManager settings for shotgun-lipidomics

Sample information		**Spray parameters**	
Sample volume	12 μL	Delivery time	17 min
Vent headspace	On	Trigger acquisition when input signal received	Off
Aspirate air after sample	Off		

Advanced parameters			
Air gap before chip	No	Aspiration delay	0 s
Contact closure	After	Contact closure delay	2 s
Voltage timing	After	Voltage timing delay	0 s
Equalization delay	3 s	Aspiration depth	1 mm
Pre-piercing	Mandrel	Pre-piercing depth	9 mm
Pre-wetting	Yes	Pre-wetting mix repeat	1

Temperature		**Spray sensing**	
Halt run if temp out of range	Off	Use spray sensing	On
Cooler is turned on in Interface settings		Begin Spray Sensing	30 s after
		Move to next nozzle when spray	5 nA
		current drops below or goes above	7000 nA
		For	10 s
		Only move to next nozzle	2 times

Gas pressure	0.9 psi
Voltage to apply	1.25 kV
Positive ion	On
Output Contact Closure	Rel3
Duration	480 s (*see* **Note 33**)

guiding values. It is suggested to change them both and monitor the spray current. With the optimal settings, it should be possible to switch polarities back and forth without losing the spray.

9. Save the method.
10. Make a data independent acquisition (DIA) method in Xcalibur.
    (a) Six different time events are necessary, two full mass spectrum-selected ion monitoring (Full MS-SIM) (one for each polarity) and four DIA (two for each polarity to increase the number of MS2 scans to 200- one DIA event is limited to 100). It is recommended to start the DIA after 0.5 min for each polarity such that a representative Full MS can be acquired for additional optical control of the spray stability (*see* **Note 34**).
    (b) Set to 0 the gas flow rates, spray voltage and current in the tune file (MS Tune) for the Q Exactive, the Capillary temperature should be set to 250°C and the S-lens RF level to 60.
    (c) Use the Xcalibur settings shown in Table 7 as a starting point to prepare the acquisition method (*see* **Note 35**).
11. Transfer a 30 μL aliquot from each sample to a 96-well plate.
12. Close the sample wells with an adhesive plate foil and place the well plate into the NanoMate.
13. Write a sample list both in Xcalibur and ChipSoft. The acquisition times and the polarity switch must correspond in both methods. A full sample description is only needed in Xcalibur as this software saves the files.
14. First save and start the Xcalibur sequence, and second the ChipSoft sequence. The start signal will be transferred via contact closure from NanoMate to Q Exactive (*see* **Note 36**).

3.4.3 Data Evaluation

1. Convert the raw files with MSConvert into mzML, following the description at the LipidXplorer Wiki (https://lifs.isas.de/lipidxplorer.html) [23].
2. LipidXplorer is a fragmentation rule based lipid identification software written in Python, which first converts the spectra in a flat-file database termed MasterScan and further identifies the lipids with a user-defined Molecular Fragmentation Query Language (MFQL). Help for installation and a short tutorial for the MFQL files are available on the LipidXplorer Wiki. Here the configuration settings and some additional help will be given, as these points are not always clear.

Table 7
Xcalibur settings for shotgun-lipidomics

Settings for 17 min	Full MS pos	Full MS neg	DIA pos	DIA neg
Runtime (min)	0–8	9–17	0.5–8	9.5–17
Polarity	Positive	Negative	Positive	Negative
In-source CID (eV)	0	0	0	0
Default charge state	–	–	1	1
Microscans	1	1	1	1
Resolution	240,000	240,000	60,000	60,000
AGC target	1.00E+06	1.00E+06	2.00E+05	2.00E+05
Maximum IT (ms)	150	150	130	130
Loop count	–	–	100	100
MSX count	–	–	1	1
MSX isochronous IT	–	–	On	On
Isolation width (m/z)	–	–	1	1
Isolation offset	–	–	0	0
Fixed first mass	–	–	–	–
nCE (%)	–	–	25	28
Number of scan ranges	1	1	–	–
Scan ranges (m/z)	250–1200	250–1200	–	–
Spectrum data type	Profile	Profile	Profile	Profile

3. Under the tab "Import Settings" different settings can be made:
   (a) Import data for each polarity separately. The time range should be similar with the measurement settings (e.g. 0.5–8 min for positive mode).
   (b) Use 0.5 Da as selection window if 1 m/z was used in Xcalibur (tolerance vs. tolerance range).
   (c) As calibration mass use a known mass in the spectrum e.g. an analyte or a known contamination (this setting is optional).
   (d) For resolution and resolution gradient calculation, a representative spectrum of the sample (*see* **Note 37**) should be opened with Xcalibur Qual Browser. In the display options under "Labels" the peaks should be labeled with mass and resolution. Afterwards the view is changed to spectrum list and the data is copied into Excel or similar

programs. Now the m/z values vs. the resolution are plotted and the slope of the trend line is calculated. This is the resolution gradient. The resolution is the resolution value of the first m/z in the spectrum.

(e) Set the tolerance higher than the instrument specification as it can be set lower later in the Run settings.

(f) Check the noise intensity in your data for a threshold value.

(g) If sample replicates are measured or different groups, a min. occupation can be set to remove features that were only present in few data files.

4. Select the files under Import source.
5. Start import (the most time consuming step).
6. Process the lipid identification under the tab "Run" (*see* **Note 38**).

   (a) Use Isotopic Correction in MS1 and MS2 for lipid quantification.

   (b) The mass tolerance can now be lower than the import setting values and changed easily by reprocessing.

7. Run LipidXplorer.
8. Use the output file for quantification.
9. The concentration of each lipid species is calculated according to Eq. 1.

$$\text{conc (analyte)} = \frac{\text{peak area (analyte)}}{\text{peak area (ISTD)}} \times \text{conc (ISTD)} \tag{1}$$

10. Quantification can now be done at the MS1 or MS2 level. If fatty acyl chains are used, quantification on the fatty acyl chain level is possible (use negative mode for phospholipids, e.g. PC 16:0_18:2 and PC 16:1_18:1 instead of sum parameter for PC 34:2). The right choice depends on the lipid class and the sample itself.

3.5 RNA Quantification

First, RNA is isolated with TRIzol or TRI reagent according to the protocol suggested by the manufacturer with small adjustments. The purified RNA is then quantified spectrophotometrically at wavelength 260 nm. As there are losses of RNA during the isolation, the samples are spiked with a defined amount of purified RNA immediately after sampling and before they are frozen. Afterwards, the results are corrected with a recovery factor. Typical recoveries that we observed were 70–90%. RNA quantities in CHO range from 13 to 23 pg per cell.

3.5.1 RNA Extraction and Quantification

Important: *Perform the isolation in the fume hood and use nitrile gloves. Before you start working, wipe the surfaces, pipettes, tip boxes etc. with RNaseZap wipes to remove RNases.*

1. For each biological replicate, take two samples of 5×10^6 cells (one will be spiked with RNA) into nuclease-free 1.5 mL tubes. If the volume of the cell suspension it too big, use a bigger tube and transfer the samples into 1.5 mL tubes after resuspension in TRIzol/TRI reagent in **step 3**.
2. Centrifuge 8 min at 200×*g*, room temperature and remove supernatant. Do not wash the cells, it might lead to degradation of mRNA (see TRIzol reagent user guide (ThermoFisher Scientific)).
3. Add 1 mL TRIzol/TRI reagent, vortex until the pellet is completely resuspended.
4. Add 30 μL of purified RNA (700–1000 ng/μL) to one of the two samples taken in **step 1**, vortex.
5. Freeze the samples at −80°C or continue with the isolation.
6. Add 0.2 mL chloroform, vortex for 30 s and incubate 3 min at room temperature.
7. Centrifuge at 12,000×*g* for 15 min at 4°C.
8. Transfer the upper aqueous phase to a fresh 1.5 mL RNase-free tube (be careful not to touch the bottom phase).
9. Add 200 μL of RNase-free water to dilute the rest of the aqueous phase that was not yet transferred to the new tubes, mix by inverting.
10. Centrifuge at 12,000×*g* for 7.5 min at 4°C.
11. Transfer the rest of the aqueous upper phase to the same tubes as in **step 8**.
12. Add 0.75 mL isopropanol, vortex and incubate for 10 min at room temperature.
13. Centrifuge at 12,000×*g* for 10 min at 4°C.
14. Decant and pipette out the supernatant.
15. Wash pellets with 70% EtOH, mix by inverting the tubes several times (until the pellets loosen).
16. Centrifuge at 12,000×*g* for 5 min at 4°C.
17. Decant and pipette out as much supernatant as possible (*see* **Note 39**).

18. Dry the pellets for approximately 10 min (until they start becoming transparent). Do not overdry them, as it might become too difficult to dissolve the RNA.
19. Dissolve RNA in 60 μL RNase-free water, vortex, incubate for 10 min at 55°C, short spin and vortex again.
20. Measure absorbance at 260 nm in triplicates (*see* **Note 40**).

3.5.2 Data Evaluation

1. If using NanoDrop, the device will calculate the RNA concentration directly. If not, calculate the concentration as $c = A \times f$, where A is the absorbance at 260 nm and $f = 40$ ng cm/μL.
2. Calculate the recovery factor = the ratio of measured:expected difference between the spiked and non-spiked samples.
3. Correct the RNA concentrations with the recovery factor.
4. Calculate the total amount of RNA in 60 μL.
5. Divide by the number of cells.

Background Information: Several other RNA quantification methods were tested. Quant-iT™ RNA Assay Kit (ThermoFisher Scientific) can be used directly in cell lysates, but strong matrix effects may lead to variation in results. RNeasy Mini Kit (QIAGEN) and Direct-zol RNA Kit (Zymo Research) are faster and easy to use, but the TRIzol/TRI reagent + chloroform extraction generated the highest yields. The purified RNA can also be quantified with a fluorescent dye instead of absorbance measurements. For instance, the Quant-iT™ RNA Assay Kit, if applied to purified RNA, generated results comparable to NanoDrop measurements. However, the kit requires additional reagents and thus cost.

3.6 DNA Quantification

DNA is quantified directly in cell lysates with the fluorescent dye PicoGreen™ (ThermoFisher Scientific), which binds specifically to dsDNA. Typical values observed for CHO cells are 5–9 pg per cell.

3.6.1 Quantification

1. Centrifuge 7.5×10^5 cells, wash with PBS, centrifuge again (*see* Subheading 3.1 for the details about sampling).
2. Store cell pellets at −80°C until analysis.
3. Resuspend the cell pellets in 100 μL PBS (*see* **Note 41**).
4. Add 10 μL proteinase K and 2 μL RNase A, vortex and incubate at room temperature for 2 min.
5. Add 100 μL buffer AL, vortex and incubate at 56°C for 10 min to lyse the cells.
6. Do a short spin and add 848 μL of 1× TE buffer, vortex.
7. Mix 20 μL of the lysate with 380 μL 1× TE, vortex.
8. Pipette 100 μL of standards and samples in triplicates into a black 96-well plate.

9. Add 100 μL 1× PicoGreen solution to each sample and cover with aluminum foil to protect from light.
10. Mix for 30s on a microplate shaker, incubate for 2 min.
11. Measure fluorescence with a microplate reader (excitation wavelength = 480 nm, emission wavelength = 520 nm, automatic gain).

3.6.2 Data Evaluation

1. Calculate the average fluorescence of each sample and each standard from the corresponding three wells.
2. Subtract the average fluorescence of the blank from all samples and standards.
3. Plot fluorescence on x-axis, concentration of the standards on y-axis and do a linear fit.
4. Use the equation of the fit to calculate the concentrations of the samples.
5. Multiply the concentrations by 0.4 to get the total amount per 0.4 mL.
6. Multiply by 53 (the sample was initially resuspended in 1060 μL and 20 μL were used, 1060/20 = 53).
7. Divide the values by the number of cells (7.5×10^5).

3.6.3 Optional Step: Matrix Correction

As the assay is performed directly from a cell lysate, other cellular components apart from DNA might interfere with the assay. To check this, samples can be spiked with a defined amount of DNA. The ratio of expected:measured difference between the spiked and non-spiked samples can be used to correct the final data with this protocol. We observed no matrix effects for our samples, therefore this step is listed as optional. If you want to check for matrix effects, add the following steps to the protocol:

1. Make another dilution in addition to the one described in Subheading 3.6 **step** 7 as follows: Mix 20 μL of the lysate with 280 μL 1× TE and 100 μL of 2 μg/mL λ DNA and continue with the procedure in the same way as for the non-spiked samples.
2. Calculate the amount of DNA in 0.4 mL as in Subheading 3.6.2, **step 5** and subtract the values of the non-spiked samples from the corresponding spiked samples.
3. Calculate the ratio of the expected vs. measured difference. In this case, the expected difference is 200 ng. If the measured difference is, for example, 160 ng, it means the concentrations in the samples are most likely underestimated. To correct the data, multiply the values of the non-spiked samples by the correction factor 200/160 = 1.25.
4. Proceed with the calculation in **step 6**.

3.7 Carbohydrate Quantification

Carbohydrates are quantified with a colorimetric assay using anthrone and sulfuric acid. The strong acid and heat hydrolyze glycolytic bonds in polysaccharides and convert monomeric sugars to furfuraldehyde derivatives, which react with anthrone and produce colored compounds. The method quantifies predominantly hexoses with only a small interference from pentoses [24, 25]. The pentoses are assumed to be mostly part of nucleic acids, which are quantified independently.

The assay is performed in a 96-well format, adapted from [26] and [27]. Because the carbohydrates are quantified directly in the cell lysates, all samples are spiked with a defined amount of glucose to correct for possible matrix effects. Typical values for CHO cells are 2–6 pg per cell.

3.7.1 Quantification

Important: *When working with sulfuric acid, always wear protective goggles, lab coat and appropriate gloves.*

1. Take 10^7 cells, centrifuge, wash with PBS, centrifuge again (*see* Subheading 3.1, *see* **Note 42**) in 1.5 mL tubes. If the volume of the cell suspension it too big, a bigger tube can be used and the suspension can be transferred to 1.5 mL tube after resuspension in PBS in **step 3**.
2. Store cell pellets at −80°C until analysis.
3. Resuspend the cell pellets in 400 μL PBS (*see* **Notes 43** and **44**).
4. Place 0.2 mL PCR tubes on ice or in a cooling block. Pipette 40 μL of the standards into the tubes in triplicates (*see* **Note 45**).
5. Mix 20 μL PBS with 20 μL of a sample (non-spiked sample) and 20 μL of 130 μg/mL glucose with 20 μL of a sample (spiked sample). Prepare the dilutions in technical triplicates.
6. Add 100 μL of 2 mg/mL anthrone (*see* **Note 46**), mix by tapping or inverting the tubes and short spin.
7. Incubate 3 min at 92°C and place on ice/cooling block or cool the samples directly in the thermal cycler.
8. Pipette 110 μL of the samples into a transparent 96-well microplate.
9. Measure absorbance at wavelength 630 nm.

3.7.2 Data Evaluation

1. Calculate an average absorbance from the technical replicates.
2. Subtract the average absorbance of the blanks from the average absorbances of all samples and standards.

3. Plot absorbances of the standards on the x-axis and concentrations on the y-axis, do a linear fit.
4. Use the equation from the fit to calculate the concentrations of the samples.
5. Subtract the concentration of the non-spiked samples from the spiked samples.
6. Calculate the ratios of the expected difference (65 μg/mL) and measured difference to get a matrix correction factor.
7. Multiply the concentrations of the non-spiked samples with an average correction factor.
8. Multiply by 2 (dilution factor).
9. Multiply by 0.4 (the initial sample volume).
10. Divide by the number of cells.

3.8 Data Conversion to Biomass Equations

To integrate the measured biomass composition into the model, the data has to be converted into the appropriate units. First, equations and average molecular masses for the macromolecules (proteins, lipids, DNA, RNA) are calculated. Then the final biomass equation is calculated using the mass percent of each macromolecule per cell dry mass and their respective molecular masses.

3.8.1 Macromolecular Equations

1. For each macromolecule, calculate the mole percent of each component, based on the available composition. These will be the stoichiometric coefficients in the equations.
   (a) For proteins, use the measured AA composition. Trp is not measured in the method presented in this chapter, therefore a value from literature has to be used. The pairs Glu + Gln and Asp + Asn are measured as a sum. Assume that their ratio is 1:1, unless other values are available in the literature. Use molecular masses of AAs as they are found in peptides, i.e. molecular mass of AA minus molecular mass of 1 H_2O molecule.
   (b) For lipids, use the sum for each lipid class (cardiolipin (CL), diacylglycerol (DG), lysophosphatidylethanolamine (LPE), phosphatidic acid (PA), phosphatidylcholine (PC), phosphatidylcholine (ether-linked) (PC-O), phosphatidylethanolamine (PE), phosphatidylcholine (ether-linked) (PE-O), phosphatidylglycerol (PG), phosphatidylinositol (PI), phosphatidylserine (PS), sphingomyelin (SM), cholesterol (ST), triglyceride (TG)).
   (c) RNA composition can be taken from literature or calculated from RNAseq data (but only if it includes rRNA quantities as well, because these form the majority of RNA).

   (d) If the genome sequence is available, DNA composition can be simply calculated from the guanine-cytosine content.

2. Add energy requirements for the synthesis of proteins, DNA and RNA, for example as in [28]. For the synthesis of 1 mole of protein, 4.306 energy bonds are hydrolyzed on average. To represent the cellular process more accurately, this was modified to 2 GTP (hydrolyzed to GDP) and 1.306 ATP (1 is hydrolyzed to AMP and 0.306 to ADP).
3. Make equations for the synthesis of the macromolecules, *see* Table 8 for examples.

3.8.2 Average Molecular Masses of Macromolecules

For each macromolecule, an average molecular mass, M^{macro}, has to be calculated, using molecular masses of the reactants, M^{react}, and products, M^{prod}, and their respective stoichiometric coefficients, s, (*see* Eq. 2).

$$M^{\text{macro}} = \sum_i s_i M_i^{\text{react}} - \sum_j s_j M_j^{\text{prod}} \quad (2)$$

3.8.3 Biomass Equations

1. Express the quantities of proteins, lipids, DNA, RNA and carbohydrates as g per g of CDM (mass percent).
2. Divide the values by the average molecular masses calculated as described in Subheading 3.8.2 to obtain mol/g. For carbohydrates, use the molecular mass of glucose, unless a specific carbohydrate composition is available.
3. Multiply by 1000 to get mmol/g. See an example of a biomass equation in Table 8.

4 Notes

1. Arginine decomposes during derivatization forming ornithine. No arginine derivative can be detected. Hence arginine is quantified via ornithine. This approach can only be followed using the isotopically labeled internal standard for arginine, which shows the same decomposition rate and derivatization efficiency as ^{12}C arginine.
2. Pipetting steps should not transfer volumes lower than 40 μL to ensure precision and accuracy.
3. As AF is not easily dissolvable in the solvent mixture, it is common practice to prepare a 1 M AF solution in water and add the necessary amount to the final solution. Ammonium acetate can be also used instead of AF.

Table 8
Composition of the macromolecular species

Protein
0.078 Ala + 0.063 Arg + 0.086 Gly + 0.051 Val + 0.083 Leu + 0.037 Ile + 0.054 Pro + 0.067 Ser + 0.051 Thr + 0.034 Phe + 0.051 Asp + 0.063 Glu
+ 0.018 MetS + 0.072 Lys + 0.02 CysA + 0.019 His + 0.029 Tyr + 0.008 Trp + 0.063 Gln + 0.051 Asn + 1.306 ATP + 2 GTP + 2.306 H_2O
→ 1 Protein + 0.306 ADP + 1 AMP + 2 GDP + 2.306 Pi + 1 PPi + 3.306 H
Lipid
0.01 CL + 0.007 DG + 0.002 LPE + 0.023 PA + 0.288 PC + 0.013 PC-O + 0.167 PE + 0.035 PG + 0.019 PI + 0.045 PS + 0.275 SM + 0.107 ST + 0.009 TG → 1 Lipid
RNA
0.264 GTP + 0.219 UTP + 0.249 CTP + 0.668 ATP + 0.4 H_2O → 1 RNA + 0.4 ADP + 0.4 Pi + 1 PPi + 0.4 H
DNA
0.294 dATP + 0.294 dTTP + 0.206 dCTP + 0.206 dGTP + 1.372 ATP + 1.372 H_2O → 1 DNA + 1.372 ADP+ 1.372 Pi + 1 PPi + 1.372 H
Total biomass
5.062 Protein + 0.209 Lipid + 0.281 RNA + 0.102 DNA + 0.110 Glucose → 1 Biomass

4. This solution has to be as clean as possible. Therefore, it should be prepared in 10 mL clean glass flasks that were heated and rinsed with 2-propanol. It is also useful to fill the solvents themselves into 10 mL flasks to facilitate the handling. To reassure the quality of this solvent mixture, the final solution should be analyzed prior to use via spray parameter optimization in ChipSoft.
5. Other LIPIDOMIX products are also available and may be cheaper or better for specific samples. An ISTD-Mix can be also prepared from single ISTD stocks to increase flexibility and reduce costs. Consult literature of the sample to find the best ISTD.
6. The Thermomixer from Eppendorf is able to mix and cool (pre-cooling before use is suggested). If such a device is not available a shaker or rotator can be placed in a cooling room to enable constant temperature.
7. If the centrifuge cannot handle 5 mL tubes, the total volume of the extraction mixture and the sample amount need to be adapted to 2 mL.
8. The purified RNA can be prepared with the RNA extraction protocol described in Subheading 3.5, but skipping the spiking

in **step 4**. Afterwards it can be diluted with nuclease-free water if necessary.

9. You will need around 40 mL of 1× TE for one 96-well plate.
10. Prepare the 2 μg/mL λ DNA solution on precision balances. Then recalculate the concentrations of the standards if needed.
11. If you use a multichannel pipette, prepare around 1 mL more of the 1× PicoGreen solution than needed. This makes pipetting from the multichannel pipette reservoir easier.
12. Take the PicoGreen stock solution out of the fridge when you start the procedure, because it is solid at 4°C and needs to warm up to thaw. Start preparing the standards and samples and by the time you need to add the PicoGreen reagent, it will have thawed. It is also good to aliquot the solution into several tubes (wrapped in aluminum to protect from light).
13. Prepare several aliquots of the 130 μg/mL standard, because it is used as a spike-in and will be used up faster than the other standards.
14. Prepare around 12 mL of the anthrone reagent for one 96-well plate assay.
15. You can measure the cell concentration in duplicates or triplicates to have a more accurate estimate of cell number. However, some cell lines get more shear sensitive the longer they are suspended in PBS and thus the measured cell number concentration decreases with each measurement. Hence, just one measurement immediately after washing can be more accurate. This is a specific property of each cell line.
16. When using a ViCell XR autosampler, do not load more than 2–3 samples at a time. The cells tend to sediment quickly and we have observed that the mixing done by ViCell before the measurement is not effective enough to obtain a representative sample. If you load more samples we recommend that you gently shake the cups before they are measured in order to counteract sedimentation.
17. Mix the cell suspension before aliquoting by gently inverting the tube. The cells sediment quite quickly.
18. Do not go over 90°C. Otherwise the evaporation process can be too violent and material may be lost. Furthermore, cell components might oxidize and thus change the result.
19. Weigh the beakers immediately after taking them out of the desiccator, because they start to absorb water and the mass increases.
20. Tryptophan cannot be quantified in the acidic hydrolysates. Base hydrolysis is recommended [29].

21. Histidine sensitivity is low and could not be further improved by changes in derivatization parameters. For better sensitivity and hence repeatability of histidine, quantification samples can be measured via HPLC-MS.
22. Quantitative transfer of cell samples into Sarstedt vials is tedious and time consuming. Sample directly into Sarstedt tubes.
23. Non-polar organic residues or cell debris in the aliquots for GC-MS/MS analysis lead to severe contamination of the liner and hence loss in sensitivity. Take care to take aliquots of the aqueous middle layer of the reconstituted hydrolysates only.
24. In order to achieve good derivatization efficiency traces of water must be removed from the protein hydrolysates by drying the samples directly before derivatization.
25. Handling of pyridine and silylation reagent has to be done in a fume hood while wearing protective clothing.
26. Prepare a method blank sample by filling a tube with cultivation medium, decant, add PBS, decant and add 1 mL MeOH. The method blank can now be treated as a sample.
27. As lipids can be oxidized and degraded, the storage time should be as short as possible. If this is not feasible, measures should be taken to avoid such degradation processes (e.g. low storage temperature).
28. The protocol can be easily adapted to a different cell number or volumes of the reagents. *See* [20] for an example with a lower starting amount of cells and volumes. This is important if 5 mL tubes cannot be centrifuged.
29. Change gloves regularly during the extraction to prevent cross-contamination of the samples.
30. Pre-wet the tip before pipetting MTBE or other organic solvents, otherwise it will drip from the pipette tip.
31. Take 2 aliquots of the upper phase as a backup. If the concentration of lipids is too high or low, the amount of solvent can be adapted, as well as the resuspension solvent.
32. For a good spray stability and signal intensity, it is important that the Nanomate is central to the MS and in a distance of 3–5 mm. This is a crucial step and should be done with care. As a rule of thumb, the reproducibility is increased by a bigger distance and the intensity is increased by a smaller distance.
33. Duration of the output closure indicates the time period until polarity switching (positive to negative mode, 8 min). Normally done at half time but an additional time for spray stabilization is highly recommended (analysis in negative mode starts at 9 min until 17 min).

34. In DIA, each mass is fragmented in chronological order independent of the intensity in the MS1 spectrum. The scan time should be long enough to fragment each mass twice. The selection of the masses is done by an inclusion list, in which the values of the m/z masses are chosen according to the expected mass defect at a specific mass. Empirically useful values start at 100.000 m/z and increase 1.001 per 1 m/z. (101.001; 102.002, etc.)
35. The MS1 mass range can be split in more overlapping ranges to increase the dynamic range and detection sensitivity by decreasing the ion package in the Orbitrap. If this is necessary, please follow the protocol of Southam et al. [30].
36. The NanoMate can perform polarity switching, but a software update and an additional straight through cable is necessary. Ask Advion for firmware and further assistance.
37. An averaged spectrum can be chosen for MS1, for MS2 a known analyte spectrum is suggested with as many fragments as possible. Use a medium concentrated sample. If the concentration differs over the sample batch, calculate the resolution gradient for low, medium and high concentrated samples and split the batch for LipidXplorer.
38. The advantage of the MFQL queries is their high flexibility. New fragmentation rules, more constraints and different reporting styles can be easily adapted. However, this flexibility needs some training to be used to its full potential. Example files can be found on the LipidXplorer Wiki. In general, it is suggested to try each lipid class one after the other and if possible, standards should be used to avoid wrong identification. Literature about lipid fragmentation can also help. A consistent labeling in the REPORT section is also important as already small typing errors will lead to a new column in the output file. A useful tool for quantification is the report option ISOBARIC=pr.Lipidclass.isobaric, which indicates whether other lipids are isobaric with this analyte.
39. Try to remove as much ethanol as possible. Decant it and pipette it out with a 200 μL tip. Then do a short spin and remove the rest with a 10 μL tip.
40. It might happen that if you measure the concentration three times, you get very different results. This may indicate that the RNA was not completely dissolved. Try vortexing the samples and remeasure them three times again.
41. The cell pellets often contain substantial amounts of residual PBS from washing. You can estimate it by weighing and then adding PBS to reach the final volume of 100 μL.

42. You can reduce the amount of cells and then adjust the volume of PBS added, e.g. if you take 5×10^6 cells, resuspend the pellet in 200 μL.
43. The cell pellets often contain substantial amounts of residual PBS from washing. You can estimate this by weighing and then adding PBS to reach the final volume of 400 μL.
44. The most effective way to resuspend the pellet is by pipetting up and down. Mix each sample just before making the dilutions, because the cells quickly sediment.
45. It is convenient to perform the assay in PCR strips, because you can use a multichannel pipette to add the anthrone reagent and to pipette the samples into the 96-well plate. The incubation at 92°C and subsequent cooling down can be performed in a PCR thermal cycler.
46. The anthrone reagent is very viscous. Pipette slowly and always pre-wet pipette tips or use reverse pipetting.

Acknowledgements

DS, SH, NB, JZ, and DER acknowledge support by the Federal Ministry for Digital and Economic Affairs (bmwd), the Federal Ministry for Transport, Innovation and Technology (bmvit), the Styrian Business Promotion Agency SFG, the Standortagentur Tirol, Government of Lower Austria and ZIT—Technology Agency of the City of Vienna through the COMET-Funding Program managed by the Austrian Research Promotion Agency FFG. DS received additional funding from the PhD program BioToP (Biomolecular Technology of Proteins) of the Austrian Science Fund (FWF Project W1224). The funding agencies had no influence on the conduct of this research.

References

1. Kyriakopoulos S, Kontoravdi C (2013) Analysis of the landscape of biologically-derived pharmaceuticals in Europe: dominant production systems, molecule types on the rise and approval trends. Eur J Pharm Sci 48:428–441
2. Mathias S, Fischer S, Handrick R, Fieder J, Schulz P, Bradl H, Gorr I, Gamer M, Otte K (2018) Visualisation of intracellular production bottlenecks in suspension-adapted CHO cells producing complex biopharmaceuticals using fluorescence microscopy. J Biotechnol 271:47–55
3. Lewis NE, Nagarajan H, Palsson BO (2012) Constraining the metabolic genotype–phenotype relationship using a phylogeny of in silico methods. Nat Rev Microbiol 10:291–305
4. Rejc Ž, Magdevska L, Tršelič T, Osolin T, Vodopivec R, Mraz J, Pavliha E, Zimic N, Cvitanović T, Rozman D, Moškon M, Mraz M (2017) Computational modelling of genome-scale metabolic networks and its application to CHO cell cultures. Comput Biol Med 88:150–160

5. Calmels C, McCann A, Malphettes L, Andersen MR (2019) Application of a curated genome-scale metabolic model of CHO DG44 to an industrial fed-batch process. Metab Eng 51:9–19
6. Hefzi H, Ang KS, Hanscho M, Bordbar A, Ruckerbauer D, Lakshmanan M, Orellana CA, Baycin-Hizal D, Huang Y, Ley D, Martinez VS, Kyriakopoulos S, Jiménez NE, Zielinski DC, Quek L-E, Wulff T, Arnsdorf J, Li S, Lee JS, Paglia G, Loira N, Spahn PN, Pedersen LE, Gutierrez JM, King ZA, Lund AM, Nagarajan H, Thomas A, Abdel-Haleem AM, Zanghellini J, Kildegaard HF, Voldborg BG, Gerdtzen ZP, Betenbaugh MJ, Palsson BO, Andersen MR, Nielsen LK, Borth N, Lee D-Y, Lewis NE (2016) A consensus genome-scale reconstruction of chinese hamster ovary cell metabolism. Cell Syst 3:434–443.e8
7. Baart GJE, Martens DE (2012) Genome-scale metabolic models: reconstruction and analysis. In: Christodoulides M (ed) Neisseria meningitidis: advanced methods and protocols. Methods in molecular biology. Humana Press, Totowa, pp 107–126
8. Dikicioglu D, Kırdar B, Oliver SG (2015) Biomass composition: the "elephant in the room" of metabolic modelling. Metabolomics 11:1690–1701
9. Széliová D (2019) Manuscript in preparation
10. Selvarasu S, Ho YS, Chong WPK, Wong NSC, Yusufi FNK, Lee YY, Yap MGS, Lee D-Y (2012) Combined in silico modeling and metabolomics analysis to characterize fed-batch CHO cell culture. Biotechnol Bioeng 109:1415–1429
11. Pan X, Dalm C, Wijffels RH, Martens DE (2017) Metabolic characterization of a CHO cell size increase phase in fed-batch cultures. Appl Microbiol Biotechnol 101:8101–8113
12. Zhang Y, Baycin-Hizal D, Kumar A, Priola J, Bahri M, Heffner KM, Wang M, Han X, Bowen MA, Betenbaugh MJ (2017) High-throughput lipidomic and transcriptomic analysis to compare SP2/0, CHO, and HEK-293 mammalian cell lines. Anal Chem 89:1477–1485
13. Chen A, Leith M, Tu R, Tahim G, Sudra A, Bhargava S (2017) Effects of diluents on cell culture viability measured by automated cell counter. PLoS One 12:e0173375
14. Fountoulakis M, Lahm H-W (1998) Hydrolysis and amino acid composition analysis of proteins. J Chromatogr A 826:109–134
15. Sandra K, Vandenheede I, Sandra P (2014) Modern chromatographic and mass spectrometric techniques for protein biopharmaceutical characterization. J Chromatogr A 1335:81–103
16. Hoofnagle AN, Whiteaker JR, Carr SA, Kuhn E, Liu T, Massoni SA, Thomas SN, Townsend RR, Zimmerman LJ, Boja E, Chen J, Crimmins DL, Davies SR, Gao Y, Hiltke TR, Ketchum KA, Kinsinger CR, Mesri M, Meyer MR, Qian W-J, Schoenherr RM, Scott MG, Shi T, Whiteley GR, Wrobel JA, Wu C, Ackermann BL, Aebersold R, Barnidge DR, Bunk DM, Clarke N, Fishman JB, Grant RP, Kusebauch U, Kushnir MM, Lowenthal MS, Moritz RL, Neubert H, Patterson SD, Rockwood AL, Rogers J, Singh RJ, Eyk JV, Wong SH, Zhang S, Chan DW, Chen X, Ellis MJ, Liebler DC, Rodland KD, Rodriguez H, Smith RD, Zhang Z, Zhang H, Paulovich AG (2016) Recommendations for the generation, quantification, storage, and handling of peptides used for mass spectrometry-based assays. Clin Chem 62:48–69
17. Weiss M, Manneberg M, Juranville J-F, Lahm H-W, Fountoulakis M (1998) Effect of the hydrolysis method on the determination of the amino acid composition of proteins. J Chromatogr A 795:263–275
18. Poole CF (2013) Alkylsilyl derivatives for gas chromatography. J Chromatogr A 1296:2–14
19. Rampler E, Dalik T, Stingeder G, Hann S, Koellensperger G (2012) Sulfur containing amino acids – challenge of accurate quantification. J Anal At Spectrom 27(6):1018
20. Coman C, Solari FA, Hentschel A, Sickmann A, Zahedi RP, Ahrends R (2016) Simultaneous metabolite, protein, lipid extraction (SIMPLEX): a combinatorial multimolecular omics approach for systems biology. Mol Cell Proteomics 15:1453–1466
21. Schuhmann K, Almeida R, Baumert M, Herzog R, Bornstein SR, Shevchenko A (2012) Shotgun lipidomics on a LTQ orbitrap mass spectrometer by successive switching between acquisition polarity modes. J Mass Spectrom 47:96–104
22. Yang K, Han X (2011) Accurate quantification of lipid species by electrospray ionization mass spectrometry – meets a key challenge in lipidomics. Metabolites 1:21–40
23. Herzog R, Schuhmann K, Schwudke D, Sampaio JL, Bornstein SR, Schroeder M, Shevchenko A (2012) LipidXplorer: a software for consensual cross-platform lipidomics. PLoS One 7:e29851
24. Trevelyan WE, Harrison JS (1952) Studies on yeast metabolism. 1. Fractionation and microdetermination of cell carbohydrates. Biochem J 50:298–303

25. Joseph H (1955) The determination of sugar in blood and spinal fluid with anthrone reagent. J Biol Chem 212:335–343
26. Laurentin A, Edwards CA (2003) A microtiter modification of the anthrone-sulfuric acid colorimetric assay for glucose-based carbohydrates. Anal Biochem 315:143–145
27. Beck A, Hunt K, Carlson R (2018) Measuring cellular biomass composition for computational biology applications. Processes 6:38
28. Sheikh K, Förster J, Nielsen LK (2008) Modeling hybridoma cell metabolism using a generic genome-scale metabolic model of Mus musculus. Biotechnol Prog 21:112–121
29. Rutherfurd SM, Gilani GS (2009) Amino acid analysis. Curr Protoc Protein Sci 58 (1):11.9.1–11.9.37
30. Southam AD, Weber RJM, Engel J, Jones MR, Viant MR (2017) A complete workflow for high-resolution spectral-stitching nanoelectrospray direct-infusion mass-spectrometry-based metabolomics and lipidomics. Nat Protoc 12:255–273

Chapter 8

Quantifying Nitric Oxide Flux Distributions

Darshan M. Sivaloganathan, Xuanqing Wan, and Mark P. Brynildsen

Abstract

Nitric oxide (NO) is a radical that is used as an attack molecule by immune cells. NO can interact and damage a range of biomolecules, and the biological outcome for bacteria assaulted with NO will be governed by how the radical distributes within their biochemical reaction networks. Measurement of those NO fluxes is complicated by the low abundance and transience of many of its reaction products. To overcome this challenge, we use computational modeling to translate measurements of several biochemical species (e.g., NO, O_2, NO_2^-) into NO flux distributions. In this chapter, we provide a detailed protocol, which includes experimental measurements and computational modeling, to estimate the NO flux distribution in an *Escherichia coli* culture. Those fluxes will have uncertainty associated with them and we also discuss how further experiments and modeling can be employed for flux refinement.

Key words *Escherichia coli*, Nitric oxide, Metabolic flux, Nitric oxide dioxygenase, Nitric oxide reductase

1 Introduction

Nitric oxide (NO) is a reactive, highly diffusible, nonpolar molecule that is used in humans as a vasodilator, neurotransmitter, and antimicrobial [1–3]. Inducible NO synthase (iNOS) produces NO within cells of the innate immune response to combat pathogens [1, 4]. Upon phagocytosis by macrophages, pathogens are internalized in membrane-bound compartments (phagosomes) where they are bombarded with a range of antimicrobial stressors, which include NO [5]. To counter NO, many bacteria have evolved defense systems, that when disrupted greatly attenuate their virulence [6–9]. Those defense systems represent antiinfective targets that if impaired would exert selective pressure only in conjunction with immunity, since they are not essential for bacterial growth

Electronic supplementary material: The online version of this chapter (https://doi.org/10.1007/978-1-0716-0159-4_8) contains supplementary material, which is available to authorized users.

Darshan M. Sivaloganathan and Xuanqing Wan contributed equally to this work.

Deepak Nagrath (ed.), *Metabolic Flux Analysis in Eukaryotic Cells: Methods and Protocols*, Methods in Molecular Biology, vol. 2088, https://doi.org/10.1007/978-1-0716-0159-4_8, © Springer Science+Business Media, LLC, part of Springer Nature 2020

[9, 10]. Such focused selective pressure, within a host at an infection site, has been predicted to slow the development of resistance to treatments [11–13]. Further, NO released from probiotic patches, catheter coatings, topical creams, and nanoparticles have been explored as an antibacterial treatments [14–17]. Inhibition of the same bacterial NO defenses that would synergize with immunity are also projected to potentiate the activities of these NO-releasing materials [9]. Unfortunately, an ability to therapeutically inactivate those defenses has yet to be realized, because known inhibitors are either toxic to humans or poorly transported into bacteria [18, 19]. Greater knowledge of how pathogens process NO stress and how different genetic or environmental perturbations modulate NO flux within bacterial reaction networks will facilitate means to therapeutically interfere with bacterial NO defenses [9].

NO reacts with O_2 and superoxide (O_2^-), directly damages iron–sulfur clusters, and reversibly binds heme, whereas its autoxidation products (e.g., N_2O_3, $ONOO^-$) can damage DNA bases, thiol groups, tyrosine, and lipid residues [20, 21]. NO is readily lost to the gas phase, it reacts both inside and outside of cells, many of its reaction products are transient intermediates, and its biochemical reaction network only has a few stable end products (e.g., NO_2^-, NO_3^-). These factors necessitate the use of computational modeling to calculate NO fluxes through its multitude of reaction pathways from experimental measurements of NO and other species [9, 22–24]. Once a computational model can capture experimental data and its predicted flux distributions have been validated to be accurate with independent experiments, the model represents a valuable tool to understand how variables of the system impact NO flux distributions. For example, in a previous study a parametric analysis of such a model identified a kinetic regime of NO delivery rates that rendered bacterial NO defenses dispensable [25]; or in other words, conditions were identified where cultures of bacteria with or without the major NO detoxification system exhibited similar NO flux distributions.

Here we describe a methodology to estimate NO flux distributions within a wild-type *E. coli* culture. Experimental measurements of NO, O_2, and NO_2^- are collected in different bioreactor configurations (e.g., with and without cells). That data is then used to optimize uncertain parameters within a mathematical model. The resulting model is then used to calculate NO flux distributions. In addition, we discuss follow-up experiments to validate model-calculated distributions, and how uncertainty in the flux distributions can be estimated and reduced. These approaches have been used previously to analyze NO fluxes in wild-type and mutant cultures of *E. coli* under different environmental conditions [26–28], and we have previously demonstrated that the methodology can be translated to other bacterial species [29, 30]. We anticipate that our approaches can be applied to higher organisms, especially given the information yielded by previous works that adopted reaction engineering methodologies to interrogate NO fluxes in mammalian systems [23, 31–33].

2 Materials

Aqueous solutions should be prepared with Milli-Q water (deionized water purified to achieve a resistivity of 18.2 MΩ cm at 25 °C). Autoclaved Milli-Q water is used to rinse stir bars and submerge the ISO-NOP NO sensor before and after use.

2.1 Bacterial Strains

E. coli K-12 MG1655. The method described here has been used previously for *E. coli* K-12 MG1655 [25]; however, it has also been applied to different strains and species [29, 30].

2.2 Media

1. LB media: Dissolve 25 g of LB broth powder (BD Difco™ Dehydrated Culture Media: LB Broth, Miller) in 1 L Milli-Q water, and autoclave for 30 min at 121 °C and 100 kPa gauge pressure.
2. MOPS minimal media: Add 100 mL of 10× MOPS Modified Rich Buffer (in MOPS Minimal Medium Kit, Teknova, Inc.), 10 mL of 0.132 M K_2HPO_4, and 10 mL of 1 M glucose to 880 mL of Milli-Q water and sterilize with a 0.22 μm bottle-top filter (Nalgene™ Rapid-Flow™ Sterile Disposable Bottle Top Filters with PES Membrane).

2.3 Plastics, Glassware, and Miscellaneous

1. Tubes and glassware: Test tube (glass), 250 mL baffled flask (glass), falcon tube (50 mL polypropylene), microcentrifuge tubes (1.5 mL polypropylene), and 150 mL beaker (glass). Autoclave glassware before use. Plastic disposables are bought sterile.
2. Magnetic stirring hotplate (Fisher Scientific International, Inc.).
3. 0.5″ and 2″ magnetic stir bars (Fisher Scientific International, Inc.) (keep 0.5″ suspended in 70% ethanol when not in use for sterility).
4. Thermometer (Fisher Scientific International, Inc.).
5. Stand and clamp (Fisher Scientific International, Inc.).

2.4 SNAP Calibration and NO Treatment Assay

1. 0.1 M Copper (II) chloride solution: 8.5 g of $CuCl_2 \cdot 2H_2O$ (Fisher Scientific International, Inc.) dissolved in 500 mL of Milli-Q water. Store at room temperature.
2. *S*-Nitroso-*N*-acetyl-DL-penicillamine (SNAP) solution: dissolve 5 mg of ethylenediaminetetraacetic acid (EDTA) in

25 mL of Milli-Q water, and then add 5.6 mg SNAP (Cayman Chemical, Inc.) to the solution. Wrap the solution in aluminum foil and store at 4 °C. It can be used within a month.

3. Dipropylenetriamine (DPTA) NONOate solution: On the day of the experiment, add 726 μL of 10 mM NaOH solution to 10 mg of DPTA NONOate (Cayman Chemical, Inc.) to make 72 mM DPTA NONOate stock, and keep on ice. It can be used within 24 h.
4. Methylamine hexamethylene methylamine (MAHMA) NONOate solution: On the day of the experiment, add 1 mL of 10 mM NaOH solution to 10 mg of MAHMA NONOate (Cayman Chemical, Inc.) to make 48.9 mM MAHMA NONOate stock, and keep on ice. It can be used within 2 h.
5. ISO-NOP NO sensor—2 mm (World Precision Instruments, Inc.).
6. Four-Channel Free Radical Analyzer (TBR4100, World Precision Instruments, Inc.).
7. *(Optional)* Nitrate/Nitrite colorimetric assay kit (Cayman Chemical, Inc.).

2.5 O_2 Consumption Assay

1. FireStingO_2—Fiber-Optic Oxygen and Temperature meter (PyroScience GmbH).
2. Robust Oxygen Probe (PyroScience GmbH).
3. Submersible Temperature Sensor TSUB21 (PyroScience GmbH).

2.6 Software

1. Windows OS (Windows 7 or later).
2. Microsoft Excel.
3. MATLAB (2016 or later).
4. LabScribe—a recording software that is compatible with ISO-NOP.
5. FireStingO_2 software—a recording software for O_2 probe.

3 Methods

3.1 Experimental Procedures

3.1.1 ISO-NOP Calibration

Here, we describe the use of SNAP as an NO donor for daily calibration of the ISO-NOP sensor. SNAP is a *S*-nitrosothiol that spontaneously releases NO in an ill-defined ratio when decomposed by heat, light, or catalyst [34]. It is generally believed that Cu(I) formed from Cu(II) through reduction by thiol rapidly catalyzes this reaction [35]. SNAP is preferred over other NO donors for the calibration process because of its stability and rapid release of NO. In this procedure, increasing doses of SNAP are added to a solution of $CuCl_2$ to generate proportional amounts of

NO. This is reflected in the signal profile generated in picoamp (pA). This signal profile is used to generate a calibration curve that enables the conversion between pA and [SNAP]. Ultimately, we want to generate a conversion between pA and [NO], and thus need a conversion of the form, [SNAP] = α [NO]. The inverse of α (α^{-1}) can be thought of as the average number of NO molecules released per parent molecule of SNAP. This is accomplished by using a second NO-donating compound at varying concentrations and incorporating α as a fitted variable in an optimization procedure with a cell-free model of aqueous NO reactivity. The second NO-donating compound in this procedure should be different from NO donors used later on to ensure an independent calibration of the NO sensor. We have elected to use MAHMA NONOate, an NO donor that releases two molecules of NO per donor molecule with a half-life on the order of magnitude of 1 min at 37 °C and pH 7.4, as the second NO-donating compound for the calibration. Please note that SNAP calibration of the probe is done daily, whereas estimation of α with MAHMA NONOate in this case should be conducted when a new SNAP solution is used or the pA vs. [SNAP] deviates significantly from previous calibrations (*see* **Note 1**).

ISO-NOP Calibration Using SNAP and $CuCl_2$

1. Pipet 10 mL of $CuCl_2$ solution in a 50 mL falcon tube and prewarm it in an incubator at 37 °C.
2. Use a 2″ stir bar, 150 mL beaker, heating plate, stand and clamp to set up the bioreactor (Fig. 1). Set the temperature of the heating plate so that the temperature of the water bath remains at 37 °C. Set the spin rate at 450 rpm (*see* **Note 2**).
3. Use tweezers to remove a small magnetic stir bar from a microcentrifuge tube filled with 70% EtOH, rinse with autoclaved Milli-Q water, and then transfer to the prewarmed $CuCl_2$ solution. Fix the falcon tube on to the clamp and lower the tube into the water bath.
4. Very carefully, lower the NO probe into the $CuCl_2$ solution so that the tip is approximately 2–4 mm below the liquid surface (*see* **Note 3**).
5. Begin recording data from the NO probe in the LabScribe software.
6. Once the NO probe signal has stabilized for at least 5 min, take a 300 μL aliquot of the prepared SNAP solution from 4 °C storage. Deliver the first volume (e.g., 32 μL) of SNAP solution to the $CuCl_2$ solution. The pA signal should increase immediately, and then drop to baseline within 2 min.
7. Once the signal approaches baseline and is flat, add the second dose of SNAP solution.

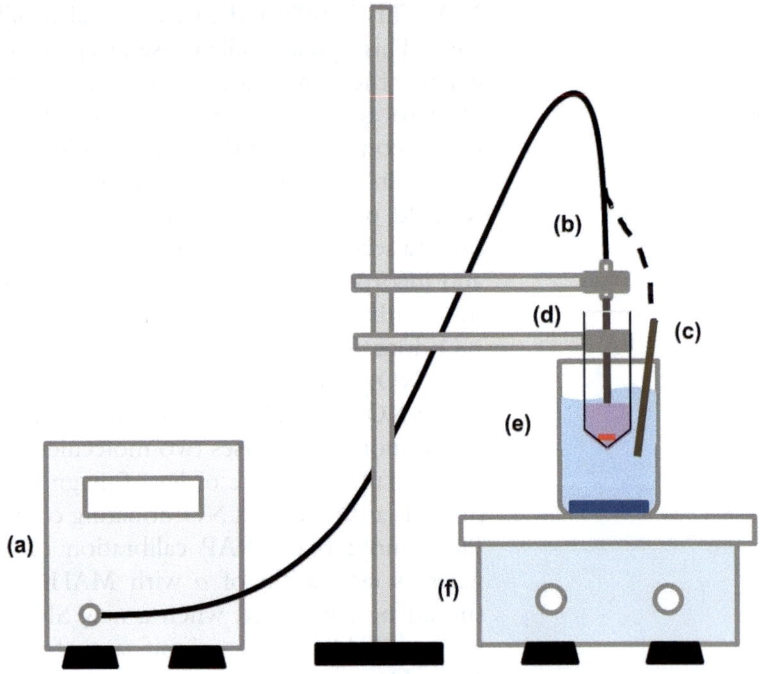

Fig. 1 Bioreactor schematic. (**a**) Free radical analyzer and/or fiber-optic O_2 and temperature meter. (**b**) NO sensor and/or O_2 probe. (**c**) Temperature sensor. (**d**) 50 mL falcon tube containing 10 mL of 0.1 M $CuCl_2$ or MOPS minimal media with a 0.5″ magnetic stir bar. (**e**) 150 mL glass beaker containing 100 mL of water and a 2″ magnetic stir bar. (**f**) Hot plate stirrer

8. Repeat until at least three points are obtained to generate a calibration curve (e.g., 32, 64, 128 μL of SNAP solution) (*see* **Note 4**).
9. Stop recording and save the data (*see* **Note 5**).
10. Raise the NO probe and submerge it into a falcon tube with autoclaved Milli-Q water. Remove the $CuCl_2$ solution from the clamp. Remove and rinse the stir bar with autoclaved Milli-Q water using tweezers and return to a microcentrifuge tube with 70% EtOH (*see* **Note 6**).
11. Open the data file and calculate the difference between the peak pA value and base pA value (ΔpA) for each dose of SNAP delivered (Fig. 2a).
12. Calculate the concentration of SNAP added with each dose (*see* **Note 7**).
13. Make a scatter plot of [SNAP] against ΔpA.
14. Use a linear regression tool to fit a line through the data that passes through the origin and calculate the slope k and R^2 value (Fig. 2b) (*see* **Notes 8** and **9**).

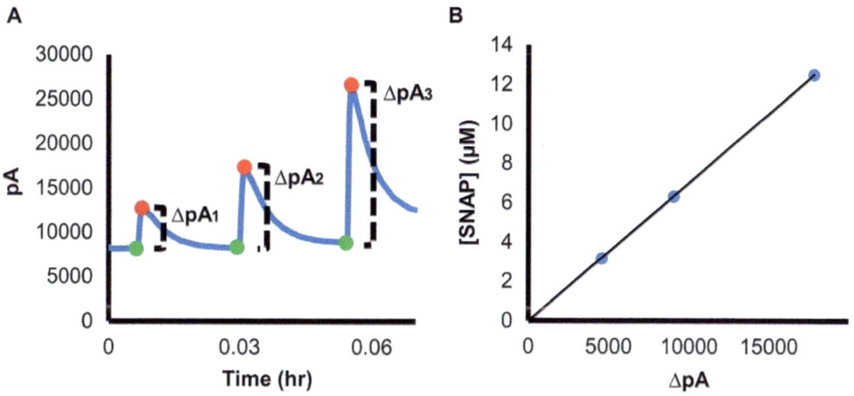

Fig. 2 SNAP calibration. (**a**) pA profile following delivery of 32, 64, and 128 μL of SNAP standard to 0.1 M $CuCl_2$. The green point and red point represent the base and peak pA value for each dose of SNAP delivered. (**b**) The difference between each pair of points (ΔpA) can be plotted against [SNAP] and fitted to a regression line

15. Divide the slope k by α (estimated in Subheading "Experiments to Determine the Conversion Rate of SNAP to NO") to convert ΔpA to [NO]. k/α is a proportionality constant between pA and [NO].

Experiments to Determine the Conversion Rate of SNAP to NO

1. Prepare two 10 mL MOPS minimal medium solutions in two separate 50 mL falcon tubes and prewarm them in an incubator at 37 °C.
2. Calibrate the NO probe following Subheading "ISO-NOP Calibration Using SNAP and $CuCl_2$". Following calibration, maintain the water bath temperature at 37 °C and spin rate at 450 rpm.
3. Fix one of the falcon tubes on to the clamp and lower the tube into the water bath. Rinse the stir bar with autoclaved Milli-Q water and place it into the media with tweezers. Lower the NO probe into the MOPS media, so that the tip is approximately 2–4 mm below the liquid surface.
4. Begin recording data from the NO probe in the LabScribe software.
5. Add 1 mL of 10 mM NaOH to MAHMA NONOate to make 48.9 mM MAHMA NONOate stock. Keep on ice.
6. Once the probe signal has stabilized for at least 5 min, add 5.1 μL of MAHMA NONOate stock to the media to reach a concentration of 25 μM.
7. *(Optional)* After 30 min, take a 300 μL sample and store at 4 °C (this can be used to measure the terminal nitrite concentration using a nitrate/nitrite colorimetric assay kit).
8. Stop recording and save the data.
9. Remove and rinse the probe in autoclaved Milli-Q water and the stir bar in 70% EtOH followed by autoclaved Milli-Q water.

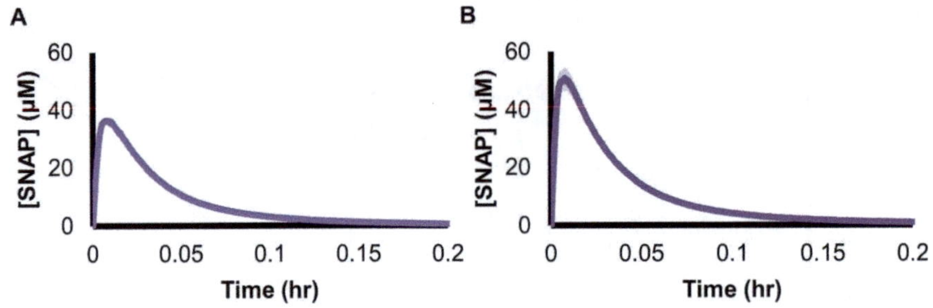

Fig. 3 [SNAP] profile following delivery of MAHMA NONOate. (**a**) 25 μM and (**b**) 50 μM of MAHMA NONOate was delivered to a cell-free bioreactor containing 10 mL of MOPS minimal media. The dark purple line represents the mean of three replicates, whereas the light purple shading represents the standard error of the mean

Replace the falcon tube with the second prepared MOPS solution. Add the stir bar into the new MOPS media. Lower the NO probe to 2–4 mm below the liquid surface.

10. Begin recording NO probe data on LabScribe software.
11. Repeat **steps 6–8** but double the volume of MAHMA NONOate delivered (10.2 μL to reach a concentration of 50 μM).
12. Raise the NO probe and submerge it into a falcon tube with autoclaved Milli-Q water. Remove the MOPS solution from the clamp. Remove and rinse the stir bar with autoclaved Milli-Q water using tweezers and return to microcentrifuge tube with 70% EtOH.
13. Truncate each data set so that all data collected before MAHMA treatment is removed.
14. Rescale the time points so that the first time point is $t = 0$.
15. For each data set: subtract all pA values by the pA value at $t = 0$. This generates a ΔpA value at each time point.
16. Multiply the ΔpA values by the slope k obtained from **step 14** in Subheading "ISO-NOP Calibration Using SNAP and $CuCl_2$". This generates a [SNAP] value at each time point.
17. Plot [SNAP] vs. time for each MAHMA NONOate concentration (Fig. 3a, b).
18. Refer to Subheading 3.2.1 to estimate the conversion factor α.

3.1.2 Measuring O_2 Transport in the Absence of the Cells

O_2 Probe Calibration

1. Prepare 10 mL MOPS media in a 50 mL falcon tube and prewarm it in an incubator at 37 °C.
2. Approximately 20 min before use, attach the prewarmed falcon tube with MOPS media to the clamp and lower it into the water bath. Rinse the stir bar with autoclaved Milli-Q water and place it into the media with tweezers.
3. Rinse and clean the O_2 probe with 70% EtOH and allow it to dry.
4. Lower the O_2 probe into the bioreactor, so that the tip is approximately 2–4 mm below the liquid surface. Place the

temperature sensor outside the bioreactor but within the heated water bath, submerged approximately 4–5 cm below the liquid surface.

5. Open the FirestingO_2 software and press the calibrate button. Wait approximately 5 min before the signal reaches a steady-state value and press the steady-state button. The O_2 saturation percentage should read 100% (*see* **Note 10**).

Oxygen Transport Measurement

1. Following calibration, bubble 100% nitrogen (N_2) gas into the MOPS media (*see* **Note 11**).
2. Once the O_2 saturation percentage drops below 2%, turn off and remove the N_2 source, and start recording data on the FirestingO2 software.
3. After 10 min, stop recording. Remove and clean the O_2 probe with 70% EtOH. Rinse the stir bar with autoclaved Milli-Q water using tweezers and return to a microcentrifuge tube with 70% EtOH.
4. Extract time and percentage of O_2 saturation data from the saved file.
5. Convert the time points from units of seconds to hours.
6. Multiply the percentage of O_2 saturation values by 210 and divide by 100. This converts percentage of O_2 saturation to units of micromolar (*see* **Note 12**).
7. Plot $[O_2]$ vs. time and $-\ln([O_2]_{sat} - [O_2])$ vs. time, where $[O_2]_{sat}$ is 210 μM (Fig. 4a, b).
8. Use a linear regression tool to fit a line through the data in Fig. 4b and calculate the slope.

   The slope obtained is the capacity coefficient for O_2 transport, $k_L a_{O2}$, where k_L is the mass transfer coefficient, a is the surface area between gas and liquid phases divided by the liquid volume, and the O2 subscript indicates that it is for O_2.

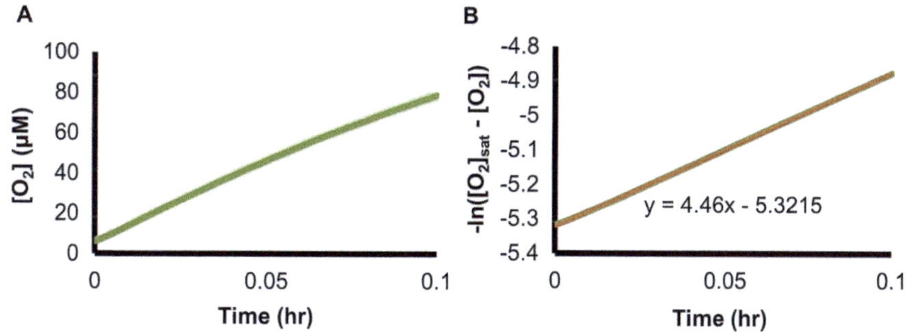

Fig. 4 Estimating the capacity coefficient for O_2 mass transfer ($k_L a_{O2}$) (**a**) $[O_2]$ profile after purging a cell-free bioreactor with 100% nitrogen. The dark green line represents the mean of three replicates, whereas the light green shading represents the standard error of the mean. (**b**) Plot of $-\ln([O_2]_{sat} - [O_2])$ vs. time. The slope of the best fit line (in red) equals $k_L a_{O2}$

3.1.3 NO Delivery and Measurement in the Absence of Cells

1. Prepare 10 mL MOPS media in a 50 mL falcon tube and prewarm it in an incubator at 37 °C.
2. Calibrate the NO probe following Subheading "ISO-NOP Calibration Using SNAP and $CuCl_2$". Following calibration, maintain the water bath temperature at 37 °C and spin rate at 450 rpm.
3. Fix the falcon tube on to the clamp and lower the tube into the water bath. Rinse the stir bar with autoclaved Milli-Q water and place it into the media with tweezers. Lower the NO probe into MOPS media, so that the tip is approximately 2–4 mm below the liquid surface.
4. Begin recording data from the NO probe in the LabScribe software.
5. Once the NO probe signal has stabilized for at least 5 min, add 69.5 μL of DPTA NONOate stock to the media to reach a final concentration of 500 μM.
6. After 2 h, stop recording and save data (*see* **Note 13**).
7. Raise the NO probe and submerge it into a falcon tube with autoclaved Milli-Q water. Remove the MOPS solution from the clamp. Remove and rinse the stir bar with autoclaved Milli-Q water using tweezers and return to microcentrifuge tube with 70% EtOH.
8. Truncate the data set so that all data collected before DPTA treatment is removed.
9. Rescale the time points so that the first time point is $t = 0$.
10. Subtract all pA values by the pA value at $t = 0$. This generates a ΔpA at each time point.
11. Multiply the ΔpA values by k/α from Subheading "ISO-NOP Calibration Using SNAP and $CuCl_2$". This generates an [NO] value at each time point.
12. Plot [NO] vs. time (Fig. 5).

3.1.4 Measuring Cellular O_2 Consumption

Cell Growth and Preparation

1. Before the day of experiment, prepare a pregrowth culture by inoculating cells from a 25% glycerol, −80 °C stock into 1 mL of LB media in a test tube, and then incubate it at 37 °C with shaking (250 rpm) for 4 h. Please note that the −80 °C permanent stock should have originated from a single colony.
2. Inoculate 1 mL of fresh MOPS media in another test tube with 10 μL of the LB preculture (1% inoculum) and incubate at 37 °C with shaking (250 rpm) for 16 h.
3. On the day of experiment, measure the OD_{600} of overnight culture. Add an appropriate volume of overnight culture into

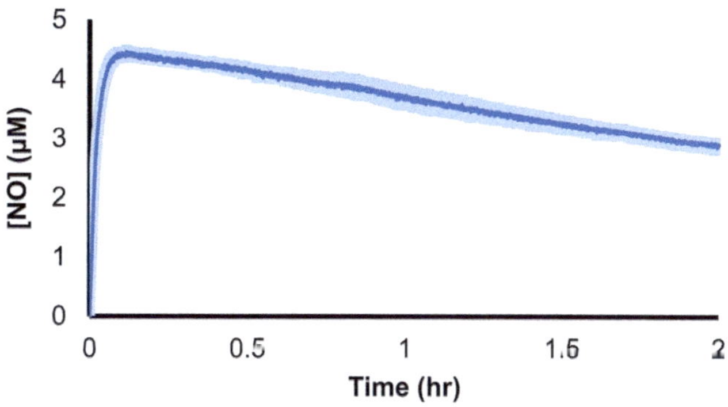

Fig. 5 [NO] profile following delivery of DPTA NONOate to a cell-free system. 500 μM of DPTA NONOate was delivered to a cell-free bioreactor containing 10 mL of MOPS minimal media. The dark blue line represents the mean of three replicates, whereas the light blue shading represents the standard error of the mean

20 mL MOPS media in a 250 mL baffled flask to obtain a final OD_{600} of 0.01.

4. Allow the flask culture to grow in the incubator (37 °C, 250 rpm) until it reaches an OD_{600} of 0.2 (*see* **Note 14**).
5. Prepare 10 mL of MOPS media in a 50 mL falcon tube and prewarm it in an incubator at 37 °C.
6. Turn on the heating plate and set the temperature to maintain the water bath at 37 °C and spin rate at 450 rpm.

O_2 Consumption Assay

1. Before the flask culture reaches an OD_{600} of 0.2, calibrate the O_2 probe following Subheading "O_2 Probe Calibration".
2. Once the flask culture reaches an OD_{600} of 0.2, transfer 8 mL of culture to the eight microcentrifuge tubes (1 mL per tube).
3. Centrifuge the eight tubes for 3 min at 15,000 rpm (21130 × *g*).
4. Remove the supernatant from all the tubes and resuspend all the pellets in 1 mL of prewarmed MOPS media (*see* **Note 15**).
5. Measure OD_{600} of the concentrated cell suspension. Calculate the volume (V) needed to reach a final OD_{600} of 0.05 in 10 mL of MOPS media.
6. Begin recording data with the FirestingO2 software.
7. Remove V of media from the 10 mL of media that is being stirred in the falcon tube and add V of concentrated culture to the media.
8. Allow the FirestingO2 software to record data for 10 min.
9. Stop recording. Remove and clean the O_2 probe with 70% EtOH and return the stir bar to a microcentrifuge tube with 70% EtOH.

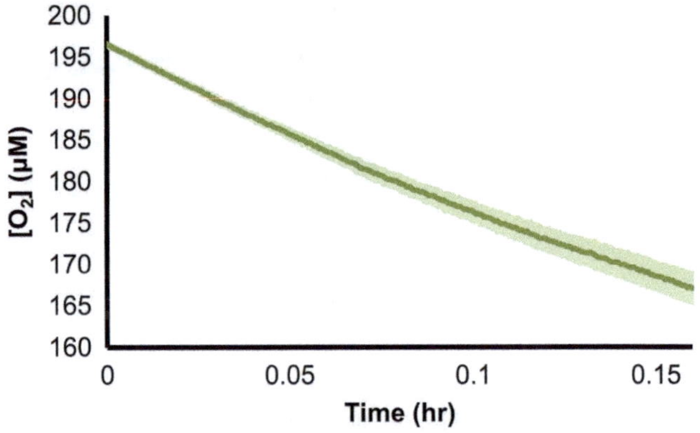

Fig. 6 [O_2] profile. An aerobic bioreactor was inoculated with *E. coli* at an OD_{600} of 0.05. The dark green line represents the mean of three replicates, whereas the light green shading represents the standard error of the mean. At the initial time point, [O_2] is not at saturation (210 μM), because the [O_2] in the volume of cell suspension added, which is generally about 300 μL, was much lower than saturation

10. Process the O_2 data following **steps 4–6** in Subheading "Oxygen Transport Measurement". After **step 5** and preceding **step 6**, truncate the data set so that all data collected before the addition of cells is removed. Rescale the time points so that the first time point is $t = 0$.
11. Plot [O_2] vs. time (Fig. 6).

3.1.5 Measuring [NO] in Cell Cultures

1. Before the day of the experiment, prepare *E. coli* MG1655 following steps in Subheading "Cell Growth and Preparation".
2. On the day of the experiment, before the culture reaches an OD_{600} of 0.2, calibrate the NO probe by following step in Subheading "ISO-NOP Calibration Using SNAP and $CuCl_2$". Following calibration, maintain the water bath temperature at 37 °C and spin rate at 450 rpm.
3. Approximately 30 min before the culture reaches an OD_{600} of 0.2, attach the prewarmed falcon tube with MOPS media to the clamp and lower it into the water bath. Rinse the stir bar with autoclaved Milli-Q water and place it into the media with tweezers.
4. Lower the NO probe into MOPS media, so that the tip is approximately 2–4 mm below the liquid surface.
5. Once the flask culture reaches an OD_{600} of 0.2, transfer 8 mL of the culture to eight microcentrifuge tubes (1 mL per tube).
6. Centrifuge the eight tubes for 3 min at 15,000 rpm (21130 × *g*).
7. Remove the supernatant from all the tubes and resuspend all the pellets in 1 mL of prewarmed MOPS media.

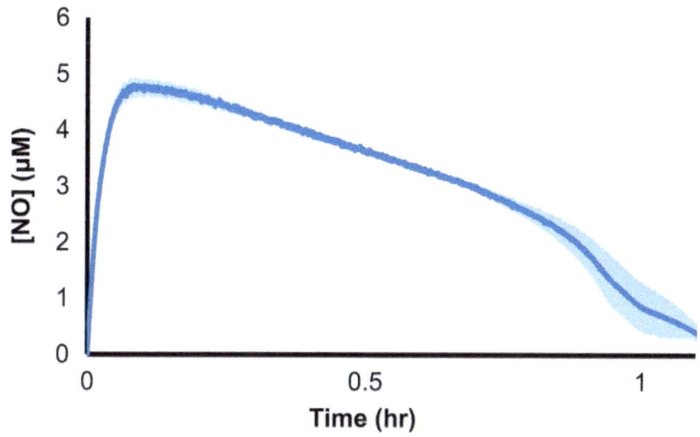

Fig. 7 [NO] profile following delivery of DPTA NONOate to cell culture. 500 μM of DPTA NONOate was delivered to a bioreactor with *E. coli* at an OD_{600} of 0.05. The dark blue line represents the mean of three replicates, whereas the light blue shading represents the standard error of the mean

8. Measure OD_{600} of the concentrated cell suspension. Calculate the volume (V) needed to reach an OD_{600} of 0.05 in 10 mL of media.
9. Begin recording data from the NO probe in LabScribe.
10. Remove V of media from the 10 mL of media that is being stirred in the falcon tube and add V of concentrated culture to the media (*see* **Note 16**).
11. Within 1 min, deliver 69.5 μL of DPTA NONOate stock to the culture to reach a final concentration of 500 μM. Monitor OD_{600} if desired.
12. The pA signal should rise from the baseline value at the beginning of the recording. Continue recording data until the signal returns to baseline. At that point, stop recording and save the data (*see* **Note 17**).
13. Raise the NO probe and submerge it into a falcon tube with autoclaved Milli-Q water. Remove the MOPS solution from the clamp. Remove the stir bar with tweezers, rinse it with autoclaved Milli-Q water, and return it to microcentrifuge tube with 70% EtOH.
14. Follow **steps 8–12** in Subheading 3.1.3 to process and plot the acquired data (Fig. 7).

3.2 Data Analysis and Computational Modeling

We note that a basic understanding of linear algebra, differential equations, and MATLAB are helpful for the following section. However, we have provided all the necessary code to analyze the data generated in the previous section (*see* **Note 18**). In this section, we will begin by estimating a value for the conversion factor α, where [SNAP] = α[NO]. Following this, we will optimize uncertain parameters in a mathematical model of NO stress in *E. coli* cultures. The model is a system of ordinary differential equations that simulate the biochemical reactions that are relevant to NO, within the cell and the extracellular environment. For a deeper understanding of the model, we refer the reader to a selection of papers [9, 25–27, 29, 36]. Due to the compartmentalization of the model, different sets of unknown parameters can be estimated independently using specific experimental conditions. The uncertain parameters fall into three categories: parameters relevant to NO reactivity in the extracellular environment, parameters relevant to cellular respiration, and parameters relevant to NO reactivity in the cellular environment. Each parameter set will be estimated from the relevant data acquired in Subheadings 3.1.1 and 3.1.3–3.1.5, using a nonlinear least square optimization algorithm (*see* **Note 19**). In the supplementary materials, we have provided all the necessary files for this section, and in Supplemental Table S1 we provide the optimized values we used in generating the figures portrayed here. An additional parameter relevant to the NO reaction network is the capacity coefficient for O_2 transport (k_La_{O2}), which was identified in Subheading 3.1.2. Before proceeding to Subheading 3.2.1, open the Excel files entitled: "AlphaModel1," "AlphaModel2," and "EcoliNOmodel" provided in the Supplementary Materials. Locate the cells (highlighted in red) and input the value of k_La_{O2} obtained in Subheading 3.1.2.

We note that the final model with optimized parameters comprise a computational tool to gain greater understanding of NO flux distributions in bacterial cultures. There will be uncertainty associated with parameters that were both estimated from measurements and adopted from previous studies, and we will discuss how that uncertainty can be assessed and reduced in the next section. Because some experimental variations are unavoidable (e.g., each probe and magnetic stirrer is a physical instrument with a finite lifespan), we perform these measurements and estimate uncertain parameters for each independent investigation we conduct. With that practice we have successfully confirmed predicted flux distributions with independent measurements [25], and uncovered the mechanisms of how genetic and environmental perturbations impact bacterial NO stress networks.

3.2.1 Estimating α

In this section, we wish to determine the conversion factor needed for Subheading "ISO-NOP Calibration Using SNAP and $CuCl_2$" of the ISO-NOP calibration. In essence, using the two [SNAP] profiles obtained for 25 and 50 μM MAHMA NONOate in Subheading "Experiments to Determine the Conversion Rate of SNAP to NO", we will optimize four parameters in a kinetic model of the NO reaction network in cell-free MOPS media. The four parameters we will train are the NO autoxidation rate, rate of NO lost to the gas phase, MAHMA NONOate degradation rate, and the conversion factor α, which is a proportionality constant between NO and SNAP.

1. Open MATLAB and set the directory to the folder entitled "Supplementary Materials."
2. Enter the command line: *model1 = LoadEcoliNOmodel()*; A search bar will appear. Select the Excel file "AlphaModel1." Input the command line: *model2 = LoadEcoliNOmodel();* and select the file "AlphaModel2." We have now loaded the contents of AlphaModel1 and AlphaModel2 into the workspace. (AlphaModel1 and AlphaModel2 contain the set of relevant species, parameters, reactions and initial conditions for experiments with 25 and 50 μM MAHMA NONOate respectively). Enter the command line: *model = [model1, model2];* to combine both model structures for simultaneous optimization of parameters using the 25 and 50 μM MAHMA NONOate data sets .
3. Import the data collected from Subheading "Experiments to Determine the Conversion Rate of SNAP to NO". Specifically, there should be a variable labeled expX for the time points measured, and a variable labeled expY for [SNAP] at each time point. Each variable should have two columns, one for each MAHMA concentration (*see* **Note 20**).
4. Enter the command line: *normFlag = 1;* and *variance = ones (length(expX),2);* (*see* **Note 21**).
5. Open the folder "Alpha Conversion." Open the subfolder "SNAP only." Import the file "fitParams" (which contains the parameter names being fit), "paramBounds" (which contains the upper and lower bound for each uncertain parameter) and "fitSpecies" (which contains the species being fit).
6. Enter the command line*: n = 1000;* (the number of optimizations we wish to perform).
7. Enter the command line: *[fitValMat,resNorms]= RandomInitParamOpt(model,expX,expY,fitParams,paramBounds,n,fitSpecies,variance,normFlag);* The function will now run 1000 optimizations with each optimization selecting a random value for each uncertain parameter within the set of parameter bounds.
8. When the function is finished running, two new output vectors should be present in the workspace. resNorms represents the sum of the squared residuals for each optimization. fitValMat represents the parameter values obtained from each optimization.
9. Running the command line: *[~,location] = min(resNorms);* will give the location of the best fit parameter set. After this run the command line: *fitValMat(location,:)*. The parameter set yielding the smallest residual error will be displayed.
10. The fourth entry is the conversion factor α.

11. Locate the parameters and input the values from **step 9** in the Excel files, "AlphaModel1" and "AlphaModel2" (highlighted in purple). Save the files.
12. Reload the newly updated model structures (**step 2**) and run the command line: *[t1,X1] = RunEcoliNOmodel(model(1), max(expX(:,1))); [t2,X2] = RunEcoliNOmodel(model(2), max(expX(:,2)));* (*see* **Note 22**).

    One way to confirm the accuracy of the conversion factor α is to compare the simulated terminal NO_2^- concentrations with the experimental data obtained from **step 10** in Subheading "Experiments to Determine the Conversion Rate of SNAP to NO". If the simulated NO_2^- concentration is not within a 95% confidence level of the measured NO_2^- concentration, we then include the measured NO_2^- data in the optimization process (*see* **Note 23**).
13. Plot the experimentally measured NO data and simulated NO curve for each concentration of MAHMA NONOate, making sure to divide both sets of data by α before plotting (Fig. 8a, b).

3.2.2 Estimating Uncertain Parameters from Cell-Free Experiments

Parameters related to NO autoxidation rate, NO gas phase mass transfer, and DPTA NONOate degradation rate can be estimated in the absence of cells. In these optimizations the cellular compartment in our model will be removed, by eliminating all relevant reactions, as we are only concerned with NO delivery and reactivity in the absence of cells. In total, we will be fitting three parameters: NO autoxidation rate, rate of NO lost to the gas phase, and DPTA NONOate degradation rate. We note that parameters associated with NO autoxidation and gas phase transport were also estimated from experiments with MAHMA NONOate (Subheading 3.2.1). Since DPTA-containing media differ chemically from MAHMA-containing media (e.g., different decomposition products), we estimate these parameters from cell-free DPTA treated media for use with DPTA-treated cell cultures. In practice, the values obtained from cell-free MAHMA and cell-free DPTA data are in reasonable agreement with one another. Further, keeping the autoxidation and gas phase transfer parameters the same (e.g., from the MAHMA NONOate optimization) while adjusting the NONOate decomposition rate will often provide good agreement between simulations and data (e.g., from the DPTA NONOate experiments) [37].

1. Open the Excel file entitled, "EcoliNOmodel." This file contains the set of reactions, rate constants, and initial species concentrations. In the top left corner, there is a cell (A7) labeled culture OD_{600}. Set the value in the adjacent cell (B7) to 0 (doing this eliminates all reactions relevant to the cellular compartment). Save and close the file.

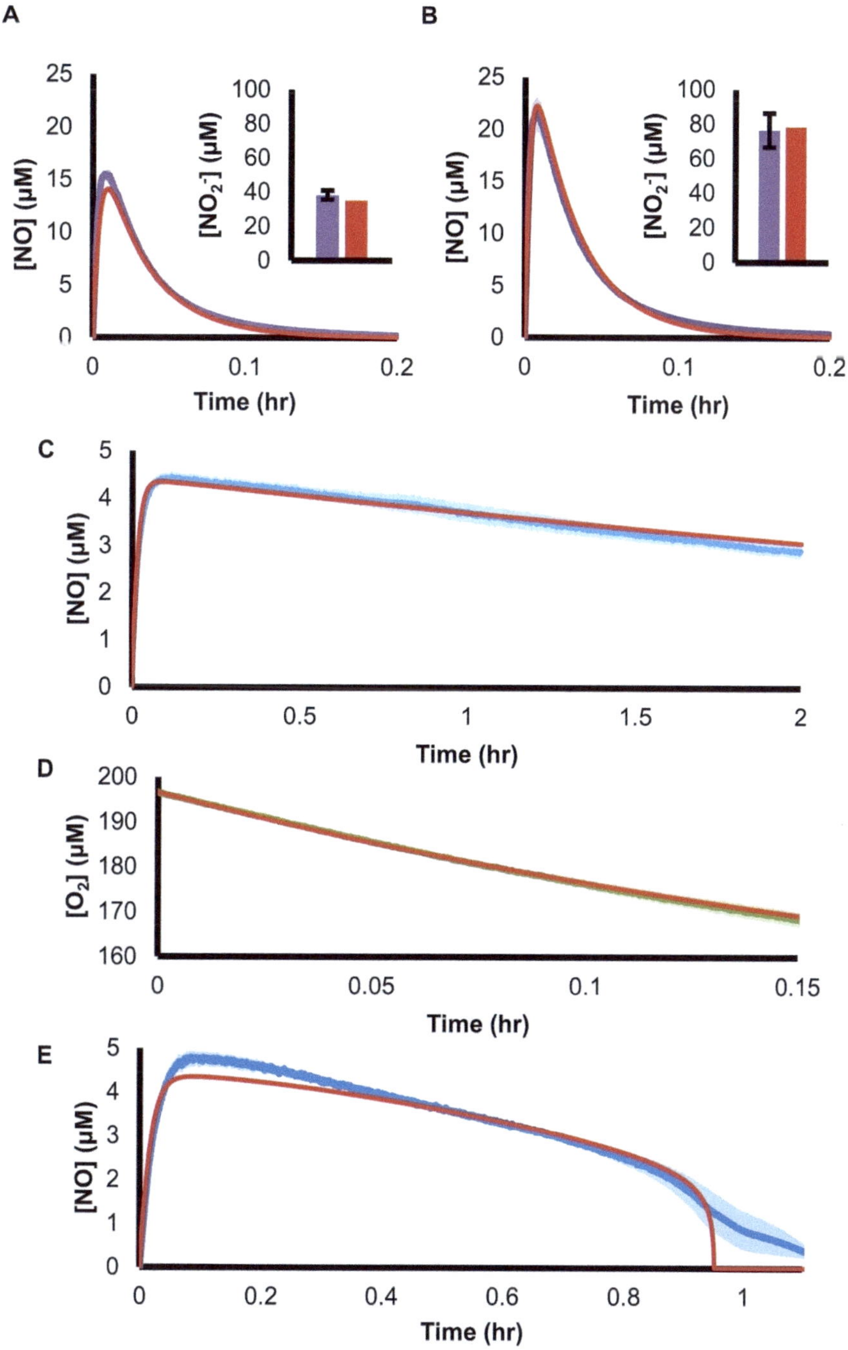

Fig. 8 Comparison between simulation and experimental results. The red line represents a model simulation using the optimal parameter set obtained (*see* Supplemental Table S1 for optimized parameter values). NO profile for (**a**) 25 μM and (**b**) 50 μM MAHMA NONOate. Parameter optimizations were performed using both [NO] and $[NO_2^-]$ data, because optimizations performed just with [NO] data provided $[NO_2^-]$ estimates that were outside the 95% confidence intervals of the measured levels. The insets show the measured (purple bars) and predicted (red bars) NO_2^- concentrations 0.2 h after MAHMA NONOate delivery. At that time the vast majority of MAHMA NONOate, whose half-life is on the order of 1 min, had decomposed. Error bars represent 95% confidence intervals. (**c**) [NO] dynamics for addition of 500 μM DPTA NONOate into a cell free bioreactor. (**d**) $[O_2]$ for cells in a bioreactor at an OD_{600} of 0.05. (**e**) [NO] dynamics for addition of 500 μM DPTA NONOate into a bioreactor with cells at an OD_{600} of 0.05

2. Open MATLAB and set the directory to the folder entitled "Supplementary Materials." Clear the workspace.
3. Enter the command line: *model = LoadEcoliNOmodel();* A search bar will appear. Select the Excel file entitled, "EcoliNO-model," which you just modified and saved. We have now loaded the contents of EcoliNOmodel into our workspace.
4. Open RunEcoliNOmodel.m. Disable lines 80 and 100, by adding a "%" character in front and enable line 103 by deleting the "%" character in front. Save RunEcoliNOmodel.m.
5. Import the data collected from Subheading 3.1.3. Specifically, you should have a vector labeled expX for the time points measured, a vector labeled expY for [NO] at each time point and a vector labeled variance for the variance in [NO] computed at each time point (if multiple experimental replicates were performed). (*see* **Note 24**).
6. Follow **steps 5–9** in Subheading 3.2.1 but setting *normFlag = 0;* and using the files located in the folder entitled "Cell Free" instead of "Alpha Conversion."

We now have obtained optimized parameter values for NO autoxidation, NO gas transport, and DPTA NONOate degradation. Please *see* Supplemental Table S1 for specific optimized values used here.

7. Locate the parameters and input the values in the Excel file, "EcoliNOmodel" (highlighted in yellow). Save the file.
8. Reload the newly updated model structure (**step 3**) and run the command line: *[t,X] = RunEcoliNOmodel(model,max(expX));*
9. Plot the experimentally measured NO data and simulated NO curve (Fig. 8c).

3.2.3 Estimating Uncertain Parameters Related to Cellular Respiration

Given the capacity of NO to react with O_2, an estimation of O_2 dynamics is important for calculating NO fluxes. In *E. coli* cultures, respiration is a major O_2 consumption pathway and respiratory inhibition has been used as an indicator of NO-induced bacteriostasis [28, 38, 39]. To account for respiration, the model contains four reactions: ubiquinone reduction by NADH dehydrogenase I and II and O_2 consumption by cytochrome *bo* and *bd*. With these four reactions there are 10 uncertain parameters that are optimized.

1. Open the Excel file entitled "EcoliNOmodel." In the top left corner, there is a cell (A7) labeled culture OD_{600}. Set the value in the adjacent cell (B7) back to 0.05 (doing this sets the initial cell density in our model to 0.05). Set the value in N126 to 0 (doing this sets the initial concentration of DPTA NONOate to 0). Save and close the file.

2. Open MATLAB and set the directory to the folder entitled "Supplementary Materials." Clear the workspace.
3. Enter the command line: *model = LoadEcoliNOmodel();* A search bar will appear. Select the Excel file entitled "EcoliNO-model." We have now loaded the contents of EcoliNOmodel into our workspace.
4. Import the data collected from Subheading 3.1.4. Specifically, you should have a vector labeled expX for the time points measured, a vector labeled expY for $[O_2]$ at each time point, and a vector labeled variance for the variance in $[O_2]$ data computed at each time point (if multiple experimental replicates were performed).
5. Follow **steps 5–9** in Subheading 3.2.1 but setting *normFlag = 0;* and using the files located in the folder entitled "Resp Compartment" instead of "Alpha Conversion."

We now have obtained optimized parameter values for parameters related to cellular respiration. Please *see* Supplemental Table S1 for specific optimized values used here.

10. Locate the parameters and input the values in the Excel file, "EcoliNOmodel" (highlighted in blue). Save the file.
11. Reload the newly updated model structure (**step 3**) and run the command line: *[t,X] = RunEcoliNOmodel(model,max(expX));*
12. Plot the experimentally measured O_2 data and simulated O_2 curve (Fig. 8d).

3.2.4 Estimating Uncertain Parameters Related to NO Detoxification and Cytochrome Inhibition

The remaining parameters necessary to calculate NO fluxes are related to the enzymatic detoxification of NO, as well as NO cytochrome inhibition. *E. coli* contains two dominant enzymatic detoxification systems where the activity of the system depends on the availability of O_2 [40, 41]. In aerobic conditions, Hmp, a flavohemoglobin, catalyzes the reduction of NO to NO_3^- [42]. We model the catalytic activity of Hmp as a subnetwork of reactions [9, 25]. Hmp synthesis is also induced by NO, and therefore, modeled (transcription, translation, transcript and protein degradation). In anaerobic conditions, NO reductase (NorV) is the dominant detoxification system where it catalyzes the conversion of NO into N_2O [43, 44]. Similarly, NorV synthesis is induced by NO exposure, and we have included it in the model. Further, NO has the ability to bind the heme centers of the terminal respiratory cytochromes *bo* and *bd* inhibiting respiration and this is also accounted for in the model. From the reactions involving Hmp, NorV, and cytochrome inhibition, there are 23 uncertain parameters that will be optimized. In addition, there are a range of other reactions in the model capturing biologically relevant NO targets, of which a few examples are: Fe–S cluster damage, DNA deamination, thiol nitrosation, and their associated repair mechanisms. Values for those and other kinetic parameters have been obtained from the literature and are not optimized here. Under conditions we have studied, NO flux through those pathways comprises far less than 1% of the total NO delivered [25, 28], and thus we have not delved deeper into improving the accuracy of flux estimates for those reactions.

1. Open MATLAB and set the directory to the folder entitled "Supplementary Materials." Clear the workspace.
2. Open the Excel file entitled "EcoliNOmodel." Change the value in N126 back to 500 (doing this sets the initial concentration of DPTA NONOate to 500 μM). Save and close the file.
3. Enter the command line: *model = LoadEcoliNOmodel();* A search bar will appear. Select the Excel file entitled, "EcoliNOmodel." We have now loaded the contents of EcoliNOmodel into our workspace.
4. Import the data collected from Subheading 3.1.5. Specifically, you should have a vector labeled expX for the time points measured, a vector labeled expY for [NO] at each time point and a vector labeled variance for the variance in [NO] data computed at each time point (if multiple experimental replicates were performed).
5. Follow **steps 5–9** in Subheading 3.2.1 but setting *normFlag = 0;* and using the files located in the folder entitled "Cell Compartment" instead of "Alpha Conversion."

We now have obtained optimized parameter values for parameters related to cellular detoxification and cytochrome inhibition. Please *see* Supplemental Table S1 for specific optimized values used here.

6. Locate the parameters and input the values in the Excel file "EcoliNOmodel" (highlighted in green). Save the file.
7. Reload the newly updated model structure (**step 3**) and run the following command line: *[t,X] = RunEcoliNOmodel(model, max(expX));*
8. Plot the experimentally measured NO data and simulated NO curve (Fig. 8e).

3.2.5 Computing and Plotting NO Flux Distributions

With a model capable of capturing [NO] dynamics, it is now possible to simulate and predict the distribution of NO among its reaction pathways. This can be done by tracking the cumulative NO flux through each consumption pathway (e.g., autoxidation, gas phase transport, enzymatic detoxification).

1. Open MATLAB and set the directory to the folder entitled "Supplementary Materials." Clear the workspace.
2. Enter the command line: *model = LoadEcoliNOmodel();* A search bar will appear. Select the Excel file entitled "EcoliNO-model." We have now loaded the contents of EcoliNOmodel into our workspace.
3. Enter the command line: *[t,X, dX,dV] = RunEcoliNOmodel (model,a);* where "a" is the length of time, in hours, of the recorded data in Subheading 3.1.5.
4. Enter the command line: *PathwayFluxEcoli(t,dV,model);* The function should output two plots. The first plot represents a simulation of the cumulative distribution of total NO consumption (Fig. 9a), whereas the second plot represents the cumulative distribution of intracellular NO consumption (Fig. 9b).

We have now generated an NO flux distribution profile for *E. coli* cells seeded in a bioreactor at an OD_{600} of 0.05, upon treatment with 500 μM DPTA NONOate.

3.3 NO Flux Validation and Uncertainty Estimates

Several experimental validation methods including gene disruption and final product measurements have been applied previously to confirm NO flux distributions [25, 26, 29, 30]. For example, good agreement between model simulations of NO dynamics in Δ*hmp* or Δ*norV* cultures and experimental measurements from those strains can lend confidence to the predictive power of the model. Another way to validate the model is to measure the stable end products of NO, such as NO_2^- and NO_3^-. We have done this previously with Griess assays on sterile-filtered samples, though we note that any residual nondecomposed NONOate and the capacity of cells to consume NO_2^- and NO_3^- must be considered for these measurements [25, 30].

Our group has also expanded on the optimization process described in Subheading 4 by applying an out-of-equilibrium adaptive Metropolis Markov Chain Monte Carlo (MCMC) method to further explore the parameter space and ensemble modeling to more rigorously account for parametric uncertainty [26, 30, 37, 45]. Optimal and near-optimal parameter sets, as identified using evidence ratios (ER) (*see* **Note 25**), obtained from the nonlinear least squares algorithm are used as initial points for the MCMC method. We often perform upward of 10,000 iterations of the MCMC procedure [26]. Parameter sets with ER $\leq$ 10 are retained as solutions with high likelihoods. Confidence intervals (CIs) for optimized parameters are generated from the sets with ER $\leq$ 10, and that uncertainty is carried forward with simulations and subsequent optimizations [26, 29, 30]. Simulations are conducted with ensembles of models, where each individual model can take on

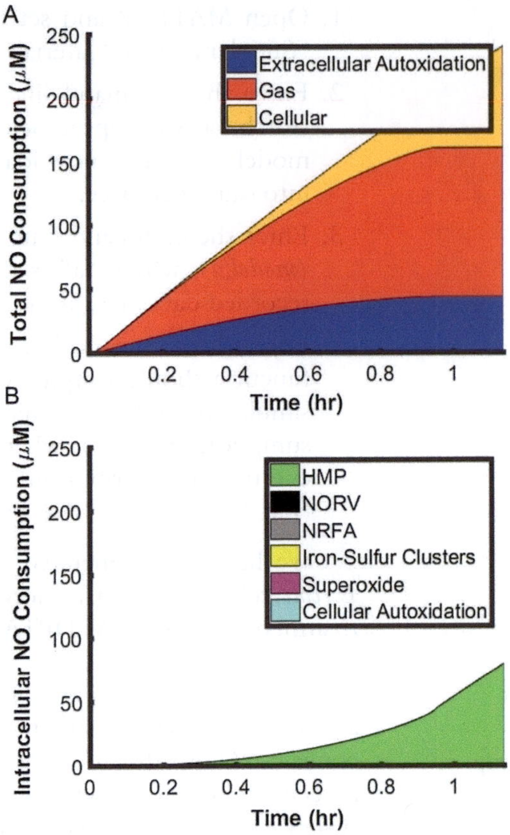

Fig. 9 Predicted cumulative NO flux distribution following delivery of 500 μM of DPTA NONOate to an *E. coli* culture. The colored regions represent the cumulative NO consumed by each pathway. (**a**) Distribution of total NO consumption. (**b**) Distribution of intracellular NO consumption. Fluxes were calculated until the time at which the cell culture had cleared NO (reduced [NO] to near baseline levels)

one set of parameters from the group obtained with ER ≤ 10. For certain conditions, the ensemble of simulations will agree and if an experiment were performed, we would have high confidence that the experimental results would confirm the model predictions. For other conditions, the ensemble of simulations will disagree and thus represent experiments that if performed would decrease the uncertainty of the model. In this way parametric uncertainty can be readily accounted for and reduced. Uncertainty in the model structure is also present (e.g., functional form for transcriptional regulation for different genes) and we have dealt with that uncertainty when we have found that parametric variation cannot recapitulate experimental measurements [26].

4 Notes

1. We previously calibrated the NO probe with a solution of 0.1 M H_2SO_4, in which we added increasing volumes of aqueous nitrite standard (spontaneously generates NO in a 1:1 ratio). However, with recent ISO-NOP electrode membranes from the manufacturer, we found that the highly acidic calibration solution was damaging the probe, which prompted us to change calibration procedures to the one that uses SNAP.
2. Achieving a stable temperature during the assay is important, as it affects the NO donor decomposition rate, probe signal, and cell phenotype. Setting the spin rate at 450 rpm ensures that the system is well mixed.
3. The ISO-NOP probe is extremely sensitive to physical perturbations; make sure that the gas permeable membrane at the tip does not touch the sides of the falcon tube.
4. The dosages of SNAP depend on the desired NO detection range. 32 μL of the prepared SNAP solution will generate a pA signal that is equivalent to approximately 2 μM NO. The calibration points should be carefully chosen, so that any data generated in subsequent steps lie within the calibration range.
5. We record 20 data points/s and take the mean of every 10 data points when we save the data in order to save memory and reduce noise. This option can be selected when saving the file in LabScribe.
6. The membrane of ISO-NOP probe can be damaged by 70% EtOH, we generally clean and store the probe in Milli-Q water only.
7. We assume that when a subsequent dose of SNAP is added, the previous dose delivered has completely decayed and does not interfere with subsequent measurements.
8. During calibration, a twofold increase in SNAP volume delivered should approximately double the value of ΔpA. If the R^2 value from this process is less than 0.95, reject the calibration and repeat the procedure.
9. Several signs can suggest a decrease in sensitivity or damage to an ISO-NOP sensor. Such as large changes in the value of the calibration slope, little to no pA response when SNAP doses are introduced, and large fluctuations in pA signal during an assay. Depending on the cause, the issue can be fixed by a polarization, a membrane change or a replacement of the ISO-NOP. Please refer to the manual from the manufacturer for ISO-NOP troubleshooting.

10. This initial 1-point calibration is recommended in order to accurately measure O_2 concentrations. This calibration relies on a preset 0% O_2 saturation value installed by the manufacturer. For more details on different calibration options refer to the FiresStingO2 user manual.
11. We use a pipet or syringe tip attached to the end of a connecting tube to a N_2 gas source. Sealing the top of the falcon tube while flushing with N_2 will accelerate the process, but that seal must be removed during the measurements of O_2 transport.
12. 210 μM represents 100 percent of O_2 saturation at 37 °C in aqueous salt solution [46], which is the same as what we have measured in MOPS minimal media [45].
13. The pA reading for a 500 μM DPTA NONOate cell free experiment can take more than 5 h to reach baseline. We find that the profile obtained over the first 2 h is sufficient for computational modeling purposes.
14. For wild-type MG1655 *E. coli* in MOPS glucose media, it takes roughly 5 h to grow aerobically from OD_{600} of 0.01–0.2 in a baffled flask.
15. Do not let cells sit in a pellet for too long, this will lead to a low O_2 levels in the cell pellet, which can alter the cellular physiology.
16. Do not leave the cells in the bioreactor or in the microcentrifuge tube for too long before adding DPTA NONOate, as the cells will consume O_2 and physiological changes can result.
17. At times, we find that pA values do not return completely to baseline and can exhibit nonzero tails. This could be an artifact of delayed electrode response or reflect residual NO. To assess whether the data reflect residual NO, we suggest adding an NO scavenger such as carboxy-PTIO while monitoring the pA signal. If there is residual NO, the pA signal should decrease.
18. Supplementary files are available on the website of the corresponding author, and can also be accessed at the web URL: https://github.com/BrynildsenLab/QuantifyingNOFluxDistributions.git.
19. We apply a nonlinear least squared optimization algorithm provided in MATLAB (lsqcurvefit). Optimized parameters are allowed to vary between specified bounds. Optimal parameter sets minimize the sum of the squared residuals between experimentally measured data and model simulations.
20. The expX variable should be in units of hours. expX and expY should be imported as numeric arrays.
21. In this optimization, we are optimizing parameters against two concentrations of MAHMA simultaneously. To ensure that

both data sets are accounted for equally, we normalize both sets by the maximum value in each data set. Furthermore, variance can be used to weight data sets during an optimization, the larger the variance of a measured data point, the smaller its weight during an optimization. Adding variance can overweight the tails of the SNAP curves, which can diminish the significance of the peak, which is the most prominent feature of the experimental data. In Subheading 3.2.1, we chose not to weight the data by variance and hence we set the variance at each point to one.

22. RunEcoliNOmodel is a differential equation solver that simulates the reaction network (specified by the first input) for a given time (specified by the second input). The two outputs represent a time vector and a matrix of species concentrations. Each column represents a concentration time course for an individual species in the model. Consult the Excel sheets AlphaModel1, AlphaModel2, or EcoliNOModel to determine the column number for specific species.

23. Repeat **steps 3–12**. However, in **step 3**, expX and expY should contain two additional columns staggered between the previous two columns. For example, the first and second columns of expY should contain [SNAP] and NO_2^- data collected for 25 μM MAHMA NONOate. Similarly, the third and fourth columns should contain [SNAP] and NO_2^- data collected for 50 μM MAHMA NONOate. It is important to note that the columns of expX and expY containing NO_2^- data should only have entries in the first row because NO_2^- is only measured once at the end point. In **step 4**, make sure to open and load the contents of the subfolder "SNAP and Nitrite" as opposed to "SNAP only."

24. The variance variable must be a numeric array and have all nonzero entries. If variance of any point is 0 (e.g., at the initial time point), change it to a value that is orders of magnitude smaller than other variance values.

25. Evidence ratio (ER) is a metric computed from the Akaike information criterion. The ER represents the likelihood of a given parameter set relative to the best fitting parameter set. An $ER \leq 10$ represents a parameter set that is 10% or more as likely as the optimal set [47].

Acknowledgments

This work was supported by National Science Foundation grant CBET-1453325 (MPB), Natural Sciences and Engineering Research Council of Canada Postgraduate Scholarship (DMS), and the generosity of Helen Shipley Hunt *71 through the Shipley Hunt Fund (MPB). We would like to thank Weng Kang Chou and Annabel S. Lemma for their assistance. The funders had no role in the preparation of the manuscript or decision to publish, and this content is solely the responsibility of the authors and does not necessarily represent the views of the funding agencies.

References

1. Bogdan C (2015) Nitric oxide synthase in innate and adaptive immunity: an update. Trends Immunol 36:161–178
2. Jin RC, Loscalzo J (2010) Vascular nitric oxide: formation and function. J Blood Med 2010:147. https://doi.org/10.2147/jbm.s7000
3. Esplugues JV (2002) NO as a signalling molecule in the nervous system. Br J Pharmacol 135:1079–1095
4. Aktan F (2004) iNOS-mediated nitric oxide production and its regulation. Life Sci 75:639–653
5. Haas A (2007) The phagosome: compartment with a license to kill. Traffic 8:311–330
6. Svensson L, Poljakovic M, Säve S, Gilberthorpe N, Schön T, Strid S, Corker H, Poole RK, Persson K, Corker H, Säve S, Poole RK, Strid S, Gilberthorpe N, Poljakovic M, Svensson L, Schön T (2010) Role of flavohemoglobin in combating nitrosative stress in uropathogenic *Escherichia coli* – implications for urinary tract infection. Microb Pathog 49:59–66. https://doi.org/10.1016/j.micpath.2010.04.001
7. Stern AM, Hay AJ, Liu Z, Desland FA, Zhang J, Zhong Z, Zhu J (2012) The NorR regulon is critical for *Vibrio cholerae* resistance to nitric oxide and sustained colonization of the intestines. MBio 3:e00013–e00012. https://doi.org/10.1128/mBio.00013-12
8. Darwin KH, Ehrt S, Gutierrez-Ramos JC, Weich N, Nathan CF (2003) The proteasome of *Mycobacterium tuberculosis* is required for resistance to nitric oxide. Science (80) 302:1963–1966. https://doi.org/10.1126/science.1091176
9. Robinson JL, Adolfsen KJ, Brynildsen MP (2014) Deciphering nitric oxide stress in bacteria with quantitative modeling. Curr Opin Microbiol 19:16–24
10. Chou WK, Brynildsen MP (2016) A biochemical engineering view of the quest for immune-potentiating anti-infectives. Curr Opin Chem Eng 14:82–92. https://doi.org/10.1016/J.COCHE.2016.08.018
11. Escaich S (2008) Antivirulence as a new antibacterial approach for chemotherapy. Curr Opin Chem Biol 12:400–408
12. Rasko DA, Sperandio V (2010) Anti-virulence strategies to combat bacteria-mediated disease. Nat Rev Drug Discov 9:117–128. https://doi.org/10.1038/nrd3013
13. Russo TA, Spellberg B, Johnson JR (2016) Important complexities of the antivirulence target paradigm: a novel ostensibly resistance-avoiding approach for treating infections. J Infect Dis 213:901–903. https://doi.org/10.1093/infdis/jiv533
14. Neidrauer M, Ercan UK, Bhattacharyya A, Samuels J, Sedlak J, Trikha R, Barbee KA, Weingarten MS, Joshi SG (2014) Antimicrobial efficacy and wound-healing property of a topical ointment containing nitric-oxide-loaded zeolites. J Med Microbiol 63:203–209. https://doi.org/10.1099/jmm.0.067322-0
15. Bryan NS (2015) Nitric oxide enhancement strategies. Fut Sci OA 1:FSO48. https://doi.org/10.4155/fso.15.48
16. Singha P, Locklin J, Handa H (2017) A review of the recent advances in antimicrobial coatings for urinary catheters. Acta Biomater 50:20–40
17. Schairer DO, Chouake JS, Nosanchuk JD, Friedman AJ (2012) The potential of nitric oxide releasing therapies as antimicrobial agents. Virulence 3:271–279. https://doi.org/10.4161/viru.20328
18. Stevanin TM, Ioannidis N, Mills CE, Kim SO, Hughes MN, Poole RK (2000) Flavohemoglobin Hmp affords inducible protection for *Escherichia coli* respiration, catalyzed by cytochromes bo' or bd, from nitric oxide. J Biol Chem 275:35868–35875. https://doi.org/10.1074/jbc.M002471200
19. Helmick RA, Fletcher AE, Gardner AM, Gessner CR, Hvitved AN, Gustin MC, Gardner PR (2005) Imidazole antibiotics inhibit the nitric oxide dioxygenase function of microbial flavohemoglobin. Antimicrob Agents Chemother 49:1837–1843. https://doi.org/10.1128/AAC.49.5.1837-1843.2005

20. Toledo JC, Augusto O (2012) Connecting the chemical and biological properties of nitric oxide. Chem Res Toxicol 25:975–989
21. Thomas DD, Ridnour LA, Isenberg JS, Flores-Santana W, Switzer CH, Donzelli S, Hussain P, Vecoli C, Paolocci N, Ambs S, Colton CA, Harris CC, Roberts DD, Wink DA (2008) The chemical biology of nitric oxide: implications in cellular signaling. Free Radic Biol Med 45:18–31
22. Kavdia M (2012) Mathematical and computational models of oxidative and nitrosative stress. Crit Rev Biomed Eng 39:461–472. https://doi.org/10.1615/critrevbiomedeng.v39.i5.60
23. Chang HL, Dedon PC, Deen WM (2008) Kinetic analysis of intracellular concentrations of reactive nitrogen species. Chem Res Toxicol 21:2134–2147. https://doi.org/10.1021/tx800213b
24. Lancaster JR (2006) Nitroxidative, nitrosative, and nitrative stress: kinetic predictions of reactive nitrogen species chemistry under biological conditions. Chem Res Toxicol 19:1160–1174. https://doi.org/10.1021/tx060061w
25. Robinson JL, Brynildsen MP (2013) A kinetic platform to determine the fate of nitric oxide in *Escherichia coli*. PLoS Comput Biol 9: e1003049. https://doi.org/10.1371/journal.pcbi.1003049
26. Robinson JL, Brynildsen MP (2016) Discovery and dissection of metabolic oscillations in the microaerobic nitric oxide response network of *Escherichia coli*. Proc Natl Acad Sci 113: E1757–E1766. https://doi.org/10.1073/pnas.1521354113
27. Robinson JL, Brynildsen MP (2015) An ensemble-guided approach identifies ClpP as a major regulator of transcript levels in nitric oxide-stressed *Escherichia coli*. Metab Eng 31:22–34. https://doi.org/10.1016/j.ymben.2015.06.005
28. Robinson JL, Miller RV, Brynildsen MP (2014) Model-driven identication of dosing regimens that maximize the antimicrobial activity of nitric oxide. Metab Eng Commun 1:12–18. https://doi.org/10.1016/j.meteno.2014.08.001
29. Robinson JL, Brynildsen MP (2016) Construction and experimental validation of a quantitative kinetic model of nitric oxide stress in enterohemorrhagic *Escherichia coli* O157: H7. Bioengineering 3:9. https://doi.org/10.3390/bioengineering3010009
30. Robinson JL, Jaslove JM, Murawski AM, Fazen CH, Brynildsen MP (2017) An integrated network analysis reveals that nitric oxide reductase prevents metabolic cycling of nitric oxide by *Pseudomonas aeruginosa*. Metab Eng 41:67–81. https://doi.org/10.1016/j.ymben.2017.03.006
31. Lewis RS, Tamir S, Tannenbaum SR, Deen WM (1995) Kinetic analysis of the fate of nitric oxide synthesized by macrophages in vitro. J Biol Chem 270:29350–29355. https://doi.org/10.1074/jbc.270.49.29350
32. Nalwaya N, Deen WM (2003) Analysis of cellular exposure to peroxynitrite in suspension cultures. Chem Res Toxicol 16:920–932. https://doi.org/10.1021/tx025664w
33. Chin MP, Deen WM (2010) Prediction of nitric oxide concentrations in melanomas. Nitric oxide. Biol Chem 23:319–326. https://doi.org/10.1016/j.niox.2010.09.003
34. Anthony AR, Rhodes P (1997) Chemistry, analysis, and biological roles of S-nitrosothiols. Anal Biochem 249:1–9. https://doi.org/10.1006/ABIO.1997.2129
35. Dicks AP, Al-Sa'doni HH, Swift HR, Cox BG, Williams DLH, Butler AR (2004) Identification of Cu+ as the effective reagent in nitric oxide formation from S-nitrosothiols (RSNO). J Chem Soc Perkin Trans 2(0):481. https://doi.org/10.1039/p29960000481
36. Robinson JL, Brynildsen MP (2016) Ensemble modeling enables quantitative exploration of bacterial nitric oxide stress networks. In: de Bruijn FJ (ed) Stress and environmental regulation of gene expression and adaptation in bacteria. John Wiley & Sons, Inc., Hoboken, NJ, pp 1009–1014
37. Chou WK, Brynildsen MP (2019) Loss of DksA leads to multi-faceted impairment of nitric oxide detoxification by *Escherichia coli*. Free Radic Biol Med 130:288–296. https://doi.org/10.1016/j.freeradbiomed.2018.10.435
38. Yu H, Sato EF, Nagata K, Nishikawa M, Kashiba M, Arakawa T, Kobayashi K, Tamura T, Inoue M (1997) Oxygen-dependent regulation of the respiration and growth of *Escherichia coli* by nitric oxide. FEBS Lett 409:161–165. https://doi.org/10.1016/S0014-5793(97)00494-8
39. Mason MG, Shepherd M, Nicholls P, Dobbin PS, Dodsworth KS, Poole RK, Cooper CE (2009) Cytochrome bd confers nitric oxide resistance to *Escherichia coli*. Nat Chem Biol 5:94–96. https://doi.org/10.1038/nchembio.135
40. Gardner AM, Gardner PR (2002) Flavohemoglobin detoxifies nitric oxide in aerobic, but not anaerobic, *Escherichia coli*: evidence for a novel inducible anaerobic nitric oxide-scavenging activity. J Biol Chem 277:8166–8171. https://doi.org/10.1074/jbc.M110470200

41. Membrillo-Hernández J, Stewart V, Anjum MF, Hughes MN, Kim SO, Poole RK (2016) Nitric oxide, nitrite, and Fnr regulation of hmp (flavohemoglobin) gene expression in *Escherichia coli* K-12. J Bacteriol 178:5487–5492. https://doi.org/10.1128/jb.178.18.5487-5492.1996
42. Gardner PR, Gardner AM, Martin L, Salzman AL (1998) Nitric oxide dioxygenase: an enzymic function for flavohemoglobin. Proc Natl Acad Sci U S A 95. https://doi.org/10.1073/pnas.95.18.10378
43. Gardner AM, Helmick RA, Gardner PR (2002) Flavorubredoxin, an inducible catalyst for nitric oxide reduction and detoxification in *Escherichia coli*. J Biol Chem 277:8172–8177. https://doi.org/10.1074/jbc.M110471200
44. Gardner AM, Gessner CR, Gardner PR (2003) Regulation of the nitric oxide reduction operon (norRVW) in *Escherichia coli*. J Biol Chem 278:10081–10086. https://doi.org/10.1074/jbc.m212462200
45. Adolfsen KJ, Brynildsen MP (2015) A kinetic platform to determine the fate of hydrogen peroxide in *Escherichia coli*. PLoS Comput Biol 11:e1004562. https://doi.org/10.1371/journal.pcbi.1004562
46. Montgomery HAC, Thom NS, Cockburn A (2010) Determination of dissolved oxygen by the winkler method and the solubility of oxygen in pure water and sea water. J Appl Chem 14:280–296. https://doi.org/10.1002/jctb.5010140704
47. Wagenmakers EJ, Farrell S (2004) AIC model selection using Akaike weights. Psychon Bull Rev 11:192–196

Chapter 9

SWATH: A Data-Independent Tandem Mass Spectrometry Method to Quantify ^{13}C Enrichment in Cellular Metabolites and Fragments

Damini Jaiswal and Pramod P. Wangikar

Abstract

Recently, the sequential windowed acquisition of all theoretical fragment ion mass spectra (SWATH) method coupled with liquid chromatography has been demonstrated for the quantification of isotopic ^{13}C enrichment in a large number of cellular metabolites and fragments. SWATH, a data–independent acquisition (DIA) method, alleviates the need for data deconvolution and shows greater accuracy in the quantification of low abundance isotopologs of fragments thereby resulting in a lower systematic error. Here we provide a detailed protocol for the design of Q1 mass isolation windows and the post–acquisition data analysis with emphasis on the untargeted nature of SWATH.

Key words Mass isotopolog distribution, ^{13}C metabolic flux analysis, Multiple reaction monitoring, Parallel reaction monitoring, Liquid chromatography–mass spectrometry

1 Introduction

Studies involving isotopic ^{13}C carbon enriched substrates as tracers have provided useful insights into cellular metabolism in a variety of biological systems [1]. For quantitative approaches such as ^{13}C metabolic flux analysis (^{13}C-MFA), accurate measurement of mass isotopolog distribution (henceforth MID) of metabolites is a key requirement [2]. Among the techniques available for quantification of MIDs, those based on liquid or gas chromatography coupled with mass spectrometry (LC/MS or GC/MS) are being widely used. Additionally, the MIDs of metabolite fragments obtained via tandem MS (MS/MS) technique provide information on positional labeling and in turn yield greater precision in flux estimates [3–5]. A tandem mass spectrometer typically comprises of three quadrupoles or mass analyzers, Q1, Q2, and Q3 (Fig. 1). Q1 is a quadrupole where precursor ions are scanned and depending on the acquisition method either all or a specific set of precursors can

Deepak Nagrath (ed.), *Metabolic Flux Analysis in Eukaryotic Cells: Methods and Protocols*, Methods in Molecular Biology, vol. 2088,
https://doi.org/10.1007/978-1-0716-0159-4_9,

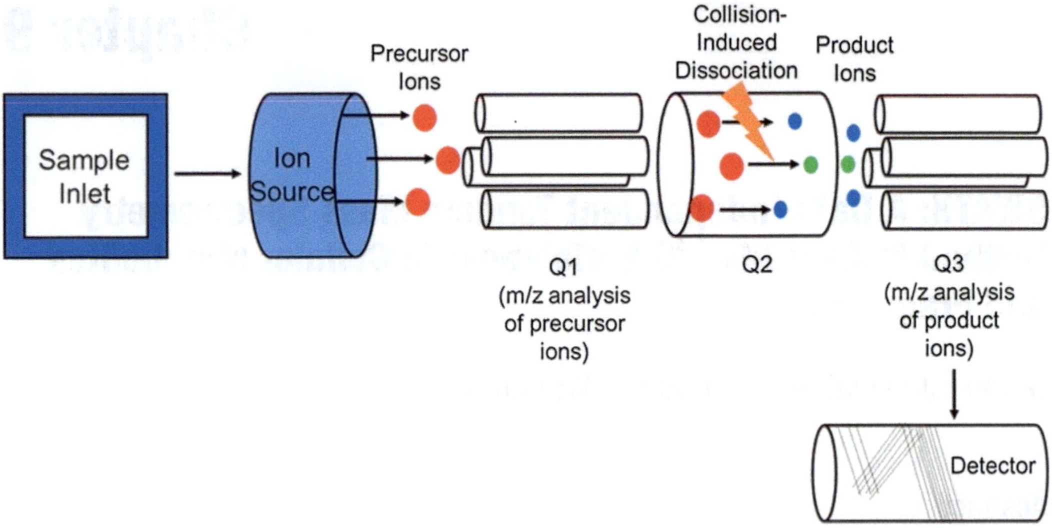

Fig. 1 The general principle of a tandem mass spectrometer. The sample is ionized and charged precursor ions are generated. In Q2, specific set or all precursor ions are fragmented resulting in the formation of fragment ions that are scanned in Q3 and coupled to a detector

be transferred to Q2, where collision-induced dissociation of precursor ions takes place resulting in fragment ions. Q3 is the third quadrupole that scans data for fragment ions and is coupled to a detector.

Multiple reaction monitoring (MRM) and parallel reaction monitoring (PRM) are the commonly used narrow pass Q1 methods for acquisition of MIDs of metabolites and fragments (Fig. 2) [4, 6]. MRM is usually configured on low-resolution triple quadrupole systems and was originally developed to overcome the matrix effects of the samples. Here, MS/MS is collected for predefined precursor–product ion pairs, typically referred to as transitions. For example, ADP-glucose shows the fragmentation pattern as shown in Fig. 3a. To collect the data for two fragments of *m/z* 241 and 346 using MRM, individual transitions need to be defined (Fig. 3b). Further, additional transitions need to be defined to account for the possible isotopologs of a given metabolite and the resultant fragments [4, 5]. Quantification of MIDs then involves a data deconvolution step [5]. In the case of PRM, MS/MS data is collected for all product ions obtained by fragmentation of a particular precursor ion (Fig. 3c) and thus requires fewer transitions to acquire the MIDs (Table 1). The use of wide-isolation window PRM may further reduce the number of transitions but may not completely eliminate data deconvolution. Although these techniques are still quite popular due to their extreme specificity, there are various limitations for their application in the quantification of MIDs. The methods show poor sensitivity for low abundance isotopologs thereby adding a systematic error in the quantification of

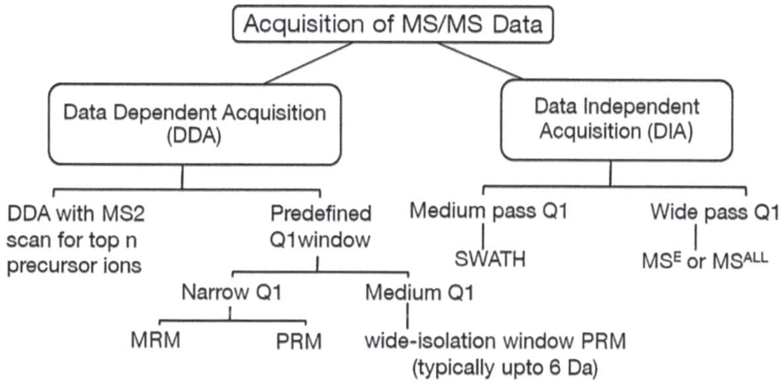

Fig. 2 Methods that are commonly used to collect tandem MS Data using high resolution LC/MS/MS. There are two broad categories, data-dependent acquisition (DDA) and data-independent acquisition (DIA). Sequential Windowed Acquisition of all Theoretical Fragment Ion Mass Spectra (SWATH), a DIA method is the focus of this book chapter

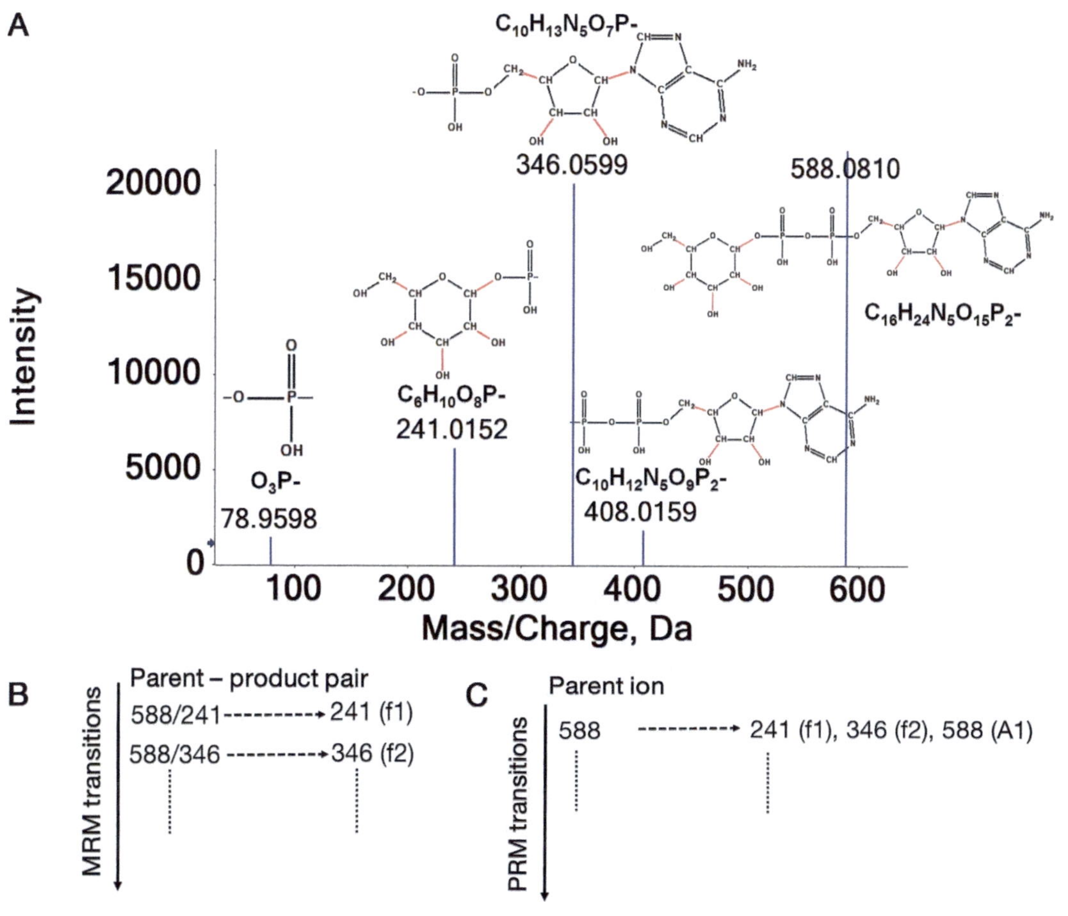

Fig. 3 The mass fragmentation pattern of ADP-glucose (**a**), schemes for acquisition of fragment data of ADP-glucose using MRM (**b**) and PRM (**c**) methods

Table 1
Comparison of data-dependent acquisition (DDA) scan, multiple reaction monitoring (MRM), parallel reaction monitoring (PRM), and sequential windowed acquisition of all theoretical fragment ion mass spectra (SWATH) for the acquisition of mass isotopologue distribution (MID) of metabolites and fragments

Approach	DDA with MS2 for top n peaks Untargeted	MRM Targeted	PRM Targeted	SWATH Untargeted
Number of transitions/windows required to obtain MID of a metabolite	None	$(n - m + 1)(m + 1)$[a]	$(n + 1)$ and $<(n + 1)$[a] for wide-isolation-window PRM	1 (in most cases) or rarely 2
XIC of precursor	Yes	No	Yes	Yes
Fragmentation pattern of the base mass	Yes	No	Yes	Yes
Precursor MID	Yes	Yes	Yes	Yes
Fragment MID	No	Yes	Yes	Yes
Deconvolution needed to obtain MIDs	No	Yes	Yes	Needed in rare instances[b]
Sensitivity for detection of low-abundant isotopologs	NS	Low	Low	High
Decoupling of data acquisition and analysis	Yes	No	No	Yes

NS not studied
[a] n and m represent the number of carbons in metabolite and fragment respectively
[b] Deconvolution, if required, will be only from two adjacent Q1 windows

MIDs (Table 1). Further, both MRM and PRM are targeted approaches and do not provide an option for decoupling of data acquisition and analysis [5] (Table 1).

The currently available high-resolution (HR) tandem MS methods fall into two broad categories, namely, data-dependent acquisition (DDA) and data-independent acquisition (DIA) (Fig. 2). DDA methods collect MS/MS data only when a preset criterion is satisfied while DIA methods allow untargeted data acquisition. To exemplify, the DDA method can be programmed to trigger the MS/MS scan for n most abundant ions in each cycle. While this may work reasonably well for unlabeled samples with monoisotopic peaks, the presence of isotopologs of each metabolite in the ^{13}C-enriched samples may pose a challenge. Some of the isotopologs of a particular metabolite may not meet the preset criteria thereby resulting in intermittent loss in the MS/MS data. Thus, DDA methods may be unsuitable for the quantification of fragment MIDs (Table 1).

The data-independent acquisition methods can be broadly categorized as wide pass and medium pass Q1 isolation window methods (Fig. 2). MS^{ALL}, an example of the former category permits fragmentation of all the precursor ions at all times. This method may not be suitable for the quantification of MIDs as the relationship between precursor and its fragments is lost (*see* **Note 1**) [7]. The latter category methods such as SWATH (Sequential Windowed Acquisition of all Theoretical Fragment Ion Mass Spectra), the data for both the precursor as well as the fragment ions are collected continuously in the MS1 and MS2 scans, respectively, while maintaining the relationship between the Q1 mass window and the fragment ions. Reconstruction of the precursor-product relationship has been shown successfully by aligning the MS1 and MS2 chromatograms, the latter drawn from the fragmentation of the respective Q1 window [8, 9].

Recently, we showed that SWATH can be applied to acquire the MIDs of metabolites and fragments [5]. The cofragmentation of all the isotopologs of a particular metabolite in a single Q1 mass isolation window results in sensitive detection of the isotopologs and obviates the need for data deconvolution. Moreover, SWATH allows untargeted analysis and potential decoupling of data acquisition and analysis. The main objective of this book chapter is to provide a detailed protocol for acquisition of MIDs of intact metabolites and fragments from a ^{13}C-labeled sample using SWATH. We describe the design of SWATH method, both for the data acquisition and analysis steps and some of the practical aspects of this approach.

2 Materials

1. An extract of a biological sample that is fully or partially labeled with a substrate enriched in ^{13}C isotopic carbon.
2. Equipment: A HPLC/UHPLC/nano-HPLC operated with an appropriate column and buffer system and coupled with a mass spectrometer capable of collecting tandem MS data via SWATH [e.g., SCIEX Triple TOF instrument, model 5600 or 6600 (Sciex, Framingham, MA)].
3. Software: PeakView 2.0 and MasterView 1.0 software (Sciex, Framingham, MA) for manual examination of the chromatographic peak quality and baseline separation. MultiQuant 3.0.2 software to integrate the area under the peaks (Sciex, Framingham, MA).
4. mol files of the metabolite of interest for fragment annotation.
5. Information on the number of carbons in the metabolites of interest (*see* **Note 2**).
6. wiff and wiff scan files of metabolite standards (*see* **Notes 3** and **4**).

3 Methods

3.1 ^{13}C Labeling and Sample Preparation

As a first step, a dynamic or steady state ^{13}C labeling experiments needs to be performed on the biological system of interest and samples collected for analysis via LC/MS/MS. In this protocol chapter, we show the results for a nonstationary ^{13}C labeling experiment with the cyanobacterium *Synechococcus* sp. PCC 7002. The relevant protocols for labeling [5, 10], extraction of metabolites [11] and chromatography [5, 12, 13] have been described in detail elsewhere. Briefly, 1 g/l $NaH^{13}CO_3$ was added to an exponentially growing culture of *Synechococcus* sp. PCC 7002 and at predetermined intervals, 20 ml samples were filtered, quenched and extracted as described earlier [5]. The extracted samples (henceforth samples) were dried and stored in a −80 °C freezer until ready for LC/MS/MS analysis.

3.2 Liquid Chromatography Mass Spectrometry Method

While a number of methods have been reported for chromatographic separation of intracellular metabolites, we chose a method that is based on ion pairing with tributylamine as described in detail previously [13]. An accompanying chapter discusses the pros and cons of the different methods that have been reported for the analysis of intracellular metabolites. The ion pairing based method has been shown to provide excellent chromatographic separation and peak quality for a number of metabolite classes such as sugar phosphates, nucleotide sugars, nucleotides organic acids and acyl CoAs [14]. Briefly, the dried metabolite extract was redissolved in 100 μl of 50/50, v/v methanol–water mixture, filtered with 0.2 μm syringe filter and 4 μl injected on LC/MS/MS. A Synergi Hydro-RP LC column 150 × 2 mm with particle size of 4 μm (Phenomenex Inc., Torrance, CA) was used for chromatographic separation using a gradient elution method consisting of mobile phases, 10 mM tributylamine +15 mM acetic acid in water (A) and 100% Methanol (B). The gradient method is as follows: 0% B (0.1 min), 0% B (2 min), 35% B (8 min), 35% B (10.5 min), 90% B (15.50 min), 90% B (20.5 min), 0% B (22 min), and 0% B (30 min). The column temperature and flow rate were 25 °C and 0.3 mL/min, respectively. A Triple TOF 5600+ mass spectrometer (SCIEX, Framingham, MA) interfaced with Shimadzu Ultra Performance Liquid Chromatography (UPLC) system (Shimadzu, Nexera LC -30 AD, Singapore) was used in negative ion mode for analysis. Each sample was injected in two technical replicates with each of the SWATH acquisition programs. Description of the SWATH programs follows.

3.3 Design of the SWATH Method and the Parameters

SWATH data is collected in a cyclic manner throughout the chromatographic time to cover the desired *m/z* range (Fig. 4a) [5, 15, 16]. A contour plot showing the MS1 scan of all the Q1 windows of

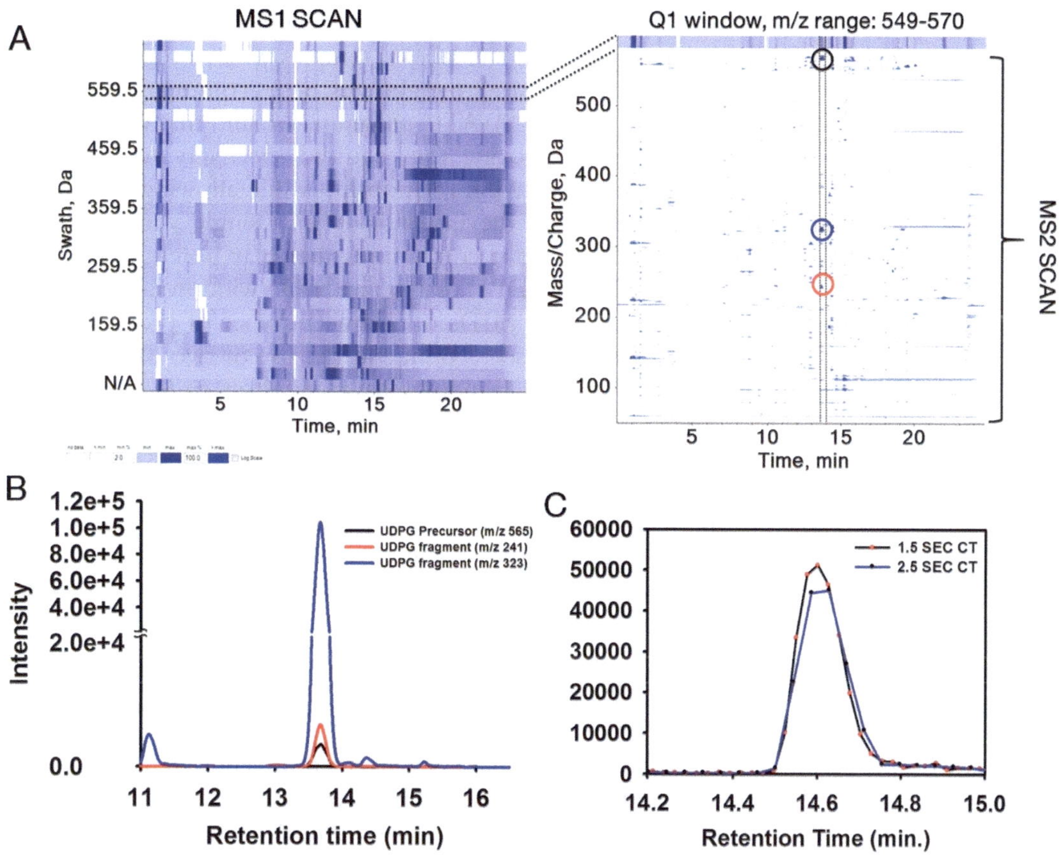

Fig. 4 (**a**) Contour plot of the MS1 scan (panel on the left) of a representative sample of cyanobacterial metabolite extract injected with reverse-phase, ion-pairing chromatography method and SWATH program # 1 (refer to Table 2 for the details of the program). The MS1 data from the different Q1 windows, of 20 Da width each, is shown in a stacked manner. Contour plot of MS2 scan for the precursor ions of Q1 window # 25 (*m*/*z* range 549–570 Da) is shown in the panel on the right. MS2 scans for each Q1 window are generated and can be processed further for data analysis. (**b**) The extracted ion chromatograms (XICs) of UDP-glucose and its fragments generated from the MS2 scans corresponding to Q1 window #25. (**c**) The effect of cycle time on peak quality of phosphoenolpyruvate for a representative sample

a SWATH program is shown in Fig. 4a. The MS2 scan of a particular window where UDP-glucose is fragmented is also shown (Fig. 4a). The XICs (extracted ion chromatograms) of UDP-glucose and its fragments obtained from that particular Q1 window show the precursor and fragment ions aligned at a particular RT (Fig. 4b). The following parameters need to be considered while designing a SWATH method for MID acquisition. Some of these parameters are interrelated:

1. *Number and width of Q1 windows*: In proteomics applications, the use of variable-width Q1 windows has been demonstrated where the windows become narrower in the analyte-dense regions of *m*/*z* [16, 17]. This may be suitable when analyzing

unlabeled samples that primarily comprise of monoisotopic peaks. To analyze ^{13}C labeled samples with isotopologs being present for almost all metabolites, we recommend the use of fixed Q1 window SWATH program. We find that Q1 windows of 18–25 Da provide the required specificity to retain the precursor-product ion relationship while allowing for cofragmentation of all the isotopologs in a single window. Many biologically relevant compounds that may potentially coelute (e.g., ATP and ADP) fall in separate Q1 windows, allowing for quantification of MIDs of a large number of precursors and fragments. We also recommend the use of two SWATH programs whose Q1 windows have an offset of 9–12 Da with respect to each other (Table 2) [5]. This is to ensure that isotopologs of all compounds fall within a single window in one of the programs. Despite this design, isotopologs of some metabolites may span two adjacent windows. MIDs of such metabolites and their fragments may be quantified by deconvoluting the data from the respective windows [5]. The number of windows may need to be restricted to achieve the desired values of the cycle time and accumulation time. This may result in the collection of the MS2 data in a more restricted Q1 range than what the instrument permits.

2. *Cycle time*: Cycle time refers to the time required to scan all the Q1 windows and therefore is the product of accumulation time and the number of Q1 windows (Fig. 4a). Cycle time also plays a crucial role in obtaining a quantifiable peak, which may be affected by the number of data points in a peak. As a rule of thumb, 8–10 data points/peak are considered satisfactory to quantify the area under the peak [18]. The number of data points in a peak is inversely inversely proportional to the cycle time and directly proportional to the peak width. Figure 4c shows the comparison of peaks of phosphoenolpyruvate (PEP) obtained using SWATH methods with a cycle time of 2.5 and 1.5 s resulting in 7 and 11 data points, respectively, for the metabolite extract of *Synechococcus elongatus* PCC 11801 [19].
3. *Accumulation time*: It is also known as the MS2 scan time or dwell time for each Q1 window (Fig. 4a). A SWATH program with a large number of Q1 windows results in a smaller accumulation time. Sufficient amount of accumulation time is required to achieve a satisfactory signal. While the minimum threshold for accumulation time may depend on the speed and sensitivity of the detector, we found that accumulation time of 40–80 ms provided satisfactory results with the hardware available with us.
4. *Collision Energy*: In tandem MS approaches such as MRM, collision energy is typically optimized for each precursor–product ion pair to maximize the intensity of the desired product

Table 2
The examples of two SWATH programs with an offset of 10 Da compared to each other

SWATH Q1 isolation window	SWATH program 1			SWATH program 2		
	Start mass (Da)	Stop mass (Da)	Collision energy (eV)	Start mass (Da)	Stop mass (Da)	Collision energy (eV)
Q1 window #1	69	90	−13.20	79	100	−13.80
Q1 window #2	89	110	−14.33	99	120	−14.90
Q1 window #3	109	130	−15.44	119	140	−16.06
Q1 window #4	129	150	−16.64	139	160	−17.27
Q1 window #5	149	170	−17.70	159	180	−18.37
Q1 window #6	169	190	−18.90	179	200	−19.52
Q1 window #7	189	210	−20.09	199	220	−20.67
Q1 window #8	209	230	−21.25	219	240	−21.80
Q1 window #9	229	250	−22.40	239	260	−22.99
Q1 window #10	249	270	−23.55	259	280	−24.12
Q1 window #11	269	290	−24.70	279	300	−25.27
Q1 window #12	289	310	−25.84	299	320	−26.43
Q1 window #13	309	330	−27.00	319	340	−27.58
Q1 window #14	329	350	−28.15	339	360	−28.73
Q1 window #15	349	370	−29.30	359	380	−29.88
Q1 window #16	369	390	−30.45	379	400	−31.03
Q1 window #17	389	410	−31.60	399	420	−32.18
Q1 window #18	409	430	−32.70	419	440	−33.33
Q1 window #19	429	450	−33.90	439	460	−34.48
Q1 window #20	449	470	−35.09	459	480	−35.63
Q1 window #21	469	490	−36.02	479	500	−36.78
Q1 window #22	489	510	−37.30	499	520	−37.93
Q1 window #23	509	530	−38.51	519	540	−39.08
Q1 window #24	529	550	−39.96	539	560	−40.23
Q1 window #25	549	570	−40.81	559	580	−41.38
Q1 window #26	569	590	−41.96	579	600	−42.53
Q1 window #27	589	610	−43.11	599	620	−43.68
Q1 window #28	609	630	−44.26	619	640	−44.83
Q1 window #29	629	650	−45.41	639	660	−45.98

This strategy allows to obtain the MIDs of metabolites, the isotopologue of which might span two adjacent windows

Table 3
The suggested parameters to design SWATH programs for acquisition of mass isotopologue distribution (MID) of metabolites and fragments

Parameters	Value
Number of Q1 windows	25–35
Cycle time	1.5–3.5 s
Accumulation time	40–80 ms
Collision energy	Rolling collision energy based on *m/z* range of Q1 window
Collision energy spread (CES)	15 eV

ion. Optimal value of CE for one fragment may be suboptimal for another fragment. Our objective is to be able to quantify both the precursor and a number of fragments for each metabolite. SWATH can be programmed to automatically select the collision energy based on the *m/z* values of the Q1 window. We find that the autoselect option provides satisfactory fragmentation pattern and permits quantification of MIDs of both the precursor and several product ions for each metabolite.

An appropriate SWATH method is obtained with a proper selection of Q1 windows, accumulation, and cycle times. These parameters need to be adjusted properly to strike a balance between sensitivity and specificity of the detection. The recommended values of the parameters are shown in Table 3.

3.4 SWATH Data Processing for MID Quantitation of Targeted Set of Metabolites

1. While the SWATH data is acquired in an untargeted manner, the data analysis is typically performed for a targeted list of metabolites. MIDs of metabolites and fragments are quantified by integrating the area under the peak for the *m/z* corresponding to each of the isotopologs using the proprietary software MultiQuant (Sciex, Farmingham, MA). We will illustrate this with the glucose containing fragment of ADP-glucose of monoisotopic mass, $F + 0$, of 241.015 Da (Fig. 3a). Incorporation of one or more ^{13}C carbons shifts the *m/z* of this fragment that would result in isotopologs $F + 1$, $F + 2$, …, $F + 6$. Area under the curve corresponding to each isotopolog *m/z* at the given RT and for the given Q1 window will allow calculation of relative abundance of each isotopolog or the MID of a given fragment. Since metabolite identification and knowledge of the Q1 window, RT, and *m/z* are prerequisites for quantification, the data analysis will require a targeted list of metabolites and fragments. The targeted list can be prepared

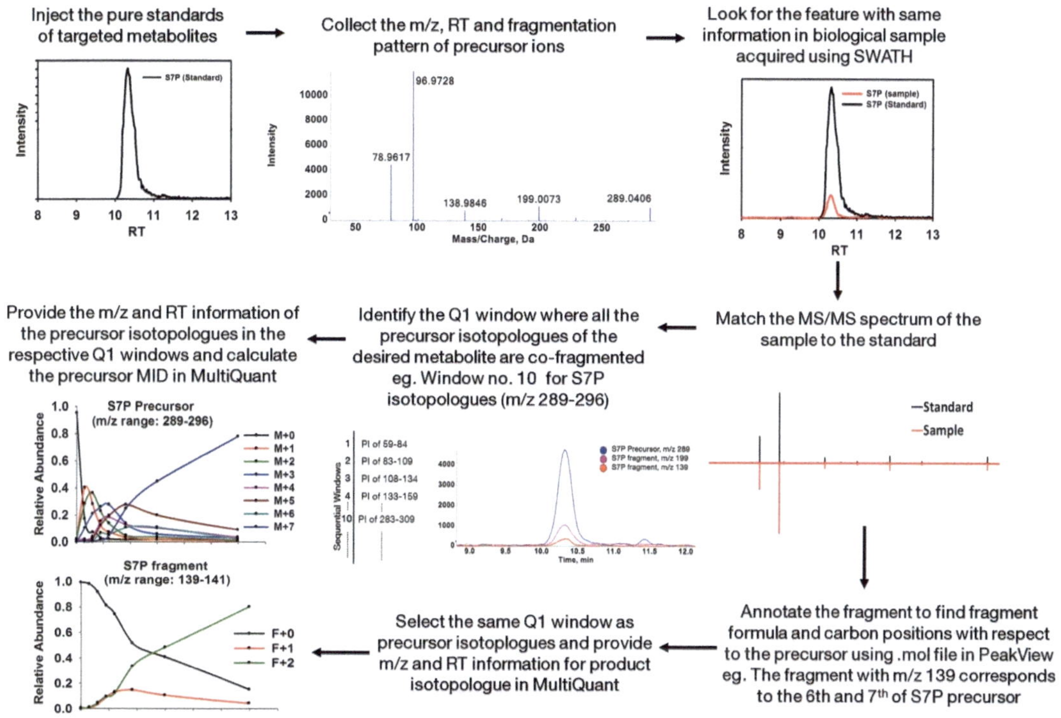

Fig. 5 The workflow for acquisition and analysis of mass isotopolog distribution of targeted metabolites and fragments using SWATH

either prior to or post–acquisition, thereby decoupling the steps of data acquisition and analysis.

2. The identity of a targeted set of metabolites present in the sample can be confirmed by injecting standard solutions of pure metabolites using the same LC/MS/MS program that is used to inject the samples (Fig. 5). This exercise provides the mass to charge ratio (m/z), retention time (RT) and MS2 fingerprint for the metabolites of interest. Proprietary software such as PeakView and MasterView (SCIEX, Framingham, MA) or open source software such as XCMS [20] can be used to visualize the peak quality, intensity, signal-to-noise ratio, and the XICs of the metabolites and the fragments [21]. This exercise is to be repeated for LC/MS/MS injections of the unlabeled samples to match the RT and m/z of the precursors and the fragments. The fragmentation pattern of the respective peak in the sample should be matched to that of the standard (Fig. 5).
3. Quantification of MIDs of the precursor and product ions begins by identifying the Q1 window that spans all the isotopologs of the precursor ion (*see* **Notes 2** and **3**). Note that the quantification of both precursor and the respective fragments can be performed from the MS2 data for the relevant Q1

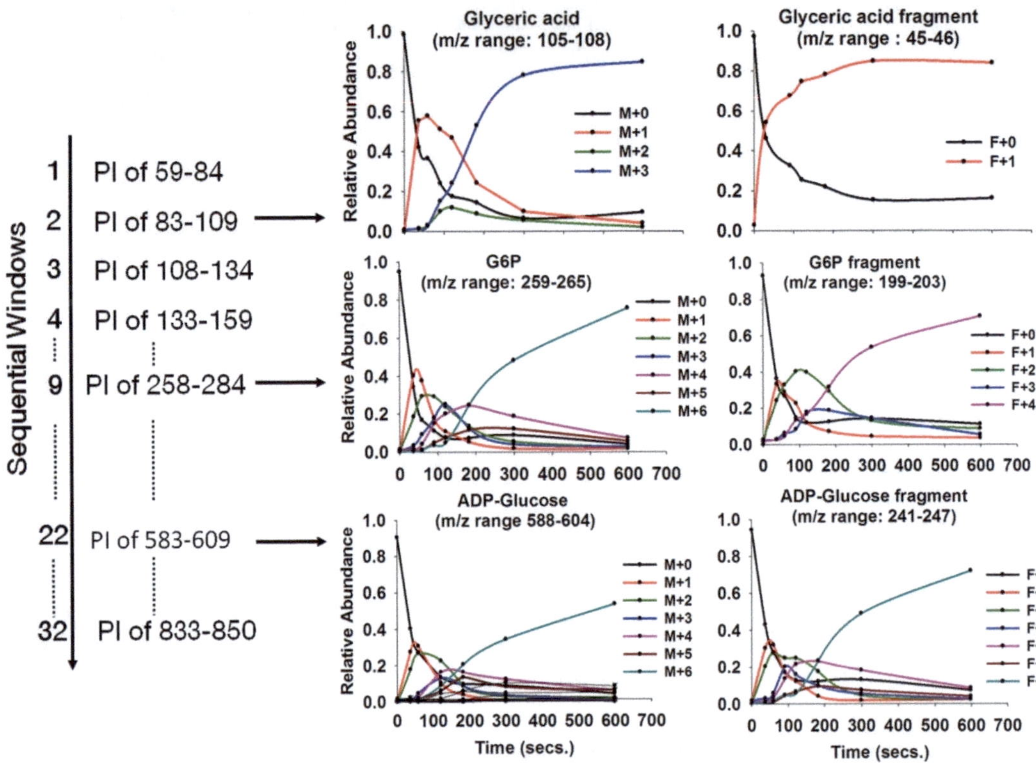

Fig. 6 Quantification of MIDs of three representative metabolites and their respective fragments from a time course labeling experiment using the SWATH method. The precursor ion (PI) windows or Q1 windows from which the respective MIDs are quantified have been indicated by arrows. The *x*-axis of the time profiles indicates the time after addition of $NaH^{13}CO_3$ to an exponentially growing culture of cyanobacteria. Samples are collected at the indicated time, quenched, extracted, dried, redissolved in the mobile phase, and injected on LC/MS/MS using the SWATH program. MID of a metabolite at a given time point is obtained from the respective SWATH injection as outlined in Fig. 5

window. While the fragments can be quantified only from the MS2 scans, our general observation is that the XICs of even the precursors tend to be cleaner in the MS2 scan from the respective Q1 window than in the MS1 scan. This may be due to the matrix effect or fragments of confounding masses resulting from other metabolites. To exemplify, the quantitation of precursor MID of seduheptulose-7-phosphate (S7P) from window number 10 is shown in Fig. 5, where all the isotopologs of S7P (m/z 289–296) are present. The MIDs of other metabolites and fragments can be similarly obtained from other Q1 windows (Fig. 6).

4. Positional labeling information can be extracted from the ^{13}C enrichment data of well-annotated fragments. Fragment annotation or assignment of chemical structure to each fragment can be performed by matching the observed MS2 fingerprint of a metabolite with the data in literature [4, 5]. Fragment

annotation can also be performed using mol file of the metabolites and the inbuilt fitting algorithm of PeakView. A user-defined metabolite and fragment library can be created in-house for routine analysis of the ^{13}C labeling data.

5. The metabolites that can be quantified from SWATH data largely depend on the chromatographic method used and the nature of the biological sample (*see* **Note 4**). The fundamentals of SWATH data collection and analysis remain unchanged regardless of the chromatographic method.
6. The MIDs of metabolites and fragments can be corrected for natural ^{13}C labeling using a software tool such as IsoCor [22].
7. We found the video tutorials at SCIEX website (https://sciex.com/support/training/course-catalog) to be very useful to understand technical aspects of SWATH data analysis using the proprietary software.

3.5 Post–acquisition Metabolite Identification and MID Quantitation from SWATH Data

1. As SWATH data is collected in an untargeted manner, the data files may contain information on a number of metabolites that the user did not intend to quantify prior to data acquisition. Thus, the MIDs of additional metabolites and fragments can be quantified post–acquisition provided their identity can be established.
2. Metabolite identification can be performed by injecting standard solutions of additional metabolites. Alternatively, one can use software tools such as MS-DIAL [8] and MetDIA [9] that rely on matching of exact mass of the metabolite and the MS2 spectra with mass spectral databases like ChemSpider and METLIN [23] (Fig. 7). This can result in an extended list of metabolites and their fragments together with the information on the Q1 window, RT, and *m/z*.
3. Once the metabolites are identified, the respective mol file can be downloaded from KEGG and fragment annotation can be performed as described above (Fig. 7).
4. The subsequent steps of quantification of MIDs of metabolites and fragments with MultiQuant can be followed as described above.
5. Post–acquisition metabolite identification and MID quantification of additional metabolites can be performed again and again, as and when additional standards become available or when the software tools and spectral libraries get updated. This will help analyze additional data from the same set of LC/MS/MS injections.

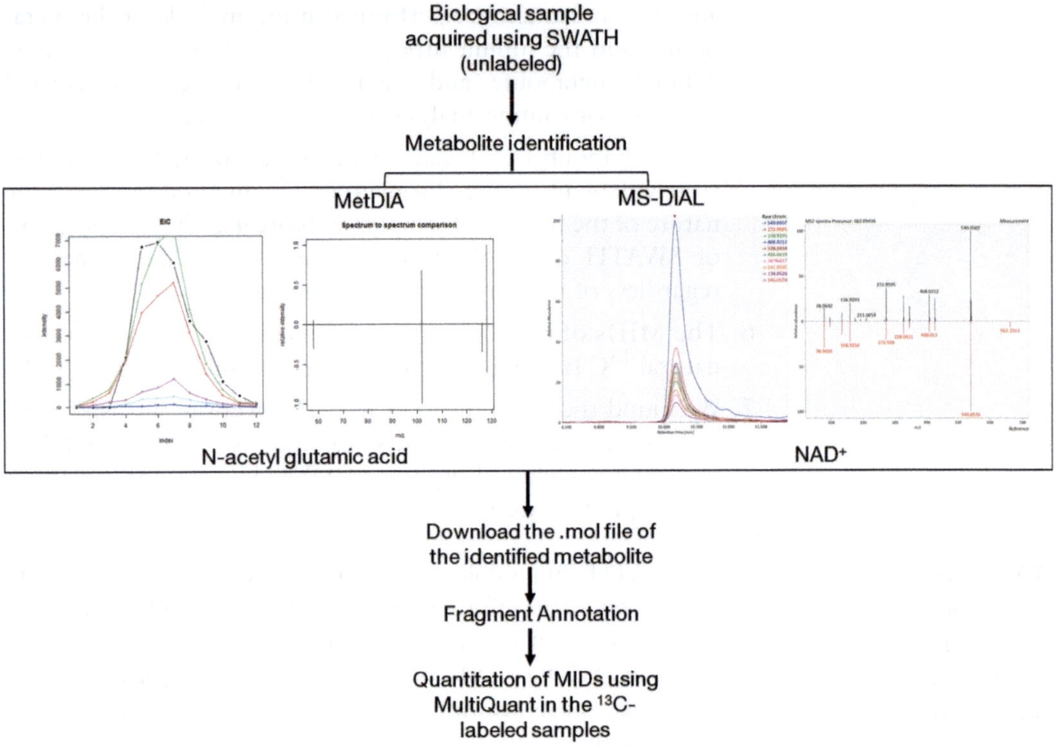

Fig. 7 The workflow for identification of additional metabolite and fragments and quantitation of their MIDs in a sample acquired using SWATH method. The software tools like MetDIA and MS-DIAL provide metabolite identification by reestablishing the link between the precursor and fragment ions via data deconvolution. This strategy can be used to enhance the repertoire of metabolites, the MIDs of which can be potentially quantitated after data acquisition

4 Notes

1. In MS^E or MS^{ALL} [24], all the precursor ions within a given m/z range are fragmented, and the data is collected for all the product ions. The MS/MS spectrum thus generated lacks specificity and is preferably used with a good chromatographic separation method [7]. Various data deconvolution algorithms have been useful to restore the link between the precursor and product ions. MSA^{LL} has not been used to acquire MIDs of metabolites owing to its lack of specificity but serves as a good method for identification of unknowns.
2. The mass isotopolog distribution (MID) is a relative fractional measurement of different isotopologs a metabolite can form. The number of isotopologs of a metabolite or fragment that is desirable upon labeling depends on the number of carbons present in it. For example, 3-phosphoglyceric acid that has three carbons and m/z of 184.98 (M) can have three isotopologs, 185.98 (M + 1), 186.98 (M + 2), and 187.98 (M + 3).

The isotopologs of a particular metabolite (M) or fragment (F) with k and l number of carbons would range from $M + 0$ to $M + k$ and from $F + 0$ to $F + l$, respectively.

3. The SCIEX instruments generate the data files as wiff and wiff scan files. Any sample acquired on these instruments thus has two files (wiff and wiff scan) with the same sample names. The software like PeakView, Master View, and MultiQuant are required to process the wiff files and wiff scan files. It is important to keep both wiff and wiff scan files of a particular sample in the same folder for analysis using these commercial software. These software at present can be installed in any Windows 7 operating system (64-bit). MasterView runs in PeakView background, and thus installation of PeakView is a prerequisite to analyze the data using MasterView.
4. To obtain the MID of a particular metabolite or fragment, the m/z and retention time of that particular analyte needs to be confirmed with standards. Although the m/z of a particular metabolite will remain constant irrespective of the method (s) or instrument(s) used to acquire the data, the retention time varies drastically depending on the type and the length of the column, chromatographic method, solvent, run time, flow rate, and so on. Therefore, one must collect this information by the injection of standard solutions using a respective chromatographic method that is to be used for SWATH acquisition. The injection of standards can be done using any mass spectrometry method.

Acknowledgments

This work was supported by a grant from Department of Biotechnology (DBT), Government of India, awarded to PPW toward DBT-Pan IIT Center for Bioenergy (Grant no. BT/EB/PAN IIT/2012).

References

1. Zamboni N, Sauer U (2009) Novel biological insights through metabolomics and13C-flux analysis. Curr Opin Microbiol 12:553–558. https://doi.org/10.1016/j.mib.2009.083
2. Sauer U (2006) Metabolic networks in motion: 13C-based flux analysis. Mol Syst Biol 2:62. https://doi.org/10.1038/msb4100109
3. Rühl M, Rupp B, Nöh K et al (2012) Collisional fragmentation of central carbon metabolites in LC-MS/MS increases precision of 13C metabolic flux analysis. Biotechnol Bioeng 109:763–771. https://doi.org/10.1002/bit.24344
4. McCloskey D, Young JD, Xu S et al (2016) MID Max: LC-MS/MS method for measuring the precursor and product mass isotopomer distributions of metabolic intermediates and cofactors for metabolic flux analysis applications. Anal Chem 88:1362–1370. https://doi.org/10.1021/acs.analchem.5b03887
5. Jaiswal D, Prasannan CB, Hendry JI, Wangikar PP (2018) SWATH tandem mass spectrometry workflow for quantification of mass

isotopologue distribution of intracellular metabolites and fragments labeled with isotopic 13C Carbon. Anal Chem 90:6486–6493. https://doi.org/10.1021/acs.analchem.7b05329

6. Li Z, Li Y, Chen W et al (2017) Integrating MS1 and MS2 scans in high-resolution parallel reaction monitoring assays for targeted metabolite quantification and dynamic 13 C-labeling metabolism analysis. Anal Chem 89:877–885. https://doi.org/10.1021/acs.analchem.6b03947
7. Zhu X, Chen Y, Subramanian R (2014) Comparison of information-dependent acquisition, SWATH, and MS all techniques in metabolite identification study employing ultrahigh-performance liquid chromatography-quadrupole time-of-flight mass spectrometry. Anal Chem 86:1202–1209. https://doi.org/10.1021/ac403385y
8. Tsugawa H, Cajka T, Kind T et al (2015) MS-DIAL: data-independent MS/MS deconvolution for comprehensive metabolome analysis. Nat Methods 12:523–526. https://doi.org/10.1038/nmeth.3393
9. Li H, Cai Y, Guo Y et al (2016) MetDIA: Targeted metabolite extraction of multiplexed MS/MS spectra generated by data-independent acquisition. Anal Chem 88:8757–8764. https://doi.org/10.1021/acs.analchem.6b02122
10. Hendry JI, Prasannan C, Ma F et al (2017) Rerouting of carbon flux in a glycogen mutant of cyanobacteria assessed via isotopically non-stationary 13 C metabolic flux analysis. Biotechnol Bioeng 114:2298–2308. https://doi.org/10.1002/bit.26350
11. Prasannan CB, Jaiswal D, Davis R, Wangikar PP (2018) An improved method for extraction of polar and charged metabolites from cyanobacteria. PLoS 2018:1–16
12. Mccloskey D, Gangoiti JA (2015) A pH and solvent optimized reverse-phase ion-paring-LC–MS/MS method that leverages multiple scan-types for targeted absolute quantification of intracellular metabolites. Metabolomics 11:1338–1350. https://doi.org/10.1007/s11306-015-0790-y
13. Luo B, Groenke K, Takors R et al (2007) Simultaneous determination of multiple intracellular metabolites in glycolysis, pentose phosphate pathway and tricarboxylic acid cycle by liquid chromatography-mass spectrometry. J Chromatogr A 1147:153–164. https://doi.org/10.1016/j.chroma.2007.02.034
14. Lu W, Clasquin MF, Melamud E et al (2010) Metabolomic analysis via reversed-phase ion--pairing liquid chromatography coupled to a stand alone orbitrap mass spectrometer. Anal Chem 82:3212–3221
15. Arnhard K, Gottschall A, Pitterl F, Oberacher H (2015) Applying "Sequential Windowed Acquisition of All Theoretical Fragment Ion Mass Spectra" (SWATH) for systematic toxicological analysis with liquid chromatography-high-resolution tandem mass spectrometry. Anal Bioanal Chem 407:405–414. https://doi.org/10.1007/s00216-014-8262-1
16. Röst HL, Rosenberger G, Navarro P et al (2014) OpenSWATH enables automated, targeted analysis of data-independent acquisition MS data. Nat Biotechnol 32:219–223. https://doi.org/10.1038/nbt.2841
17. Schubert OT, Gillet LC, Collins BC et al (2015) Building high-quality assay libraries for targeted analysis of SWATH MS data. Nat Protoc 10:426–441. https://doi.org/10.1038/nprot.2015.015
18. Oswald S, Gröer C, Drozdzik M, Siegmund W (2013) Mass spectrometry-based targeted proteomics as a tool to elucidate the expression and function of intestinal drug transporters. AAPS J 15:1128–1140. https://doi.org/10.1208/s12248-013-9521-3
19. Jaiswal D, Sengupta A, Sohoni S et al (2018) Genome features and biochemical characteristics of a robust, fast growing and naturally transformable cyanobacterium Synechococcus elongatus PCC 11801 isolated from India. Sci Rep 8:16632. https://doi.org/10.1038/s41598-018-34872-z
20. Tautenhahn R, Patti GJ, Rinehart D, Siuzdak G (2012) XCMS online: a web-based platform to process untargeted metabolomic data. Anal Chem 84:5035–5039. https://doi.org/10.1021/ac300698c.XCMS
21. Kanehisa M, Goto S (2000) KEGG: kyoto encyclopedia of genes and genomes. Nucleic Acids Res 28:27–30. https://doi.org/10.1088/1751-8113/44/8/085201
22. Millard P, Letisse F, Sokol S, Portais JC (2012) IsoCor: correcting MS data in isotope labeling experiments. Bioinformatics 28:1294–1296. https://doi.org/10.1093/bioinformatics/bts127
23. Tautenhahn R, Cho K, Uritboonthai W et al (2012) An accelerated workflow for untargeted metabolomics using the METLIN database. Nat Biotechnol 30:826–828. https://doi.org/10.1038/nbt.2348
24. Bateman KP, Castro-Perez J, Wrona M et al (2007) MSE with mass defect filtering for in vitro and in vivo metabolite identification. Rapid Commun Mass Spectrom 21:1485–1496. https://doi.org/10.1002/rcm.2996

Chapter 10

Quantifying Metabolic Transfer Mediated by Extracellular Vesicles Using Exo-MFA: An Integrated Empirical and Computational Platform

Abhinav Achreja, Noah Meurs, and Deepak Nagrath

Abstract

Extracellular vesicles (EVs) are ubiquitous nanoscale particles released from many different types of cells. They have been shown to contain proteins, DNA, RNA, miRNA, and, most recently, metabolites. These particles can travel through the intercellular space and bloodstream to have regulatory effects on distant recipients. When an EV reaches a target cell, it is taken up and degraded to release its contents for utilization within the cell. In addition to regulatory effects, EVs have been shown to supplement the high metabolic demands of recipient cells in a nutrient-deprived tumor microenvironment. We developed an integrated empirical and computational platform to quantify metabolic contribution of source cell-derived EVs to recipient cells. The versatile Exo-MFA software tool utilizes ^{13}C stable-isotope tracing data to quantify the metabolic contributions of EVs from a source cell type on a recipient cell type. This is accomplished by creating EV-depleted culture medium, producing isotope-labeled EVs from the source cells, isolating the labeled EVs from the culture supernatant, culturing the recipient cells in the presence of the labeled EVs, and measuring the resulting metabolite levels across several time points.

Key words Extracellular vesicles, Exosomes, Exo-MFA, Stable-isotope tracing, 13-Carbon metabolic flux analysis, Multicellular metabolic flux analysis

1 Introduction

Among the wide range of intercellular interaction mechanisms, microvesicles secreted by cells are an important apparatus of communication between cells. *Extracellular vesicles (EVs)* have been documented as part of fundamental biological machinery as well as being involved in disease progression of aggressive cancers [1–4]. In this chapter, we focus on EVs—also referred to here as exosomes—which are 30–150 nm in diameter and transport proteins and nucleic acids including miRNA intercellularly including distant targets via the bloodstream [5, 6]. Recently, using ^{13}C tracer

Abhinav Achreja and Noah Meurs contributed equally to this work.

Deepak Nagrath (ed.), *Metabolic Flux Analysis in Eukaryotic Cells: Methods and Protocols*, Methods in Molecular Biology, vol. 2088, https://doi.org/10.1007/978-1-0716-0159-4_10,

experiments our lab provided compelling evidence that these EVs also transport free metabolites that are directly incorporated into recipient cells' central carbon metabolite pools [4]. However, a quantified insight into the contribution of metabolite cargo toward rescuing recipient cells from nutrient deprivation was lacking at the time. To address this, we designed a novel paradigm, *exosome-mediated metabolic flux analysis (Exo-MFA)*, to predict fluxes involved in metabolite trafficking from source cells to recipient cells [7]. The protocols described here can be applied to any type of cells that exchange biological cargo via exosomes or any other type of EV.

Exo-MFA integrates a novel experimental protocol using ^{13}C-labeled substrates with an enhanced metabolic flux analysis to provide insight into metabolic cross talk. Exo-MFA utilizes the fundamentals of the *13-carbon metabolic flux analysis (^{13}C-MFA)* algorithm and tracer experiments [8, 9], traditionally used for single cell systems that only exchange metabolites with their medium, to provide an enhanced platform to analyze metabolite fluxes in multicellular systems. Exo-MFA is designed to quantify intracellular fluxes and flux of metabolites via extracellular vesicles (EVs) from *source cells* to *recipient cells* (Fig. 1). The paradigm described herein considers two important processes that occur in the system; (1) packaging of metabolite cargo into EVs and secretion of EVs within *source cells* and (2) internalization of EVs and release of cargo within *recipient cells.* Source cells are cultured in medium containing stable ^{13}C isotope-labeled tracer substrates in order to obtain EVs with ^{13}C-labeled cargo. The ^{13}C-labeled EVs are then introduced into cell medium, which release their labeled cargo into recipient cells. Exo-MFA describes a ^{13}C-MFA problem that includes packaging fluxes describing the generation of EVs and utilizes data from (1) tracer experiments within source cells and EVs, (2) extracellular fluxes of source cells and (3) growth rate measurements. Source cells are cultured in medium with the labeled tracers for 72 h to allow for sufficient production of EVs and enrichment of glycolytic and TCA cycle intermediates along with de novo synthesized amino acids. EVs enriched with labeled metabolites are isolated from the spent medium and their cargo is analyzed using GC-MS. Exo-MFA then considers the transient nature of EV internalization and utilizes data from time-series isotope-labeled EV experiments, measurements of extracellular fluxes, and composition of EV metabolite cargo to estimate intracellular fluxes in recipient cells that internalize EVs from the medium. Exo-MFA for recipient cells is set up as a time-series ^{13}C-MFA problem and solved for measurements sampled at various times such as 3, 6, 12 and 24 h (Fig. 1).

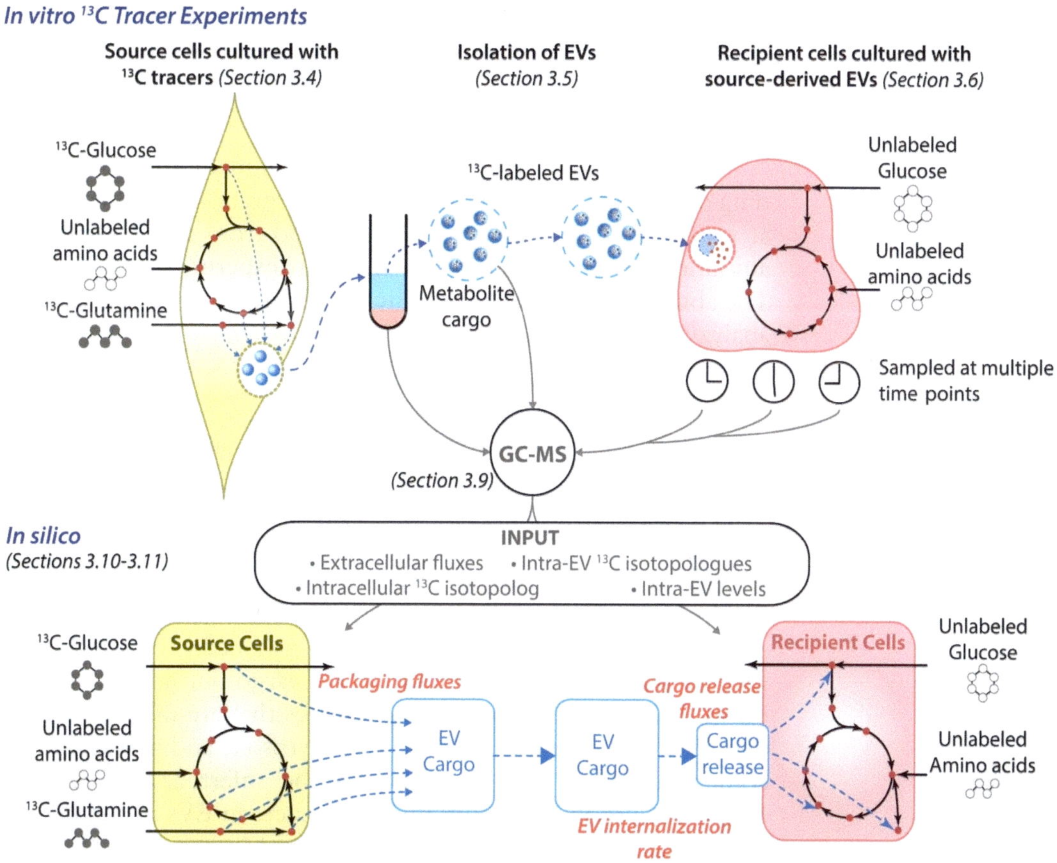

Fig. 1 Empirical and computational workflow for Exo-MFA. Recipient cells are cultured in stable-isotope labeled medium to produce EVs with isotope-labeled metabolite cargo. Upon isolation EVs are isolated and a fraction of them are introduced into culture medium for recipient cells. Metabolic analysis is performed on intracellular extracts from source and recipient cells, EVs, and spent culture medium using GC-MS. All empirical data is compiled as inputs to the Exo-MFA software to quantify intracellular fluxes in source and recipients cells as well as metabolic fluxes carried by EVs from source to recipient cells

1.1 Modeling Packaging Fluxes in Source Cells

In order to obtain sufficient information for estimating metabolite packaging fluxes into EV cargo, the rate of EV secretion along with extracellular fluxes are empirically determined. If each source cell is assumed to produce EVs at a constant rate of r (mg EV/mg protein/h), the total EVs secreted at the end of a tracer experiment is described by Eq. 1, with the knowledge that exponential growth rate of source cells is μ (h^{-1}) and seeding density of source cells is X_0 (cells):

$$\frac{d(\mathrm{Exo})}{dt} = rX = rX_0e^{\mu t} \tag{1}$$

Integrating Eq. 1 will give the expression for total number of EVs produced as given by Eq. 2

$$\text{Total exosomes [mg exosome]} = \int d(\text{Exo}) = \int rX_0 e^{\mu t} dt = \frac{rX_0 e^{\mu t}}{\mu} + C \quad (2)$$

Assuming that there are no EVs in fresh medium at the beginning of the experiment, the constant of integration can be evaluated at $t = 0$ and therefore r can be estimated from Eq. 2 once total amount of EVs produced at the end of the experiment is measured. The abundance of each metabolite in EV cargo can be estimated by analyzing EVs in the GC-MS and Eq. 3.

$$\text{Total } i\text{th exosomal metabolite [µmol]} = \left(P_i \left[\frac{\text{µmol}}{\text{mg exosome}} \right] \right) (P_i).(\text{Total exosome [mg exosome]}) \quad (3)$$

However, measurements for all metabolites that are packaged in cargo are not practical to obtain empirically. For this purpose, Exo-MFA has the ability to predict the packaging fluxes for set of metabolites not measured via targeted GC-MS analysis. This is achieved by introducing P_i as unknown parameter for metabolites that are postulated to contribute to the cargo but not measured a priori. Furthermore, even for metabolites that are measured, contribution of the same metabolite from cytosolic and mitochondrial compartments to EV cargo cannot be inferred from Eq. 3. Exo-MFA also includes isotopomer balances in EVs in addition to intracellular isotopomer balances described in the ^{13}C-MFA formulation (described in detail in [7]).

Mass balance for intracellular metabolites that are packaged into EVs within source cells is described under the steady state assumption by Eqs. 4a and 5c, where C_i is the intracellular concentration of metabolite i; S_i is the stoichiometric vector corresponding to metabolite i and the parameter introduced to represent EV packaging flux is v_i^{exo}. For metabolites that exist in multiple compartments, mass balances in Eqs. 5a–5c are defined for each subcompartment the same way as they are for metabolites in single compartments (Eqs. 4a and 4b), however the balance for EV packaging flux is modified as in Eq. 5c to include contribution from both compartments.

$$\frac{dC_i}{dt} = S_i \vec{v} - v_i^{\text{exo}} = 0, i \in M_{1C} \quad (4a)$$

$$v_i^{\text{exo}} - rP_i = 0, i \in M_{1C} \quad (4b)$$

$$\frac{dC_{i,\text{cyt}}}{dt} = S_{i,\text{cyt}} \vec{v} - v_{i,\text{cyt}}^{\text{exo}}, i \in M_{2C} \quad (5a)$$

$$\frac{dC_{i,\text{mit}}}{dt} = S_{i,\text{mit}} \vec{v} - v_{i,\text{mit}}^{\text{exo}}, i \in M_{2C} \quad (5b)$$

$$v_{i,\text{cyt}}^{\text{exo}} + v_{i,\text{mit}}^{\text{exo}} - rP_i = 0, i \in M_{2C} \tag{5c}$$

Further, the isotopomer balance for metabolites which are measured in source cells and EVs via GC-MS is described in a generalized form in Eq. 6. The isotopomer distribution of metabolite i in the EV is a combination of isotopomer distributions derived from multiple compartments that is proportional to the packaging fluxes originating from each compartment.

$$v_{i,\text{cyt}}^{\text{exo}} \cdot \begin{pmatrix} y_0 \\ \vdots \\ y_{2^n-1} \end{pmatrix}_{i,\text{cyt}} + v_{i,\text{mit}}^{\text{exo}} \cdot \begin{pmatrix} y_0 \\ \vdots \\ y_{2^n-1} \end{pmatrix}_{i,\text{mit}} - rP_i \cdot \begin{pmatrix} y_0 \\ \vdots \\ y_{2^n-1} \end{pmatrix}_{i,\text{exo}} = 0 \tag{6}$$

Here, M_{1C} and M_{2C} are set of metabolites that exist in single compartment or two compartments, respectively; $v_{i,\text{cyt}}^{\text{exo}}$, $v_{i,\text{mit}}^{\text{exo}}$ refer to packaging fluxes of ith metabolite from the single compartment, cytosolic compartment or mitochondrial compartment, respectively; $\overrightarrow{y}_{i,\text{cyt}}$, $\overrightarrow{y}_{i,\text{mit}}$, $\overrightarrow{y}_{i,\text{exo}}$ are isotopomer distribution vectors (IDVs) of ith metabolite existing in the cytosol, mitochondria or EV. Equations 4a–6 are included in the Exo-MFA algorithm along with ^{13}C-MFA mass balance and isotopomer balance constraints. The objective function for Exo-MFA in is modified from the ^{13}C-MFA objective function to consider the error residuals of additional measurements in source cells and EVs, that is, EV secretion rate, metabolite levels in EVs, and *mass isotopolog distributions (MID)* in EVs.

1.2 Modeling Cargo Release in Recipient Cells

Exo-MFA modifies the metabolic model for recipient cells to include EV internalization and release of metabolite cargo that contributes to the endogenous metabolite pools. The rate of internalization is considered to be $u(t)$ (mg EV/mg protein/h). Internalization is assumed to be time-dependent due to dependence on extracellular concentration of EVs and the transient nature of the nutrient-deprived recipient cells. The content of EVs, however, is assumed to be consistent throughout the process of packaging, transport and internalization. Release of cargo into recipient cells is slightly more complex than packaging, since (1) not all metabolites are utilized in the same way, and (2) intracellular metabolic fluxes are not at steady-state. For this purpose, the cargo is categorized according to their utilization, (1) central carbon metabolites that are incorporated directly into central carbon metabolism, (2) essential amino acids that are incorporated only into biomass. The mass balance for central carbon metabolites are formulated in Eq. 7. Cargo release fluxes are proportional to rate of internalization $u(t)$ and the intra-EV composition P_i, therefore the flux term becomes $u(t)P_i$ (μmol/mg protein/h). Equation 7 describes the mass balance of metabolite i derived from EV cargo and S_i is the stoichiometric vector corresponding to that metabolite describing the inracellular reactions it is involved in. Equation 8 represents the isotopomer balance equations as described by the Isotopomer Mapping Matrix algorithm [10, 11], but with a small modification that includes the term $u(t)P_i y_{i,\text{ exo}}$ that represents the influx of isotopomers of metabolites derived from EV cargo.

$$\frac{dC_i}{dt} = u(t)P_i + S_i\vec{v} = 0, \quad i \in M_{CCM} \tag{7}$$

$$\frac{d}{dt}\left(C_i \begin{pmatrix} y_0 \\ \vdots \\ y_{2^n-1} \end{pmatrix}_{i,\text{cell}} \right) = u(t)P_i \begin{pmatrix} y_0 \\ \vdots \\ y_{2^n-1} \end{pmatrix}_{i,\text{exo}}$$

$$+ \sum_{j=1}^{N} S_{ij}.v_j \left(\prod_{\substack{k,\ S_{ij}>0 \\ S_{kj}<0}} IMM_{k\to i}. \begin{pmatrix} y_0 \\ \vdots \\ y_{2^n-1} \end{pmatrix}_{k,\text{cell}} \right) + \sum_{j=1,\, S_{ij}<0}^{N} S_{ij}. \begin{pmatrix} y_0 \\ \vdots \\ y_{2^n-1} \end{pmatrix}_{i,\text{cell}} = 0, \tag{8}$$

C_i is the total intracellular concentration of metabolite i. The objective function of Exo-MFA in recipient cells, is modified to include residuals of metabolite levels in EVs, and MID in both recipient cells and EVs.

2 Materials

1. Fetal bovine serum (FBS).
2. Ultracentrifuge.
3. Vacuum centrifuge.
4. Gas Chromatography-Mass Spectrometer (GC-MS).
5. HP-5MS or equivalent column.
6. Nanoparticle tracking analysis (NTA) instrument.
7. 0.2 μm filter.
8. Culture medium.
9. Isotope-labeled substrates.
10. Phosphate buffered saline (PBS).
11. RIPA Buffer.
12. Bicinchoninic (BCA) Kit.
13. Wako Glucose Kit.
14. HPLC-Grade Substrate Standards.
15. Methanol.
16. Norvaline.

17. Chloroform.
18. 0.9% saline in water
19. 2% methoxyamine hydrochloride in pyridine.
20. MBTSTFA + 1% TBDMCS.
21. Desktop computer or access to computational cluster with MATLAB license.
22. MATLAB 2016a or later, with licenses for Optimization and Parallel Computing Toolboxes.
23. Artelys Knitro 10 or later (or MATLAB's in-built Optimization Toolbox and Parallel Computing Toolbox).
24. Microsoft Office Suite (or any spreadsheet software that can save files in the. XLS or. XLSX formats).
25. Source code and files can be downloaded from http://nagrath.bme.umich.edu/exomfa_sourcecode/

3 Methods

3.1 Fetal Bovine Serum Depletion

1. Load fetal bovine serum into clean, sterile ultracentrifuge tubes. Tubes should be filled completely and balanced precisely. Centrifuge at 100,000–120,000 × *g* for 19 h at 4 °C (*see* **Note 1**) (Fig. 2a).
2. Carefully remove the light-colored upper layers into fresh tubes, discarding the lower dark layer and pellet. The upper layer consists of approximately 90% of the serum.
3. Sterile filter the depleted serum.
4. Use immediately or aliquot and store at −20 °C (*see* **Notes 2** and **3**).

3.2 Preliminary Characterization of Extracellular Vesicles and Cell Lines

1. Seed donor cells and grow to 70% confluency in recommended culture medium (*see* **Note 4**).
2. Aspirate culture medium and wash cells twice with PBS.
3. Add culture medium containing EV-depleted FBS.
4. After 48 h, remove the spent medium from the flasks into centrifuge tubes for EV isolation (*see* **Note 5**).
5. Quantify donor cells via protein assay after 48 h for normalization.

3.3 Isolation of Extracellular Vesicles from Conditioned Medium via Ultracentrifugation

1. Centrifuge spent culture medium containing EVs from **step 4** of the previous section at 300 × *g* for 10 min at 4 °C to remove cell debris.
2. Transfer the supernatant to fresh centrifuge tubes, careful to avoid disturbing the pellet. Centrifuge the supernatant at 10,000 × *g* for 30 min at 4 °C to remove larger vesicles.

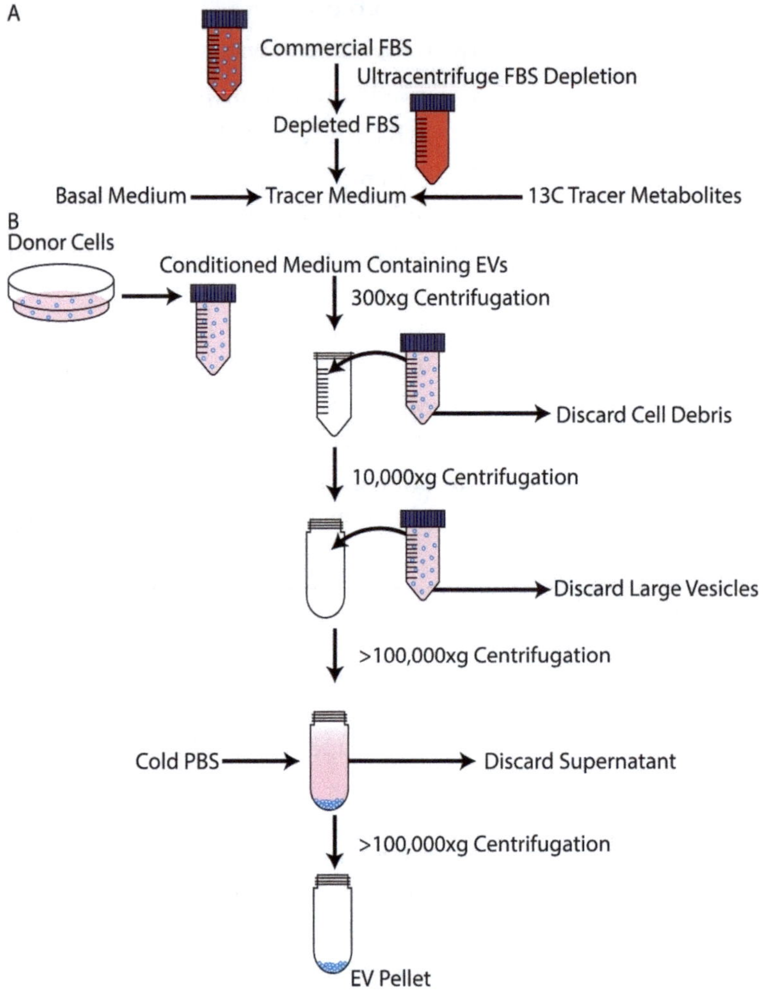

Fig. 2 (**a**) Depleted FBA is produced by ultracentrifugation of commercial FBS. The resulting solution can be used to supplement culture medium spiked with ^{13}C stable-isotope tracers to produce the stable-isotope labeled medium for production of EVs. (**b**) After culturing cells in EV-deprived stable-isotope tracer medium, EVS can be isolated via differential ultracentrifugation. After a series of increasing centrifugations, the purified EV pellet is finally collected at the bottom of the centrifuge tube where it can be resuspended and used for further analyses

3. Transfer the supernatant into clean, sterile ultracentrifuge tubes, careful to avoid disturbing the pellet. Tubes should be balanced precisely. If tubes are not full to maximum volume, dilute with PBS. Centrifuge at 100,000–120,000 × *g* for 90 min at 4 °C (*see* **Notes 1** and **6**) (Fig. 2b).
4. Aspirate the supernatant, careful to avoid disturbing the pellet. Resuspend the pellet in a full volume of PBS at 4 °C and repeat the centrifugation described in **step 3** with the tube in the same

orientation. Aspirate the supernatant again and resuspend the pellet of EVs in an appropriate volume of solution for the desired application (*see* **Note 7**).

5. For protein quantification, the pellet can be resuspended directly in 100 μL RIPA buffer and quantified via BCA assay following manufacturer instructions (*see* **Note 8**).
6. For particle number, the pellet can be resuspended in 100 μL of PBS and diluted for counting via Nanoparticle Tracking Analysis (*see* **Note 9**).
7. For treatment of cells, the pellet can be resuspended in 100 μL of culture medium and immediately applied to the recipient cells. After 24 h of treatment the recipient cells should be quantified via protein assay for normalization.

3.4 Isotopic Labeling of Donor Cells for Extracellular Vesicle Production

1. Seed donor cells and grow to 70% confluency in recommended culture medium (*see* **Note 4**).
2. Aspirate culture medium and wash cells twice with PBS.
3. Add culture medium with extracellular vesicle depleted FBS and isotope-labeled substrates of interest replacing the unlabeled at the original concentrations. Reserve 1 mL of medium for metabolic analysis and store at −80 °C (Fig. 3a).
4. After 48 h, remove the spent medium from the flasks into centrifuge tubes for EV isolation (*see* **Note 5**). Reserve 1 mL of medium for metabolic analysis and store at −80 °C.
5. Extract intracellular metabolites immediately. Wash with cold saline once and then quench the cells in cold methanol (*see* **Note 10**). Add an equal volume of water with 1 μg of norvaline and scrape cells thoroughly.
6. Pipet mixture into fresh tubes and add two volumes of chloroform. Vortex at 4 °C for 30 min and centrifuge at 5000 × *g* for 10 min at 4 °C. Remove the upper aqueous layer containing polar metabolites into a new tube. Dry via vacuum centrifugation. Store dried samples at −80 °C.

3.5 Isolation of 13-Carbon Labeled Extracellular Vesicles from Conditioned Medium via Ultracentrifugation

1. Centrifuge spent culture medium containing labeled EVs from **step 4** of the previous section at 300 × *g* for 10 min at 4 °C to remove cell debris.
2. Transfer the supernatant to fresh centrifuge tubes, careful to avoid disturbing the pellet. Centrifuge the supernatant at 10,000 × *g* for 30 min at 4 °C to remove larger vesicles.
3. Transfer the supernatant into clean, sterile ultracentrifuge tubes, careful to avoid disturbing the pellet. A separate tube containing 300 μg protein based on earlier quantification should be aliquoted for direct analysis of EV samples. Tubes should be balanced precisely. If tubes are not full to maximum

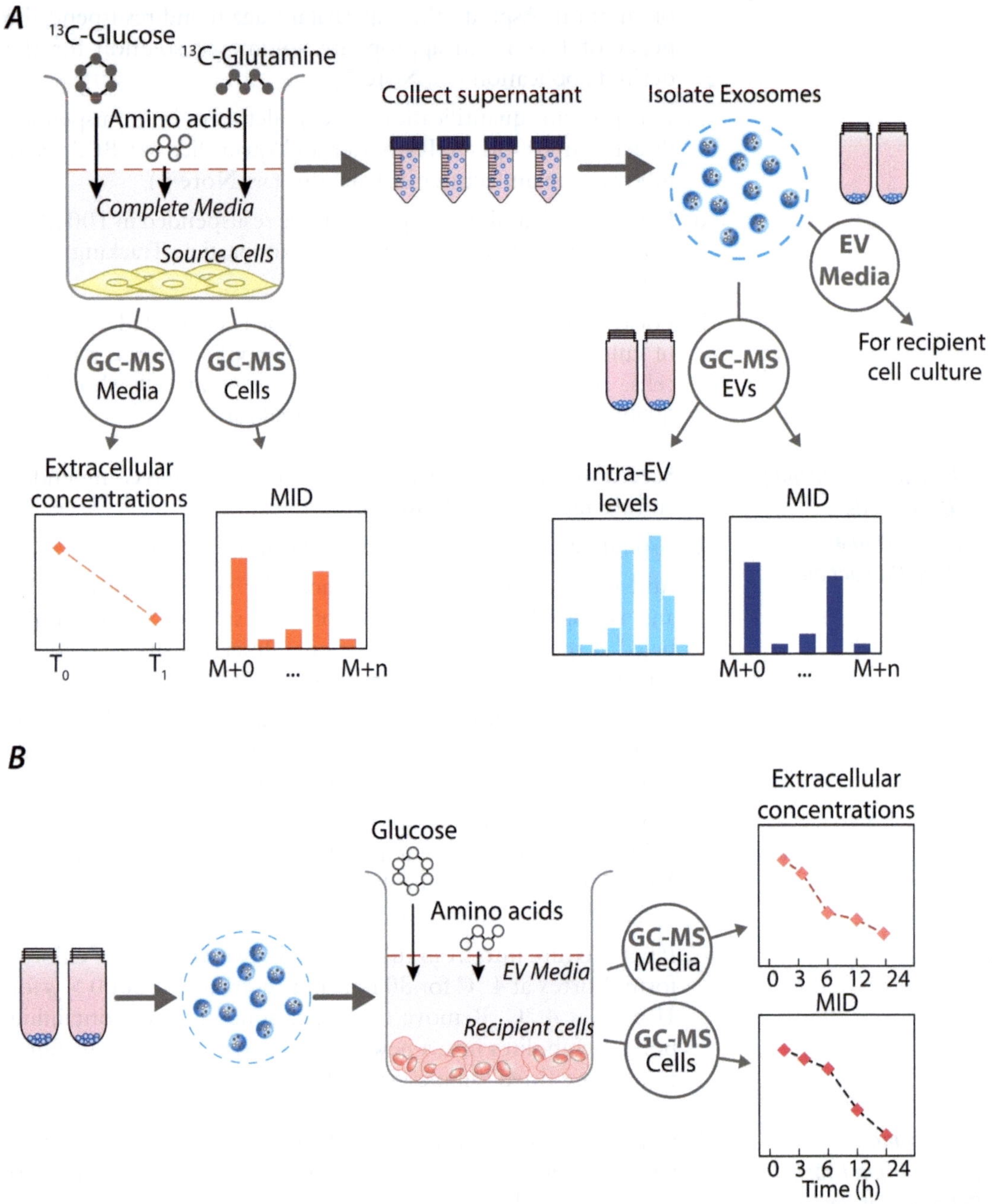

Fig. 3 (**a**) Source cells are cultured with stable-isotope labeled medium (e.g., ^{13}C-Glucose and ^{13}C-Glutamine). Spent medium with secreted EVs is collected and divided into two batches: the first batch is reserved for formulating culture medium for recipient cells and the second batch is analyzed using a GC-MS. (**b**) EVs carrying stable-isotope labeled metabolite cargo are introduced into culture medium for recipient cells. Cells are incubated in this medium and sampled at multiple time points within 24 h in order to capture the dynamic metabolic contribution of EV metabolites. Both spent culture medium and intracellular extracts are analyzed using the GC-MS

volume, dilute with PBS. Centrifuge at 100,000–120,000 × *g* for 90 min at 4 °C (*see* **Notes 1** and **6**) (Fig. 2b).

4. Aspirate the supernatant, careful to avoid disturbing the pellet. Resuspend the pellet in a full volume of PBS at 4 °C and repeat the centrifugation described in **step 3** with the tube in the same orientation.
5. For treatment of recipient cells, the pellet can be resuspended in culture medium with EV depleted FBS. The EVs should be diluted to approximately 200 μg protein/mL or 40 × 10^9 particles/mL for treatment based on earlier quantification. Reserve three replicate samples of EVs equivalent to 100 μg protein each resuspended in 150 μL water containing 1 μg norvaline for metabolic analysis, stored at −80 °C.

3.6 Treating Recipient Cells with Isotopically Labeled Extracellular Vesicles

1. Seed recipient cells at desired confluency in a 6-well plate and leave overnight. Then replace culture medium with fresh medium spiked with ^{13}C labeled EVs (Fig. 3b).
2. At 3, 6, 12, and 24 h extract intracellular metabolites. Reserve 1 mL from each well and store at −80 °C for metabolic analysis. Aspirate remaining medium. Wash with cold saline once and then quench the cells in cold methanol. Add an equal volume of water with 1 μg of norvaline and scrape cells thoroughly.
3. Pipet mixture into Eppendorf tubes and add two volumes of chloroform. Vortex at 4 °C for 30 min and centrifuge at 5000 × *g* for 10 min at 4 °C. Remove the upper aqueous layer containing polar metabolites into a new tube. Dry via vacuum centrifugation. Store dried samples at −80 °C.

3.7 Extracellular Vesicle Sample Preparation for Metabolic Analysis

1. Transfer 75 μL of cold methanol to each EV sample after thawing from −80 °C. Reserve 20 μL of the resulting solution for protein assay.
2. Add 150 μL of cold chloroform to each sample and vortex for 30 min at 4 °C. Centrifuge at 5000 × *g* for 10 min at 4 °C to separate the phases. Remove the upper aqueous layer containing polar metabolites into a new tube. Dry via vacuum centrifugation. Store dried samples at −80 °C.

3.8 Media Sample Preparation

1. Transfer 200 μL of medium collected from fresh ^{13}C tracer medium, donor cell conditioned medium, and recipient cell conditioned medium after thawing from −80 °C.
2. Add 10 μL of water containing 1 μg norvaline in water to each tube as internal standard.
3. Add 800 μL of prechilled methanol to each tube and vortex for 10 min.
4. Allow samples to deproteinize for 2 h at −20 °C.

5. Centrifuge at 14,000 × *g* for 10 min at 4 °C to collect protein at the bottom of the tube and transfer the supernatant to fresh tubes. Dry the samples via vacuum centrifugation and store the dried metabolites at −80 °C.
6. From remaining medium, quantify glucose via Wako Glucose Kit following manufacturer instructions.

3.9 GC-MS Analysis of Metabolites

1. Prepare external standard curve mixture including relevant amino acids, glycolytic intermediates, and TCA cycle metabolites via serial dilution (*see* **Note 11**).
2. Retrieve polar metabolite samples from storage at −80 °C including donor cells, EVs, recipient cells, and media samples. Dry samples and external standards via vacuum centrifugation briefly to remove condensation (*see* **Note 12**).
3. Derivatize samples by dissolving in 30 μL of 2% methoxyamine hydrochloride in pyridine, sonicating for 10 min, and incubating at 37 °C for 2 h.
4. Add 45 μL of MBTSTFA+1% TBDMCS and incubate at 55 °C for 1 h.
5. Transfer into vials containing glass inserts for GC-MS measurement.
6. The GC-MS should be equipped with an HP-5MS column or equivalent. The method parameters are as follows: helium carrier gas with flow of 1 mL/min. Injection volume of 1–2 μL at 270 °C. Oven temperature of 100 °C for 3 min raised at 5 °C/min to 300 °C over 40 min and then held for 5 min. Solvent delay of 6–10 min depending on when the initial saturating signal is observed to have dropped. MS source is set to 230 °C, and MS quadrupole is set to 150 °C. MS detector operated in scan mode from 50 to 450 m/z.

3.10 Estimating Extracellular Fluxes

1. Estimate the growth rate of source cells and recipient cells from assays performed in Subheading 3.2, **step 5** and Subheading 3.3, **step 7**. Estimate growth rate, μ, assuming exponential growth by fitting to Eq. 9.

$$X = X_0 e^{\mu t} \tag{9}$$

2. For *source cells*, using the measurement of extracellular metabolite concentrations from Subheading 3.4, **step 3** at the start time (fresh medium) and Subheading 3.4, **step 4** at time T (spent medium), estimate the extracellular flux v_{ext} from the integrated form of Eq. 10.

$$\int_{C_0}^{C_T} \mathrm{d}C = -v_{\text{ext}} X_0 \int_0^T e^{\mu t} \mathrm{d}t \tag{10}$$

3. Repeat **step 2** for all extracellular metabolites that you will consider in the Exo-MFA model.
4. For *recipient* cells, extracellular fluxes are measured soon after treatment with culture medium with EVs. Thus, their metabolism is not at steady state and their extracellular fluxes will not be constant over the course of the experiment. Their extracellular fluxes extracellular metabolites are derived analytically and integrated for a set of unknown parameters as in Eqs. 11a and 11b (*see* **Note 13**).

$$v_i(t) = a + bt \tag{11a}$$

$$\int \frac{\mathrm{d}C_i^{\text{ext}}}{\mathrm{d}t}\mathrm{d}t = -\int v_i(t) X_0 \mathrm{e}^{\mu t}\mathrm{d}t \tag{11b}$$

3.11 Using Exo-MFA to Quantify Intracellular and EV-Mediated Metabolic Flux

1. Build and curate a model defining the metabolic network in donor and recipient cells. A template for the model input files can be downloaded from supplemental files.
2. The first column is ignored by the Exo-MFA program, and is meant to help users identify reactions. The second column defines the reaction behavior: I, irreversible; R, reversible; E, exchange or boundary. The third column semantically defines the stoichiometry of each reaction. The fourth column defines the carbon atom transitions. Metabolites and compartments are defined by the user but must follow the format: Metabolite_Compartment (*see* **Note 14**).
3. The empirically measured data for *source cells* is compiled in an input Excel file, for which the template can be downloaded from supplemental files.
4. For *source cells*, enter the stable-isotope tracers used and their carbon labeling pattern in the sheet "tracer" (*see* **Note 15**). Enter measured extracellular fluxes in the first column of the sheet "flux," followed by the standard deviation and name of the metabolite (as referred to in the model file). Enter the mass isotopolog distribution data for intracellular metabolites and intra-EV metabolites in the first column of the sheet "MID," followed by the standard deviation of each measurement, name of the metabolite and the carbon chain. These are the measurements obtained in Subheading 3.9 and 3.10. In the sheet "exo," list the metabolite names (without denoting the compartment) that are detected in the EV cargo in the first column, followed by the compartments in the source cell which are possible sources of those metabolites (*see* **Note 16**). The third and fourth columns are the measured abundance and standard deviations of the metabolites in the EV cargo. In the "exo_rate" sheet, report the rate of secretion of EVs by dividing yield

of EVs (in protein or particle number) by the time over which the EVs were collected.

5. Load the "Main_ExoMFA_source.m" script in MATLAB (*see* **Note 17**) and run until all the steps are completed.
6. Exo-MFA is based on 13-carbon metabolic flux analysis (^{13}C-MFA) that utilizes Isotopomer Mapping Matrices (IMM) method to model 13-carbon atom transitions. The model is assumed to be at isotopic steady-state and is solved for intracellular fluxes, v, and mass isotopomer distributions, y.
7. Exo-MFA provides results for intracellular fluxes of source cells as well as packaging fluxes that represent metabolites packaged into EVs. The confidence intervals are also estimated by the script and published in an Excel file with the flux results. The MATLAB script also provides a file labeled "exo_IDV" that is used as an input for estimating intracellular fluxes in recipient cells.
8. The empirically measured data for *recipient* cells is compiled in an input Excel file, for which the template can be downloaded from supplemental files. There should be a separate input file for each time point at which the recipient cells were sampled.
9. For *recipient* cells, enter measured extracellular fluxes in the first column of the sheet "flux," followed by the standard deviation and name of the metabolite (as referred to in the model file). Enter the mass isotopolog distribution data for intracellular metabolites and intra-EV metabolites in the first column of the sheet "MID," followed by the standard deviation of each measurement, name of the metabolite, and the carbon chain. These are the measurements obtained in Subheading 3.9. In the sheet "exo," list the metabolite names (without denoting the compartment) that are detected in the EV cargo in the first column, followed by the compartments in the recipient cells, which are possible recipients of those metabolites (*see* **Note 18**). The sheet labeled "exo_IDV" should contain the data as published in the output file in **step 7**.
10. Load the "Main_ExoMFA_recipient.m" script in MATLAB and run until all the steps are completed. Repeat this step for every time point at which recipient cells were sampled.
11. Exo-MFA provides results for intracellular fluxes of recipient cells as well as cargo release fluxes that represent EV metabolites being internalized by cells. The confidence intervals are also estimated by the script, and an Excel file with the flux results is published. The Exo-MFA algorithm also estimates an additional parameter, $u(t)$, which represents the rate of EV internalization (*see* **Note 18**).

4 Notes

1. The precise centrifugation speed and time required to effectively precipitate a particle cannot be determined solely from the relative centrifugal force (RCF). Instead it must be determined by the geometry of the rotor as well as the speed of rotation expressed as the *k*-factor. The optimal speed for a given rotor should therefore be determined uniquely for each application. RCF quoted here is offered as a general guideline.
2. Depletion efficiency can be verified via Nanoparticle Tracking Analysis.
3. Depleted FBS can also be purchased.
4. EV production rates can vary dramatically by cell type and origin. Before beginning a new experiment, the donor cells should be characterized to select the optimal cell number. As a general guideline, high-production cells like cancer can be grown in as few as five T-75 flasks while low-production cells like fibroblasts may need as many as 15 T-160 flasks.
5. At least 48 h are necessary for EVs to accumulate in the medium. After longer than 48 h the EVs can be taken up again and degraded by the source cells, preventing increased yield.
6. The EV pellet can be difficult to observe. It is helpful to mark the centrifuge tube where the pellet is expected to assist in maintaining the orientation and avoiding disturbances.
7. EVs should be characterized according to the standards defined in the Minimal Information for Studies of Extracellular Vesicles [12]. This includes a determination of particle number and protein content as described here in addition to characterization of typical protein markers.
8. The micro BCA kit is specifically designed for low sample concentrations (0.5–20 μg/mL) and is particularly well suited for measuring low EV protein concentrations.
9. For an uncharacterized EV source, several dilutions must be tested to determine the optimal concentration for measurement. Particle number can be estimated from protein concentration using the ratio of 4.9 μg protein/10^9 particles but should be verified independently.
10. If PBS is used to wash in place of saline, the phosphate can be detected on the GC-MS and obscure the measurements.
11. The standard curve may need to be adapted based on the experimental conditions. A typical curve would span 0.1–10 nmol of each metabolite.

12. Any amount of water will prevent derivatization. It is imperative to ensure samples are completely dry before continuing with analysis.
13. For each extracellular metabolite $v_{ext}(t)$ was assumed to be a linear function of time. For our experimental design, the parameters were estimated by fitting Eq. 11a to extracellular concentrations measured at 0, 3, 6, 12, and 24 h of culture for recipient cells. The $v_{ext}(t)$ function can be a polynomial as well if the fit for a linear model is not good. However, additional parameters will need to be fitted for higher order polynomials, and users should take care that the number of measurements must be larger than the number of parameters being fitted.
14. We consider three compartments in the source cells: c, cytosol; m, mitochondria; x, extracellular; and d, a dilution source of unlabeled metabolites that contribute to EV cargo. The user can add more compartments or merge cytosolic and mitochondrial compartments, depending on the complexity of the model. The metabolites from compartment "d" only affect the 13-carbon enrichment of EV cargo and do not interfere in the estimation of packaging flux metabolites from source cells' cytosolic and mitochondrial compartments to EV cargo. This is included to account for the production of EVs when intracellular metabolites are not saturated with 13-carbon from the stable-isotope tracers.
15. Tracer labeling pattern follows a binary pattern (e.g., U-$^{13}C_6$-Glucose containing six 13-carbon atoms is represented as "111,111" and 5-$^{13}C_1$-Glutamine is represented as "00001").
16. Zhao et al. and Achreja et al. have previously shown the presence of amino acids, glycolytic intermediates, and TCA cycle intermediates which exist in multiple compartments in source cells. Further, the mass isotopolog distribution of metabolites known to be present multiple compartments did not match distributions observed in EVs, indicating that EVs can package metabolites from mitochondrial compartments in addition to cytosolic compartments. Given these observations, and the incomplete knowledge of metabolite packaging in EVs, Exo-MFA lets the user define which intracellular compartments can contribute to EV cargo.
17. Exo-MFA was developed with MATLAB 2016a and Artelys Knitro Solver 10.0. Using earlier or later versions of these software may lead to errors due to incompatibility.
18. Internalization is assumed to be time-dependent due to dependence on extracellular concentration of EVs and the transient nature of the nutrient-deprived recipient cells. The cargo of EVs is consistent throughout the process of packaging, transport and internalization. Release of cargo into recipient cells is

slightly more complex than packaging, since (a) not all metabolites are utilized in the same way, and (b) intracellular metabolic fluxes are not at steady state. Therefore, the EV cargo is categorized according to its utilization, that is, (a) central carbon metabolites that are incorporated directly into central carbon metabolism which are either in the cytosol or mitochondria, and (b) essential amino acids that are incorporated only into biomass.

Acknowledgments

This work was supported by grants from the National Institute of Health (R01-CA222251, R01-CA204969, and R01-CA227622) awarded to D.N.

References

1. Rajagopal C, Harikumar KB (2018) The origin and functions of exosomes in cancer. Front Oncol 8:66–66
2. Ruivo CF et al (2017) The biology of cancer exosomes: insights and new perspectives. Cancer Res 77(23):6480
3. Wu M et al (2019) Emerging roles and therapeutic value of exosomes in cancer metastasis. Mol Cancer 18(1):53
4. Zhao H et al (2018) The key role of extracellular vesicles in the metastatic process. Biochim Biophys Acta Rev Cancer 1869(1):64–77
5. Stoorvogel W (2012) Functional transfer of microRNA by exosomes. Blood 119(3):646
6. Yu X, Odenthal M, Fries JWU (2016) Exosomes as miRNA carriers: formation-function-future. Int J Mol Sci 17(12):2028
7. Achreja A et al (2017) Exo-MFA – a 13C metabolic flux analysis framework to dissect tumor microenvironment-secreted exosome contributions towards cancer cell metabolism. Metab Eng 43(Pt B):156–172
8. Dey P et al (2017) Genomic deletion of malic enzyme 2 confers collateral lethality in pancreatic cancer. Nature 542:119
9. Yang L et al (2016) Targeting stromal glutamine synthetase in tumors disrupts tumor microenvironment-regulated cancer cell growth. Cell Metab 24(5):685–700
10. Zupke C, Stephanopoulos G (1994) Modeling of isotope distributions and intracellular fluxes in metabolic networks using atom mapping matrices. Biotechnol Prog 10(5):489–498
11. Schmidt K et al (1997) Modeling isotopomer distributions in biochemical networks using isotopomer mapping matrices. Biotechnol Bioeng 55(6):831–840
12. Thery C et al (2018) Minimal information for studies of extracellular vesicles 2018 (MISEV2018): a position statement of the International Society for Extracellular Vesicles and update of the MISEV2014 guidelines. J Extracell Vesicles 7(1):1535750

Chapter 11

Comparative Metabolic Network Flux Analysis to Identify Differences in Cellular Metabolism

Sarah McGarrity, Sigurður T. Karvelsson, Ólafur E. Sigurjónsson, and Óttar Rolfsson

Abstract

Metabolic network flux analysis uses genome-scale metabolic reconstructions to integrate transcriptomics, proteomics, and/or metabolomics data to allow for comprehensive interpretation of genotype to metabolic phenotype relationships. The compilation of many Constraint-based model analysis methods into one MATLAB package, the COBRAtoolbox, has opened the possibility of using these methods to the many biologists with some knowledge of the commonly used statistical program, MATLAB. Here we outline the steps required to take a published genome-scale metabolic reconstruction and interrogate its consistency and biological feasibility. Subsequently, we demonstrate how mRNA expression data and metabolomics data, relating to one or more cell types or biological contexts, can be applied to constrain and generate metabolic models descriptive of metabolic flux phenotypes. Finally, we describe the comparison of the resulting models and model outputs with the aim of identifying metabolic biomarkers and changes in cellular metabolism.

Key words Constraint-based metabolic models, Genome-scale reconstruction, Flux balance analysis, Transcriptomics, Metabolomics, Systems biology, Data integration

1 Introduction

Cell metabolism is an important indicator and controlling factor of cell function in many situations. Recent advances in metabolomics measurement techniques have provided an increasing number of comprehensive data sets. This is in parallel with the increasing detail of genome annotations, and transcriptomic and proteomic data sets that provide detailed readouts of whole cell state. With the increasing availability of large-scale data sets of various data types both the need and the possibility of their integrated analysis has become apparent. An increasingly common method for this integration is the building of a genome-scale metabolic reconstruction (GEM), a way of cataloguing the metabolic genes annotated to an organism [1, 2]. This GEM may then be constrained and modeled using

Deepak Nagrath (ed.), *Metabolic Flux Analysis in Eukaryotic Cells: Methods and Protocols*, Methods in Molecular Biology, vol. 2088, https://doi.org/10.1007/978-1-0716-0159-4_11,

Constraint-based metabolic analysis (COBRA) methods [2–6]. COBRA methods can use transcriptomics, proteomics, and metabolomics data in order to recapitulate the metabolism of a particular cell type within an organism or the metabolism of an organism or cell type in a particular context [5, 7–13].

Reconstructions may not always have been curated to the extent that they are biologically and chemically feasible. For example some reactions may be listed as reversible that should in fact be able to operate in only one direction. Also transport reactions, especially nonenzymatic transport between organelles, may be missing or incorrectly assumed to exist. These sorts of problems may lead to issues for example loops of cofactors being resynthesized incorrectly. There are two approaches to addressing these problems; one is manual comparison of model reactions to the existing literature, databases of reaction information and other curated models, the other is to use various CobraToolbox functions to assess the model and check for such features. These methods should be applied in an integrated fashion and a researcher's judgment must be applied to finding a sensible reconstruction to base the rest of their modeling on. The CobraToolbox website offers various strategies for checking that a model is operating in a biochemically sensible manner in the various reconstruction tutorials. Below some of the key options are demonstrated using RECON2M2 [14] as an example. We have compiled this methodology based on the tutorials provided on the CobraToolbox website and our own experience. The following tutorials have been particularly useful in compiling this methodology and in model building in general and may be useful to find more background information on the methodology described below; "Atomically resolve a metabolic reconstruction" [15], "Proton shuttle testing with sparse flux balance analysis" [16], "Sparse flux balance analysis test for a minimal stoichiometrically balanced cycle involving ATP hydrolysis" [17], "FastGapFill Tutorial" [17], "Example use of functions listed in the Standard operating procedure for metabolic reconstruction" [18], "Test physiologically relevant ATP yields from different carbon sources for a metabolic model" [19], "Testing chemical and biochemical fidelity" [20], "Testing basic properties of a metabolic model (aka sanity checks)" [21].

To investigate context specific metabolism more constrained models may be built. The metabolic capacity of a cell is defined by which enzymes it is currently expressing, a subset of all of the enzymes encoded in its genome. It is possible to obtain a snapshot of the enzymes expressed in a particular cell type under given circumstances using transcriptomic and proteomic data. Both of these data types are somewhat correlated with enzyme activity [22, 23]. It should, however, be noted that the presence of mRNA or even protein in a data set does not necessarily guarantee enzymatic activity, and certainly does not necessarily give an absolute indication of the magnitude of enzymatic activity. Given these caveats various methods to prune a genome-scale metabolic reconstruction based on expression data, mostly transcriptomics, have been developed. Opdam et al. identified three categories of model extraction method with different characteristics [7]. The first group, "GIMME-like," includes only GIMME and relies on the minimization of flux through reactions that are associated with lowly expressed genes [5]. The group described second, "iMAT-like," includes iMAT, and INIT [24, 25]. This group optimizes the compromise between removing reactions linked to poorly expressed genes with the retention of reactions linked to highly expressed genes. The third "MBA-like" group of MBA, Fastcore, and mCADRE, define active core reaction sets and then remove as many as possible noncore reactions [8, 9, 26]. Another review by Herrgard and Machado identifies the fact that some methods including GIMME force a minimal flux through the objective, while others, such as iMAT, do not and that when no objective flux is forced the model often will not "grow" [4]. Both reviews note that the selection of parameters such as expression cutoff greatly affects the model output [4, 7]. The CobraToolbox has implemented six methods for the integration of expression data to contextualize genome-scale metabolic reconstructions in a single function createTissueSpecificModel. Others may be downloaded as separate MATLAB packages or part of the RavenToolbox [10]. Further background information on the use of this function is available in a CobraToolbox tutorial [27]. It is possible to use either data that has been produced specifically for a project or obtained from either the supplementary data of a paper or from a public database listed in Subheading 2.

The function createTissueSpecificModel takes a base model (normally a genome-scale reconstruction that has had some uptake constraints based on medium composition), and some form of expression data, either arranged per reaction or per gene or as a list of reactions to be listed as core, and creates a tissue specific model using one of the methods described in **Notes 25–30**.

COBRA models may also be improved by the application of constraints from metabolomics measurements. Metabotools is a set of functions that are now implemented within the CobraToolbox which may be used to constrain COBRA models with quantitative or semiquantitative metabolomics data such as that obtained from mass spectrometry experiments [11].

Metabotools offers a method for applying semiquantitative data based on the relative uptake and secretion rates between two different contexts (cell types or situations). This requires the relative abundances of various metabolites linked to exchange reactions in the model with a limit of detection per metabolite, although it does not require quantified data for each metabolite. The semi-quantitative method relies, to a large extent, on relative differences between two conditions and therefore is most suitable for creating pairs of related models for comparison. For a complete description of the original method and its background read the Metabotools I tutorial [28]. Metabotools may also be used to apply data from quantitative metabolomics data this was originally described in the Metabotools II tutorial [29].

Once accurate metabolic reconstructions have been built, COBRA models may be analyzed with a variety of techniques to elucidate the metabolic characteristics of the cell type or context of interest. Perhaps the most common of these techniques is flux balance analysis (FBA) and its variants [6]. FBA predicts the fluxes through the reactions of a model when the flux through an objective reaction, often a reaction known as the biomass reaction modeling cell growth, is optimal. This assumption of optimal growth simplifies the computation of fluxes; however, it is unlikely to be a biologically valid assumption for the majority of cell types from multicellular organisms [30, 31].

Other metabolic flux analysis methods do not assume optimality in the biomass reaction but are able to assess the possible fluxes through the system for a range of circumstances, which may or may not require the presence of a biomass function. One of these possible techniques is flux variability analysis (FVA). FVA produces the minimum and maximum possible fluxes through each reaction of the metabolic network [32, 33]. Another is the analysis of repeated random sampling of feasible fluxes through the metabolic network. Although the random sampling approach does not guarantee the finding of extreme flux values, it does provide more information about the likely distribution of metabolic fluxes between the extreme values [34–37]. Random sampling of the possible flux space of a metabolic model avoids the potential biases of FBA toward metabolic functions associated with the biomass objective. In order to fully explore the space sufficient samples must be made to ensure that it is uniformly covered. The CobraToolbox provides various methods to implement this sampling [34–37] which are discussed in a tutorial, this methodology focuses on one we have found straightforward to use [38].

As well as analyzing single models to capture and describe metabotypes at specific environmental or nutrient conditions, it is possible to use COBRA techniques to compare models of two different cell types or cells under different biological contexts. It is possible to do this both by comparing the outputs from the analyses described above or by applying techniques specifically designed to compare COBRA models. One of the most widely used of these techniques is minimization of metabolic adjustment (MOMA) [39]. This technique reveals the minimum set of alterations needed to make one model as metabolically similar to another as possible, without necessarily producing an optimal flux through the biomass reaction. For two given GEMs, MOMA predicts an affected phenotype by minimizing the adjustment of the original phenotype. This reveals the suboptimal flux distribution of the affected GEM which is intermediate between the original nonaffected flux distribution and the optimal one. Ultimately, the analysis reveals what reactions (and subsequently, pathways) need adjustment for one model to take on a phenotype more similar to another model. Quadratic programming is used to minimize the Euclidean distance between the original and the affected flux distributions. MOMA analysis is available in the Cobra Toolbox [40].

All of the above methods rely on the appropriate formation of one or more context-specific COBRA models based on a species-specific GEM constrained with some combination of transcriptomics, proteomics, and metabolomics data. This chapter will describe the process of constraining, checking, and curating a pair of context-specific COBRA models with transcriptomic and metabolomics data primarily using the COBRAtoolbox [3, 41, 42] in MATLAB (Mathworks). The method is accompanied by a MATLAB script that can be followed accordingly. In order to understand changes to metabolism between the pair of metabolic phenotypes, comparative analysis of the metabolic reaction fluxes within the two models is demonstrated affording deliverables similar to those shown in Fig. 1a–c. These can then be used for data presentation, to update metabolic hypotheses and mathematically interrogate metabolic differences in silico.

2 Materials

2.1 Constraint-Based Modeling Software

2.1.1 CobraToolbox

The CobraToolbox is a MATLAB based package intended to simplify the implementation of Constraint-based metabolic modeling methods [3, 41, 42]. CobraToolbox v.3 has recently been released (2019) [41]. The CobraToolbox provides a compilation of methods for building, checking, manipulating, and integrating data with GEMs and COBRA models.

This comprehensive package allows the user to build or input a network reconstruction or model. The model can then be subjected to a variety of data integration methods so that it includes information on context-specific gene expression and metabolite production and utilization. The context-specific model may then be subject to analysis by FBA, random sampling, and other methods. The CobraToolbox also provides some functions that enable the user to present the data in the form of figures and tables.

2.1.2 Other Software

See **Notes 1** and **2**.

Building GEMs and COBRA models is a means of integrating data from multiple types of large-scale experiments. This is a useful function as it allows for a more comprehensive and contextualized

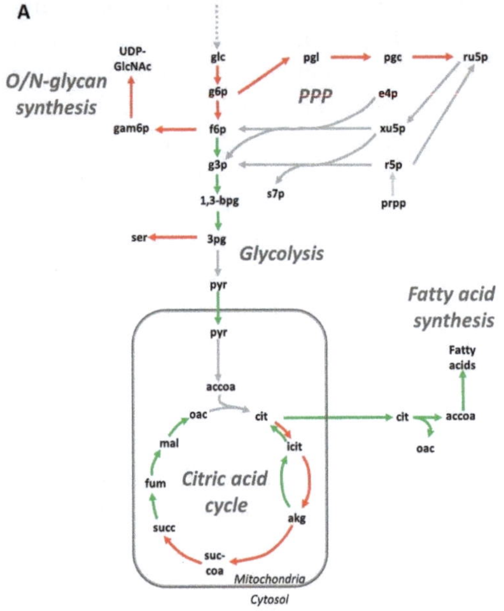

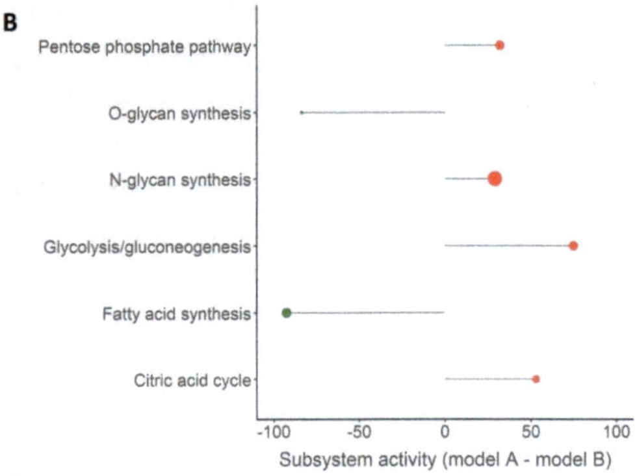

C

Genes essential for model A	
SLC16A10	Solute carrier family 16 member 10
AGXT	Serine-pyruvate aminotransferase
GSS	Glutathione synthetase
Genes essential for model B	
PDHX	Pyruvate dehydrogenase protein X
ASL	Argininosuccinate lyase
ASS1	Argininosuccinate synthase

Fig. 1 An example of output from the analysis and comparison of two metabolic network models. (**a**) A flux map showing relative reaction activity within selected reactions of central carbon metabolism. Red indicates higher activity in model A whereas green indicates higher activity in model B. The flux values are based on histograms of random sampling results (*see* Section x), where only significantly different reactions are used. (**b**) Pathway activity based on results from (**a**). *Y*-axis shows the subsystems shared by the models and the *X*-axis represents percentage, where +100% means that all reactions within a given subsystems are more active in model A, and − 100% the same for model B. Size of bubbles within graph represents size of subsystems. (**c**) Results from single gene deletion analysis (*see* Section vi and vii) which gives information about genes exclusively essential for feasibility of model A or model B. This provides cell-type specific gene targets which can be further validated. These analyses are based on methodology previously used by Johannsson et al., Halldorsson et al., and McGarrity et al. [142–144]

2.2 Databases of Publically Available Data Sets and Models

analysis of these often seemingly overwhelming data types. However, this integrated analysis requires multiple data types that it may well be outside the scope of a single project to collect. For this reason, a reliable source of complementary data is necessary, as large-scale data sets have become both richer and more common, various repositories have been set up to allow the sharing, and therefore better utilization, of these data sets. Increasingly the publication of journal articles about such large-scale studies is dependent on depositing the data in a repository and this has meant that there has been an increasing availability and improved curation of the data in recent years. *For full details see* **Notes 3–18**.

2.2.1 Array Express [43]

Array express is a database of publically available transcriptomic data [44]. Processed data is downloadable in a tab-delimited format for most experiments. Array Express data is also downloadable via a Bioconductor package in R [45].

2.2.2 Gene Expression Omnibus [46]

The Gene Expression Omnibus (GEO) is a repository for microarray and next generation sequencing data sets [47].

2.2.3 PRIDE [48]

The PRIDE (Proteomics Identifications) archive is a database of mass spectrometry proteomics data [49]. Data is available via the website as well as a stand-alone inspection tool.

2.2.4 Peptide Atlas [50]

Peptide Atlas is a database of information from high-throughput proteomics experiments [51]. Peptide information may be accessed via the database web page and is linked to ENSMBL identifiers [52].

2.2.5 MASSive [53]

A database of mass spectrometry-based proteomic data [53]. The website offers both analytical tools and the ability to download data sets.

2.2.6 iProX [54]

A repository for proteomics data [55]. Proteomic data sets may be downloaded.

2.2.7 Japan ProteOme STandard Repository [56]

A repository for proteomic mass spectrometry data in either raw or processed format [57]. It may be accessed via an online interface.

2.2.8 The Human Protein Atlas [58]

The human protein atlas provides information about the tissue, cell, organelle, and pathological condition specific expression of proteins in humans and on mouse brain protein expression [25].

2.2.9 MetaboLights [59]

Metabolights is a repository for metabolomics data sets [60].

2.2.10 Genome RNAi [61]

A repository for RNAi screens in Drosophila and human cell lines [62]. It may be useful for confirming the outputs of gene and reaction essentiality described below.

2.2.11 The Database of Genotypes and Phenotypes, The European Genome-Phenome Archive, and The Japanese Genotype-Phenotype Archive [63–65]

A repository for data sets describing genotype–phenotype interactions in human derived data [66]. Other similar databases are the European Genome-Phenome Archive [67] and The Japanese Genotype-Phenotype Archive [62]. These databases may be useful for confirming the results of models of human metabolism.

2.3 Species-Specific Genome-Scale Metabolic Reconstructions

2.3.1 Human

There are two main efforts to build, maintain, and refine a human genome-scale metabolic reconstruction Recon and the Human Metabolic Reconstruction [14, 68–71]. The current version of Recon is Recon3D [69] which attempts to account for a high number of reactions and the multiple tissues types. It uses some of the reactions of fatty acid metabolism from HMR2. HMR2 [72] is the most recent version of the human GEM called HMR or human metabolic reconstruction [73], it has been used extensively examine cancer metabolism particularly hepatocellular carcinoma. The details and downloadable versions of the Recon models may be accessed at the BIGG database website [74] and the vmh website [75, 76]. The HMR models and details may be accessed at the Human Metabolic Atlas website [77].

2.3.2 Other Mammalian

Metabolic reconstructions have been built for Chinese Hamster Ovary Cells mostly aimed at improving their productivity as cell factories there has been a recently published consensus GEM published [26]. There have also been various efforts to produce a mouse metabolic reconstruction. One largely based on the human reconstruction Recon1 [78] and a tissue-specific set by the group responsible for HMR [79]. This is important, as mice are a commonly used experimental system.

2.3.3 Plants

Metabolic reconstructions have been made for various plant species [80]. Most extensively there are multiple models of *Arabidopsis thaliana* [81, 82], rice [83], tomatoes [84], and maize [85, 86]. Class-based reconstructions are also available [87].

2.3.4 C. elegans

Several metabolic reconstructions of *Caenorhabditis elegans*, some strain specific, have been created and recently there has been an effort called WormJam to use these to produce a consensus reconstruction [88–92].

2.3.5 Bacteria

Metabolic reconstructions exist of a wide variety of bacterial species, indeed optimization of bacterial production was one of the original applications of COBRA techniques in biology. These have been applied to the synthesis of valuable chemicals and assessment of likely therapeutic targets [93–98]. Recently there has been an increasing effort to model communities of bacteria, such as gut microbiota, to better understand how metabolic interactions can support differing community compositions and interactions with hosts [99–101].

2.4 Metabolic Knowledge Databases

2.4.1 KEGG [102]

KEGG is a database of genes and genomes, as well as metabolite, metabolic pathways and drugs [52].

2.4.2 MetaCyc [103]

MetaCyc is a database of metabolic reactions, enzymes, and related chemicals [52]. The database can be searched online or has files available for download.

2.4.3 BRENDA [104]

BRENDA is a large collection of information on enzymes, their activity, inhibition, and expression [105, 106].

2.4.4 Reactome [107]

Reactome is an EBI maintained comprehensive database of biological reactions including metabolic pathways [105].

2.4.5 BioModels Database [108]

The biomodels database is a repository for biological models including COBRA models [109]. It allows the download of models in the systems biology mark-up language (SBML) [110], this is compatible with the CobraToolbox. It includes the path2models set based on metabolic databases [111].

3 Methods

This method is based on the CobraToolbox version 3 [41] in MATLAB 18b in Windows, with other MATLAB functions as necessary [112].

1. MATLAB is available from the MathWorks website and requires a licence (*see* **Note 19**) [113].
2. It is also necessary to install extra solvers (*see* **Note 20**) [114, 115].
3. The easiest way to install CobraToolbox from Windows is to download Gitbash from its website [116] choosing the "Use Git Bash and optional Unix tools from the Windows Command prompt" and "Checkout as-is, commit Unix-style line endings" options during the installation. Then in gitbash follow *line 3* of the accompanying script as described in the installation instructions [117].
4. When this is completed open MATLAB, add the CobraToolbox folder to the MATLAB path and use the function in *line 5* of the accompanying script to initiate the CobraToolbox. This function must be repeated at the beginning of each MATLAB session when the CobraToolbox is used.
5. A base model for the appropriate species should be selected and downloaded, *see* **Note 21** for considerations in making the selection.
6. Once a model has been selected and downloaded to a folder in the current MATLAB path (and the CobraToolbox has been initiated as above) it is necessary to load it into MATLAB. The function used to do this depends on the file type downloaded. For .xml files, containing models encoded in SBML use the read SBML function as in *line 11*. The variable name selected for the model may be anything but it should be distinct from any other model loaded and versions created later. Also it may be necessary to add a file path before the file name if it is not in the folder currently set as the current path, as in *line 12*.
7. For COBRA format models in .mat files use the read Cobra model function as described in *line 15*, as with the read SBML function a folder's path may be added as part of the file name. By using the "varagin" options in this function other model

formats such as correctly set up Excel files and Simpheny files may be loaded using this function.

8. First check the model has a biomass, or growth function, and that this is set as the objective. This is done most easily as shown in *line 22*. This will save the name of the biomass function as the variable name selected. The reaction name will also be displayed in the workspace, as will a table showing the composition of the objective reaction, metabolites and their stoichiometries will be displayed. Most components of this reaction will have a negative stoichiometry indicating that they are being consumed; however, ADP, inorganic phosphate, and possibly protons will be produced (positive stoichiometry) indicating energy consumption as ATP cleavage.
9. To see the reaction formula of the objective then follow *line 24*, this function may also be used to display other reactions within a model. The components and their stoichiometry should be checked against available literature to ensure they are biologically reasonable, if not then follow the procedure below to add a different biomass function.
10. If the reaction named as the current objective is not the biomass reaction, or for other analyses than those relying on cell growth, another objective is desired then follow *line 28* to select another existing reaction as the objective.
11. If necessary add a biomass reaction to the model, as in *line 26* and then use *line 28* to select this as the objective function. It may also be desirable to remove the previous biomass reaction to avoid confusion. This can be done simply by deleting the relevant reaction as in *line 30*.
12. Further exploration of the biomass function may be desirable. For example it may be interesting or necessary to check how the metabolites in the biomass reaction are produced. Most straightforwardly a function called "surfNet" allows the user to navigate a model in the MATLAB workspace beginning at any reaction or metabolite, including the biomass reaction, as in *line 32*.
13. An alternative method would be to find all of the reactions associated with one of the metabolites consumed by the biomass reaction, for example tryptophan, as in *line 34*. Sort out which of these reactions are not the biomass reactions, in *line 35*, then display the formulas of these reactions as in *line 36*.
14. Next check that all of the metabolites in the model are able to be produced and consumed and that all reactions can carry flux. Firstly find a list of the ID number of dead end metabolites, unconsumed or unproduced metabolites, as in *line 38*.

15. Then find the relevant metabolite abbreviations, shown in *line 39*.
16. Next identify the reactions linked to the dead end metabolites as in *line 40*.
17. Following this check the reaction bounds of these reactions, as in *lines 41* and *42*, as it may be that they are set to 0 or the reaction is incorrectly reversible or irreversible. A reaction is reversible if the lower bound is negative and the upper positive.
18. It is also possible to identify reactions that are unable to carry flux as in *line 43*, gaps that leave a section of the model isolated as in *line 45*.
19. Following this it may be useful to repeat lines *41 and 42* with the blocked reactions variable from *line 43*.
20. To identify the areas of metabolism affected by gaps found in *line 46* use the "surfNet" function described above in *line 32*. If there are unlinked metabolites and inactive reactions then this should addressed; either by removing metabolites or reactions that are unused, or additional reactions should be added to ensure functionality. This may be accomplished by using the same functions for addition and removal of reactions shown in *lines 26* and *30* using information from the literature or from the databases described above. *For alternative method see* **Note 22**.
21. By checking that ATPsynthase is inactive when there are no metabolites able to be taken up it is possible to identify infeasible loops of reactions created by reaction reversibility, particularly of transport reactions of cofactors. ATP production should require some metabolic input, if it does not it indicates that some metabolites, often cofactors or protons are forming infeasible loops. These may be elucidated by performing sparse flux balance analysis (FBA) with ATPsynthase as the objective function. This process is has been derived from CobraToolbox tutorials [16, 118]. A detailed explanation of the theoretical background to these ideas and their ideal implementation is also available in Fleming et al. and Schellenberger et al. [119, 120].
22. In RECON2M2 the ATPsynthase reaction is called 'ATPS4m'. In order to check the ability of the model to make ATP with no input we will first identify uptake reactions in *lines 52–54* then close them in *lines 56–58*.
23. Then set up the parameters to perform FBA with the ATPsynthase as the maximization target in *lines 61–65*.
24. SparseFBA, finding the minimum set of active reactions for an optimal objective [121], is then performed in *line 66* the expected objective of this FBA is 0, this is displayed in the workspace by *line 67*.

25. Further examination of this output, in *lines 70–74*, identifies the reactions which are carrying flux above the cutoff value 1e–6, and displays their reaction formulas.
26. It is likely that some of these are forming inappropriate loops and should be made irreversible, as in *line 76*. The following variations on the above procedure will help to identify which reactions it is necessary to alter.
27. To gain further clues as to the location of the infeasible loops the above sparseFBA should be repeated with varying combinations of open and closed fluxes.
28. First closed exchange fluxes but all internal reactions open as in *lines 79–83*.
29. It is likely some ATPsynthase flux will be present as all possible reversible reactions are present, to see where there is flux in this case repeat *lines 70–74* but with the model created in *lines 80* and *81*.
30. Next repeat the process with closed exchanges but fully open transport reactions between various compartments, the set up for this is described in *lines 86–91*, and repeat *lines 70–74* with the new model.
31. In this case it is likely that there will be flux through ATP-synthase, despite the lack of allowed uptake of metabolites. This is due to the fully open transport reactions forming infeasible loops shuttling cofactors between compartments in a way that is unlikely in biology.
32. Finally, to examine which transport reactions between compartments form loops that allow the formation of ATP with no metabolites being taken up it is necessary to run sparseFBA for ATPsynthase with exchanges closed and transport reactions between the cytoplasm and the mitochondria, then cytoplasm and golgi, endoplasmic reticulum, and peroxisome and finally between the extracellular space and the cytoplasm open. *Lines 94–97* set up which pair of compartments will be tested and will need to be rerun for each pair of compartments with *lines 94* and *95* altered to represent the correct compartments.
33. Then open the selected set of reactions in the closed exchange model and run sparseFBA in *lines 100–104*. In the version of the model with all fluxes open between each compartment pair it is likely that there will be some flux through the ATPsynthase reaction.
34. Which reactions have flux in each state can be determined by rerunning *lines 70–74* with the model of interest.
35. Once reactions that are forming active shuttles are identified by this method change their bounds to irreversible as in *line 76* but with the otherwise open model and then run the

sparseFBA with the output, eventually a state should be found with no ATPsynthase flux. *For alternative methods see* **Note 23**.

36. It is also useful to check that various other impossible metabolic functions either do or do not occur. For example water or oxygen alone should not create ATP or biomass.

37. Glucose and oxygen should, however, be able to produce ATP in the mitochondria via the electron transport chain. This is accomplished by opening and closing uptake reactions to represent different states and assessing the productions of biomass or ATP.

38. Begin by creating a closed version of the model that has had infeasible loops removed created in in *lines 109–118*, this model allows for secretion of metabolites but not uptake.

39. Then use this model to check if ATP is produced from water by changing the objective to the demand reaction for cytoplasmic ATP, *line 121*, allowing creation of ATP, *line 122*, opening only the water uptake flux, *line 123*, and running FBA in *line 124*, the .obj of the variable FBAatph2o should be 0.

40. The process should be repeated for ATP production from water and oxygen as in *lines 126–131*. "FBAatph2oo2.obj" in *line 131* should also be 0.

41. Then check if the closed model produces protons in *lines 134–138*, this should be repeated with m replaced by c in *lines 135–137* to check both cytoplasmic and mitochondrial protons. Again "FBAhm.obj" should be 0.

42. If any of the above "FBA.objs" is positive then running *lines 70–74* with the relevant FBA and model will illustrate the paths taken and suggest where reactions may need to be made irreversible or removed as in *line 76* or *30*.

43. Whilst it is important to exclude nonrealistic reactions it is also useful to check that desirable metabolic functions are possible. This is straightforward to do for over 100 common human metabolic features using the "test4HumanFctExt" functions, in *line 140*. As at this stage a generic model is being built it should be able to perform all or most of these functions. For some of them such as aerobic glucose metabolism to produce ATP it should also be noted what rate of production is occurring, 32 ATP per mole is expected from glucose under aerobic conditions. *For an alternative method for nonhuman models see* **Note 24**.

44. The Metabotools section of the CobraToolbox has a function that applies constraints on uptake exchanges in a model based on the concentration of metabolites known to be in a given medium [11]. Uptake of all of the listed compounds is allowed, the maximum rate of uptake allowed is equivalent to the rate if

all of the available metabolite in the medium was used in the time between medium changes (t). Information about medium composition is often obtainable from the medium distributor or by mass spectrometry analysis; an estimate of the number of cells used in a standard culture experiment is also needed and will depend on the standard experimental conditions of the user. The cell weight may be obtained from the literature, but weights of around 1–2 ng (1e–9 g) are typical for mammalian cells [122]. The code below is adapted from that in the Metabotools tutorial [11, 28].

45. Begin by loading the list of exchange reactions related to medium components from an Excel file in which the names of the exchanges are found in cells A1–A30 in *line 144*. The variable medium composition will appear as a cell array of 30 × 1.
46. Then load the associated concentrations from column B of the same file, *line 146*. These will form a 30 × 1 array of doubles called met_Conc_mM, it is important that these values are in mMoles per litre of medium (Fig. 2a).
47. Then various metabolites are defined as being allowed to be taken up, although not in the medium composition, *lines 147* and *148*. This allows the uptake of substances such as carbon dioxide, ammonia, and water. This list may be adapted by removing any that are listed in the experimental medium used, or where the uptake rate for a specific cell type is known.
48. Further specific constraints, for example oxygen uptake rate are defined in *lines 150–152*.
49. The medium_compounds (*lines 147* and *148*) are constrained with only a specific uptake rate while customizedConstraints (*lines 150–152*) have both a defined uptake (negative) rate and a defined secretion (positive) rate. Any estimated rates of a more specific magnitude, for example the oxygen uptake rate in this example should be expressed as mmol/g dry weight/h.
50. In order that the maximal allowable uptake rates of the medium components may be applied to the model it is necessary to define the number of cells per mL in the experimental set up *line 155*, the cell weight in g (see above) *line 156*, the time between medium changes in hours *line 157*, this is as the model is defined in mmol/g dry weight/h. Then the value for the maximum flux magnitude for the model is defined in *lines 158* and *159*. All of these are applied to the model in *line 160*.
51. It would now be wise to rerun the section on model feasibility. Particular attention should be paid to the possibility of having acquired blocked reactions or dead end metabolites (checked in *lines 40–48*) or lost important metabolic functions (checked in *line 140* or **Note 24**).

a

EX_ca2(e)	1.2
EX_chol(e)	2.4
EX_cl(e)	120
EX_pnto_R(e)	0.0004
EX_glc_D(e)	25
EX_fe3(e)	1.00E-05
EX_fol(e)	0.003
EX_gly(e)	0.4
EX_arg_L(e)	0.4

b

NM_018000	55686
NM_004211	9152
NM_004212	9153
NM_018004	55076
NM_018003	55075
NM_004210	9148
NM_078483	206358
BC089396	284029
BC054009	3174

c

A_14_P112718	AK026901	chr1:000604268-000604327
A_14_P201353		chr1:000799622-000799681
A_14_P108353	NM_024796	chr1:000801796-000801852
A_14_P129881	BC042880	chr1:000827354-000827412
A_14_P114030	NM_017891	chr1:001059676-001059720
A_14_P139527	NM_017891	chr1:001089934-001089993
A_14_P118340		chr1:001139597-001139656
A_14_P106890	NM_153254	chr1:001160693-001160740
A_14_P122641	NM_016176	chr1:001192357-001192416

d

A_14_P112718	0.54
A_14_P201353	0.59
A_14_P108353	0.71
A_14_P129881	0.52
A_14_P114030	0.61
A_14_P139527	1.00
A_14_P118340	0.50
A_14_P106890	0.96
A_14_P122641	0.09

e

EX_gthrd(e)	0.0018	0.0021	0.004	0.0045
EX_gthox(e)	0.0017	0.0019	0.004	0.0040
EX_trp_L(e)	0.0018	0.0021	0.004	0.0039
EX_cys_L(e)	0.0018	0.0020	0.004	0.0042
EX_arg_L(e)	0.0019	0.0022	0.004	0.0043
EX_fru(e)	0.0019	0.0021	0.004	0.0040
EX_glc_D(e)	0.0018	0.0021	0.004	0.0045
EX_his_L(e)	0.0019	0.0021	0.004	0.0043
EX_gua(e)	0.0019	0.0022	0.004	0.0042

f

Time	EX_gthrd(e)	EX_gthox(e)	EX_trp_L(e)	EX_cys_L(e)	EX_arg_L(e)	EX_fru(e)	EX_glc_D(e)	EX_his_L(e)	EX_gua(e)
0	0.0018	0.0017	0.0018	0.0018	0.0019	0.0019	0.0018	0.0019	0.0019
0	0.0021	0.0019	0.0021	0.0020	0.0022	0.0021	0.0021	0.0021	0.0022
0	0.0042	0.0038	0.0036	0.004	0.00399	0.0038	0.0042	0.0041	0.003885
24	0.0045	0.0040	0.0039	0.0042	0.0043	0.0040	0.0045	0.0043	0.0042
24	0.0018	0.0017	0.0018	0.0018	0.0019	0.0019	0.0018	0.0019	0.0019
24	0.0021	0.0019	0.0021	0.0020	0.0022	0.0021	0.0021	0.0021	0.0022
4	0.0042	0.0038	0.0036	0.004	0.00399	0.0038	0.0042	0.0041	0.003885
4	0.0045	0.0040	0.0039	0.0042	0.0043	0.0040	0.0045	0.0043	0.0042
4	0.0048	0.0042	0.0042	0.0044	0.0046	0.0042	0.0048	0.0045	0.0045
8	0.0018	0.0017	0.0018	0.0018	0.0019	0.0019	0.0018	0.0019	0.0019
8	0.0021	0.0019	0.0021	0.0020	0.0022	0.0021	0.0021	0.0021	0.0022
8	0.0042	0.0038	0.0036	0.004	0.00399	0.0038	0.0042	0.0041	0.003885

Fig. 2 Examples of data formats required for model construction. (**a**) Format of the Excel files named *mediumfile.xlxs* (*lines 144* and *146*), *massspec_info.xlxs* (*lines 240* and *241*). Format for exchange IDs in (**a**) and concentration values in (**b**). (**b**) Format of *gentoent.xlsx* (*line 164*). DAVID output with initial IDs in (**a**) and converted IDs (Entrez) in (**b**). (**c**) Format of Array IDs (*line 165*). The array data downloaded from Array Express (notes in top lines have been removed). (**d**) Format of *SampleTr_A.xlxs* (*line 184*). IDs in the initial format in A and expression values in (**b**). (**e**) Format of *massspec_info.xlxs* sheet 2 (*line 248* and *249*). Exchange IDs and concentrations. (**f**) Format of *massspec_infoALTERNATE.xlxs* (*line 294*). A has information about sample times, row 1 has exchange IDs and the rest is concentrations

52. It may be possible to solve these problems by altering the "mediumCompounds" in *lines 147–148* or "customizedConstraints" lists in *lines 150–152* above and rerunning the "setMediumConstraints" in *line 160*. This will have a similar effect to making the reaction bounds larger by using "changeRxnBounds" run similarly to in *line 76* but with nonzero values and the choice of "l" or "u" to indicate which bound (lower or upper) is to be changed. It should also be noted that as the model is now constrained to a specific culture medium it may be reasonable that some reactions are no longer active and some metabolic features no longer possible. It will be necessary to examine the literature to distinguish between appropriate and inappropriate restrictions.
53. The first step to applying transcriptomic or indeed proteomic data (either from your own experiments or a database listed in **Notes 3–10**) to a model using one of the methods described in **Notes 25–30** is to obtain expression scores related to the gene identifiers in the base model. This maybe done using an online conversion tool such as DAVID [123]. In this example a data

set from Array express has been downloaded [44], both the processed data relating to the cell type of interest as a .tab file (converted to an Excel file by using save as), and a .txt file of the array information (also converted to Excel).

54. The IDs in the array information have been converted to EntrezIDs [124](used in RECON2M2 for gene IDs) using DAVID [125] and the output downloaded and saved as an Excel file (after conversion from .tab) (Fig. 2b, c).
55. The array information and the conversion to Entrez IDs are loaded in to MATLAB in *lines 164* and *165* then a "for loop" is used to match the EntrezIDs and add them to the array information, *lines 167–182.*
56. Then in lines 184–197 the data set is loaded and the EntrezIDs matched to the expression values (Fig. 2d).
57. The "model_medium.genes" ("model_medium" created above) provides the list of genes for "expressionData.gene" variable in *line 199.* Then the final matching takes place with gene IDs from "model_medium" set as a variable in *line 203.*
58. But if there are any slight format differences in format then IDs in the format of the DAVID output are loaded from Excel in *line 202.*
59. Either way then matched to the data set accounting for duplicates by using the mean expression value of each occurrence, it is also possible to ignore the extra values, in *lines 204–215.* The output of this stage forms the .value section of the "expressionData" variable.
60. The per gene expression values should be mapped to per reaction values required for GIMME, iMAT, and, with an adjustment such that expressed is positive and unexpressed negative, INIT options in the "createContextSpecificModel" function, as shown in *line 217.* Reactions that are not associated with the genes identified in the data set are set to a value of −1.
61. For other options in the "createContextSpecificModel" function, specifically FastCore, and as two lists (high and medium importance) MBA, a list of core reactions is required. This may be defined based on the expressionRxns values and a cutoff number to define those reactions highly enough expressed to be classed as core as in *line 220.*
62. Other reactions may be added based on the literature or other experiments as in *line 221.*The selected cutoff is data set specific and is a parameter that may be adjusted and will affect the final model output. mCADRE requires a similar structure based on occurrences in transcriptomic and literature sources separately.

63. The next step to running the "createContextSpecificModel" function is to create a structure called options. This contains the expression data and any extra parameters needed for the desired pruning method and also defines the method (*see* Table 1 for details). The selection of parameters such as the expression cutoff value (or values) and a threshold for minimal allowable flux for a reaction to be considered to be active will affect the completed model. An example is given in *lines 223–226*. This sets up the GIMME algorithm; all the options structures for other methods are shown in Table 1. "createTissueSpecificModel" is run in *line 228*.
64. It may be that it is necessary to run the function "createTissueSpecificModel" multiple times with varying parameters. Once again it would now be wise to rerun the section on model feasibility. Particularly the loss of important metabolic functions (checked in *line 140* or in **Note 24**). It may be that rerunning the "createContextSpecificModel" with new parameters or another method is necessary. It will be necessary to examine the literature to distinguish between appropriate and inappropriate removed reactions and lost functions. Also checking that uptake reactions for medium metabolites are retained is possible as in *line 230–233*. The lengths displayed in *line 231* and *233* should be 0 if all of the metabolites in the medium and customized constraints are known to be critical, otherwise examine the variables created in *lines 230* and *232* to see if the losses are acceptable.
65. In order to create a single model using semiquantitative mass spectrometry data the following adaptation of the Metabotools method may be used [28]. To begin use either the "model_-medium" or "tissueModelM4" from *line 160* or *229* above and define it as "modelstart" in *line 239*.
66. Next define the scope of the data set and the limits of detection; this will probably be derived from mass spectrometry. First load, from an Excel file (or as a cell array), a list of the exchanges representing the compounds in the metabolic data set *line 240*.
67. Then, either load a matching list of limits of detection in mM as an array of doubles as in *line 241* (Fig. 2a), or a list of molecular weights and a list of limits of detection in ng/mL (not shown) and calculate the mM values as in *line 242*.
68. Between *lines 243* and *247* various potential lists of metabolites, to be excluded or included in lists of secreted metabolites based on prior knowledge irrespective of the data, are defined. This includes the special case of essential amino acids. They are defined as cell arrays of exchange reaction abbreviations.

Table 1
The parts of the "options" structure for "createTissueSpecificModel" needed for each of the methods from CobraToolbox help

FastCore	GIMME	iMAT	INIT	MBA	mCADRE
solver: 'fastCore'	*solver*: 'GIMME'	*solver*: 'iMAT'	*solver*: 'INIT'	*solver*: 'MBA'	*solver*: 'mCADRE'
core: indices of reactions in the model that are part of the core set of reactions	*expressionRxns*: gene expression data corresponding to model. rxns.	*expressionRxns*: gene expression data corresponding to model. rxns.	*weights*: column with positive (high expression) and negative (low expression) weights for each reaction	*medium_set*: list of reaction names with medium confidence	*ubiquityScore*: ubiquity scores, vector of the size of 'model.rxns' quantifying how often a gene is expressed accross samples
epsilon: smallest flux value that is considered nonzero	*threshold*: gene expression threshold, reactions below this are minimized	*threshold_lb*: lower bound of gene expression threshold, reactions with expression below this value are "non-expressed"	*tol*: minimum flux value for "expressed" reactions	*high_set*: list of reaction names with high confidence	*confidenceScores*: literature-based evidence for generic model, vector of the size of 'model.rxns'
printLevel: 0 = silent, 1 = summary, 2 = debug	*obj_frac*: minimum fraction of the model objective function that need to be maintained in the extracted model	*threshold_ub*: upper bound of gene expression threshold, reactions with expression above this value are "expressed"	*logfile*: name of the file to save the MILP log	*tol*: minimum flux value for "expressed" reactions	*protectedRxns*: cell array with reactions names that are manually added to the core reaction set
		tol: minimum flux value for "expressed" reactions	*runtime*: maximum solve time for the MILP		*checkFunctionality*: Boolean variable that determines if the model should be able to produce the metabolites associated with the 'protectedRxns'
		core: list of reaction names that are associated with a high confidence			*eta*: tradeoff between removing core and zero-expression reactions
		logfile: name of the file to save the MILP log			*tol*: minimum flux value for "expressed" reactions
		runtime: maximum solve time for the MILP			

69. Then in *lines 248* and *249* the exchange reaction names of measured metabolites and then the measurements are loaded (Fig. 2e). The measurements do not need to have a particular unit; they may be peak areas from a chromatogram. The measurements should be loaded in four columns. The first should represent measurements in blank medium at time one, the second in blank medium from time two, the third in spent medium from time one, and the fourth from spent medium at time two.
70. Normally the input of metabolite measurements would be repeated for a second data set and the full procedure of Metabotools followed; however, the second data set may be made to match the first as in *line 250*, although not all of the functions in Metabotools will then be able to be utilized.
71. The metabolite data is used to define lists of those metabolites that are secreted (increase between time point one and two) or are taken up (decrease between measurements) in *lines 251* and *252*, the related limits of detection are found in *lines 253* and *254*.
72. The user may then note any metabolites where the data quality is low and may be ignored in *line 255*. Metabolites which may not be measured but which are likely to be exchanged with the medium, for example salt ions are defined in *line 256*.
73. This information along with the lists of metabolites secreted and taken up is then used to constrain the model in *line 257*.
74. The current method may introduce some lower bounds that are greater than upper bounds. This problem is checked for and then corrected, with reference to the predicted uptake or secretion direction, in *lines 259–276*.
75. *Lines 277–280* then further constrain the model according to the growth rate of the cells, based on the increase in cell weight over the known doubling time (defined from experience or the literature). This forces the biomass reaction to take an optimal value that represents known doubling time.
76. If the user has previously used an expression based pruning method then they may wish to skip lines *282* and *283*. If not, then using the "expressionData" variable created between *lines 164* and *215* define a list of unexpressed genes and then set the flux through linked reaction to 0 in *lines 282* and *283*.
77. The constrained model may then be checked for blocked reactions and the necessary metabolic functions (checked in *lines 40–48* and *line 140* or **Note 24**).
78. If known functions are lost the above procedure may be repeated with a longer list of ambiguous metabolites in *line*

255 or defined uptake or secretion metabolites in *lines 234–237* or bounds may be relaxed as in *line 76*.

79. However, it may be desirable to leave only the active reactions for the defined metabolic data (if the user is confident in the data). In this case *lines 285–287* remove all of the inactive reactions, following this it would be wise to check that all necessary metabolic functions for the context remain as above and if not rerun the procedure as described above.
80. The following adaption of the Metabotools method may be used to create a model based on quantitative metabolomics data [29]. First an uptake rate must be defined, for example by taking measurements in culture a few hours apart and using linear regression to obtain uptake and secretion rates, these must be normalized for cell weight and in units compatible with the model.
81. The data should be loaded in a table with the time point the first column and the exchange reactions associated with each metabolite in the top row then measurements in nano moles per liter in the rest of the table (Fig. 2f), as shown in *line 294*.
82. The cell concentration and weight are used from the medium definition above (or may be loaded by the user) along with the experimental volume in liters as in *line 296*.
83. A regression analysis is run in *lines 303–319* for each metabolite giving a table including the exchange reaction abbreviation, the predicted flux and the 95% confidence intervals of the flux for each metabolite.
84. In *lines 321–323* this data is arranged into the input format for the Metabotools functions, a sample name is added.
85. In *line 324* the model is defined.
86. Parameters are set up to check if metabolites listed in the data set are able to be secreted and taken up. These parameters are the minimum and maximum, *lines 325* and *326*, flux to be tested for and the output directory in which to save the results of the tests, in this case the current path is used in *line 327*.
87. The test to find which of the measured uptakes and secretions is possible for the base model is performed in *line 328* and the results are saved.
88. The data lists saved in *line 328* are summarized in *line 333*, the number of possible exchanges to examine is defined as the number of uptake or secretion exchanges predicted in the initial data in *lines 330–332*.
89. In *lines 335* and *336* the solver "ibm_cplex" is set. Then a minimum growth rate and various metabolites that should be treated specially are defined, for example peroxide and carbon

dioxide should not be taken up and oxygen should not be secreted are defined between *line 337* and *343*.

90. The model to be used and minimum values for a reaction to be considered active, and for a number to be considered 0 are set in *lines 344–347*.
91. The data is then applied to the model and minimal fluxes are allowed through other exchanges, two models are the model outputs of *line 348*. Within the output structure "ResultsAll-Cell" lines are two model structures; one is equivalent to the output of *line 280*, "modelMin," and "modelPruned" on the other hand is equivalent to *line 287*.
92. Similar checks to those described above to check model functionality should be performed on the models produced above. It should also be noted that multiple data sets may be used at once to follow the tutorial exactly as shown online, see the Metabotools II tutorial [29].
93. FBA determines the optimal flux through a model to maximize the flux through one reaction [6]. When FBA is performed for a biomass reaction the maximal cell growth is tested [30]. If FBA is repeated after a reaction has been blocked by setting both bounds to 0 and there is no longer growth then the reaction is considered essential. If a gene is blocked by setting the bounds of all linked reactions to 0 and there is no longer growth then the gene is essential. These tests are equivalent to knocking out a gene or pharmacologically blocking a reaction in vitro and the results may be compared to experimental data (your own or from databases such as in **Notes 12** and **13**).
94. *Lines 350–362* alter the bounds of each reaction in the model to zero in turn. The maximized objective function for each knock out model is then recorded in a list "GrowthR."
95. The reactions that result in an objective function of 0 are listed as "LethalR" and those that reduce the objective to less than a tolerance defined in *line 351* are listed as "SubLethalR."
96. *Lines 364* to *377* produce analogous results for each gene called "GrowthG," "LethalG," and "SubLethalG," respectively. These may be presented in a Table such as shown in Fig. 1c.
97. It is more straightforward to compare two models if the two models are of the same size (contain the same reactions). Two models built from the same reconstruction will contain a largely overlapping set of reactions and therefore this process is fairly straightforward and this is the case described here. Begin by establishing all of the reactions contained in the two models, in *line 383*.
98. Next find the reactions in the original model which are absent from both and may be deleted, in *line 384*, then delete them.

99. Find the reactions that are not in each model, *line 385 and line 386.*
100. Use these lists to alter both bounds of these reactions to 0 as in *lines 389* and *395.*
101. Then use a "for loop" to change the bounds of each reaction in turn to match those in "modelA" or "modelB" *lines 390–393* and *396–399.* These models may be used for random sampling and other analyses.
102. First parameters to set up the sampling are defined, the number of warm up points, how many starting possibilities there are, the steps per point, how far should the sampling go and the maximum time allowed, to stop if the sampling takes too long. These are defined in *lines 401–403.* The models used are those defined above in create intersection models.
103. Next the files for the output are defined the points per file multiplied by the file number should be equal to the steps per point multiplied by the number of warm up points. These are defined in *lines 405–408.*
104. The file name for each model is then defined and then the sampling is run, in *lines 410* and *411* and *413* and *414.*
105. The first step to comparing two sampling outputs is to load the sampling outputs as in *lines 418–457,* assumed to be from models with the same number of reactions, although many of the functions in the CobraToolbox may be used with unequal models.
106. The next step is to establish the basic statistics such as the mean, median and standard deviation for each reaction. The mean is found in *line 459–461,* mean may be replaced by median, std., mode, skew, and kurt to calculate other basic statistics.
107. The ratio between each mean flux in "model_A" and "model_B" is worked out in *line 461–463.*
108. Then the Kolmogorov-Smirnov statistics in *line 464.*
109. Between *lines 467* and *476* the reaction IDs and abbreviations of reactions which reach a significance level of 0.01 are identified as are those either up regulated in "model_A" or upregulated in "model_B".
110. In *lines 478* and *479* the two sets of upregulated reactions are subjected to enrichment analysis to find over represented metabolic subSytems.
111. In *line 481* sample histograms of significantly altered reactions are plotted or a bubble plot showing the subsystem distribution of reactions that are different in the two models (Fig. 1b).

112. The output from the random sampling may be used to generate a map similar to Fig. 1a. One generates a file with RECON 2 reaction ids, and the corresponding median flux values for every reaction in both models.
113. This is loaded onto a previously generated flux map in [126, 127] and the relative flux difference can be viewed in all reactions in the map.
114. If the two models share an origin, or represent cells that are in different stages of differentiation, it can be useful to use MOMA. MOMA analysis is carried out in *line 485*, which outputs a solution vector B, solution vector A and total flux difference between the vectors. Solution vector B represents the adjusted flux distribution of model A, which is closer to the optimized flux distribution of model B.
115. Between lines *489* and *502*, the adjusted reactions are being ranked and sorted based on their "adjustment amount," that is, how much the reaction needs to be adjusted for A to become more like B.
116. In *lines 505* and *506*, the 100 reactions that needed the most adjustment are used for enrichment analysis, which reveals the subsystems significantly enriched in the adjusted reactions. These subsystems are the ones that need the highest amount of adjustment for A to take on a phenotype closer to B (differentiation, EMT, etc.).

4 Notes

1. CobraPy.

   COBRAPy is designed as a way of better tackling the "complexity inherent in integrated biological networks" and a better "framework for the multiomics data used in systems biology"; moreover, as it is based on the free Python and not the commercial MATLAB, it is more accessible. It is also currently possible to allow interactions between COBRAPy and the COBRAToolbox [128].
2. RavenToolbox.

   RAVEN (Reconstruction, Analysis, and Visualization of Metabolic Networks) Toolbox was originally described as "a software suite that allows for semiautomated reconstruction of genome-scale models. It makes use of published models and/or the KEGG database, coupled with extensive gap-filling and quality control features" in 2013 [129]. It has since been updated to a second version RAVEN Toolbox 2.0 additional features include "(a) de novo reconstruction of GEMs based on the MetaCyc pathway database; (b) a redesigned KEGG-based reconstruction pipeline; (c) convergence of reconstructions from various sources; (d) improved performance, usability, and compatibility with the COBRA Toolbox" [97]. A particular feature of the RAVEN Toolbox is its links to various metabolic databases to enable the reconstruction of GEMs based on the information contained in them.

3. Array Express.

   Array express is a database of publically available transcriptomic data. It is maintained by the Functional Genomics team of EMBL-EBI and contains both sequence and microarray data. It currently contains details of over 70,000 experiments either directly uploaded or submitted by the NCBI Gene Expression Omnibus [44]. Processed data is downloadable in a tab-delimited format for most experiments. Array Express data is also downloadable via a Bioconductor package in R [45]. Data in the Array Express fulfils the Minimum Information about a Microarray Experiment (MIAME) [130] or Minimum Information about a Sequencing Experiment (MIASE) [131] standard depending on the data source.

4. Gene Expression Omnibus.

   The Gene Expression Omnibus (GEO) is a repository for microarray and next generation sequencing data sets [47]. It was established by the National Center for Biotechnology Information (NCBI) and contains around 20,000 data sets. Its guidelines for uploading data mirror MIAME and MIASE standards.

5. PRIDE.

   The PRIDE (Proteomics Identifications) archive is a database of mass spectrometry proteomics data [49]. It provides expression data and protein and peptide identifications. It complies with the standards of the Proteome Exchange Consortium [51] and data is available via the website as well as a stand-alone inspection tool.

6. Peptide Atlas.

   Peptide Atlas is a database maintained by the Seattle Proteome Center [51]. Its aim is to provide a publically accessible database of information from high-throughput proteomics experiments in order to annotate eukaryotic genomes with experimentally validated expression information. Peptides are identified and may be accessed via the database web page and are linked to ENSMBL identifiers [52]. It is also a member of the Proteome Exchange.

7. MASSive.

   The Center for Computational Mass Spectrometry at the University of California, San Diego maintains a database of mass spectrometry-based proteomic data that contains over 9000 data sets [53]. This is another member of the Proteome Exchange consortium and therefore the MASSive data sets contain compatible identifiers. The website offers both analytical tools and the ability to download data sets.

8. iProX.

   The integrated proteome resources center is a Chinese run repository for proteomics data that is a member of the Proteome Exchange consortium [55]. It provides a database from which proteomic data sets may be downloaded, data sets have proteome exchange consortium identifiers and follow proteome exchange consortium guidelines. It currently contains around 500 data sets (200 public) and about 110,000 data files.

9. Japan ProteOme STandard Repository.

   The Japan Proteome Standard Repository (JPOST) is a data repository for proteomic mass spectrometry data in either raw or processed format [57]. It is a member of, and assigns identifiers for, the Proteome Exchange Consortium. It may be accessed via an online interface. It contains over 400 projects related to 82 species.

10. The Human Protein Atlas.

    The human protein atlas provides information about the tissue, cell, organelle, and pathological condition specific expression of proteins in humans [25]. This is based on a combination of antibody-based images, mass spectrometry, and transcriptomics. A more recent addition is information on mouse brain protein expression. The data from the human protein atlas has been used to parameterize tissue specific metabolic models [25].

11. MetaboLights.

    Metabolights is a repository for metabolomics data sets it is maintained by the European Bioinformatics Institute [60]. It provides data sets from multiple species from both mass spectrometry and nucleic magnetic resonance experiments. Some of the data relates to cultured cells and some to clinical samples. It is associated with the Metabolomics Standards Initiative.

12. Genome RNAi.

    Is a repository for RNAi screens in Drosophila and human cell lines [62]. It contains around 700 experiments detailing the phenotypic outcome of RNAi experiments and may be useful for confirming the outputs of gene and reaction essentiality described in Section Xi below.

13. The Database of Genotypes and Phenotypes, The European Genome-Phenome Archive, and The Japanese Genotype-Phenotype Archive.

The database of genotypes and phenotypes (dbGaP) is an NCBI maintained repository for data sets describing genotype–phenotype interactions in human derived data [66]. Similar databases are maintained in Europe, the European Genome-Phenome Archive [67], by the Center for Genomic Regulation and the European Bioinformatics Institute, and Japan, The Japanese Genotype-Phenotype Archive [62], maintained by the National Bioscience Database Center (NBDC) of Japan Science and Technology Agency. These databases may be useful for confirming the results of models of human metabolism by allowing checks to be made on the clinical features of genetic alterations predicted as important by a model.

14. KEGG.

    KEGG is a database of genes and genomes, as well as metabolite, metabolic pathways, and drugs. It is made of a collection of 18 databases representing different aspects of biology covering, information on how systems are connected together, how genes function, chemical compounds important to biology and how underlying biology causes health problems [30].

15. MetaCyc.

    MetaCyc is a database of metabolic reactions, enzymes, and related chemicals [52]. The information is arranged in metabolic pathways which are in turn collected into networks called super-pathways. The data covers eukaryotes, prokaryotes, and archaea. The database can be searched online or has files available for download.

16. BRENDA.

    BRENDA is a large collection of information on enzymes, their activity, inhibition and expression. Where possible BRENDA uses enzyme commission (EC) classification numbers. This allows for the identification of enzymes according to various levels of their function and allows for the identification of similar enzymes [105, 106].

17. Reactome.

    Reactome is an EBI maintained comprehensive database of biological reactions in the broadest sense, of any change to a biological molecule. This includes metabolic pathways. Reactome may be used to identify the scope of metabolic pathways and also to compare orthologues between humans and various other species [105].

18. BioModels database.

    The biomodels database is a repository for biological models including COBRA models [109]. It allows the download of models in the systems biology mark-up language (SBML) [110], this is compatible with the CobraToolbox. Models in the repository are generally initially uploaded uncurated and are marked as such; however, many are curated and therefore may be considered as a more solid base for future exploration. There is also a set of automatically generated metabolic models from a project called path2models based on metabolic databases [111].

19. Installing the MATLAB license.

    The MATLAB license may be free to acquire for students and academics or your institution may have a licence, if not you will need to purchase one.

20. Installing other solvers.

    Not all have been fully compatibility tested for CobraToolbox with Windows 10. Ensure that the version of the solver is compatible with the version of MATLAB. The IBM Cplex and Gurobi solvers are both available for free to academics.

21. Selecting a base model.

    There are several GEMs available for commonly researched species and many are downloadable as either .mat (MATLAB) files or .xml (systems biology mark-up language) files from either author maintained websites or repositories such as the BioModels database, *see* above in Subheading 2 for model databases and publications that frequently contain models as supplementary material. The selection of a particular model largely depends on the species chosen, and in general the most up-to-date version would be recommended. However, if a user has a particular interest in a specific area of metabolism then checking that the area of interest is extensively included and curated may result in another choice, it is best to read the peer reviewed article describing the model to determine this. Other considerations such as the use of identifiers or annotations that are consistent with other information sources may play a part in model selection, for example RECON3D [69] is compatible with AGORA [99] bacterial models allowing for easy integration into a multispecies model while Recon2m2 [14] has provided multiple versions with various labeling systems for metabolites and genes allowing easy integration with other data sets. It is also possible to select a reduced metabolic model reconstruction focusing on a narrower area of metabolism such as MitoCore [132].

22. Fast Gap Fill.

    It may also be possible to apply the automated "fastGapfill" function to add reactions, this is described in the

CobraToolbox tutorial of the same name [17, 133]. However, the user has more control of the reactions added if performed manually. This may reduce the time needed for the steps below.

23. Thermodynamic constraint.

    The above procedure checks the feasibility of producing ATP in the mitochondria and helps to ensure thermodynamic consistency, and biochemically sensible reversible reactions. However, there are also functions available that apply facts about the Gibbs free energy of metabolites found in databases in order to apply thermodynamic constraints, one matTFA is downloadable from github and includes a version of the Cobra-Toolbox [134, 135]. The option available in the CobraToolbox is not available for Windows but is, if the CobraToolbox is installed on a linux machine there is also an alternative method available [136–138] that uses CobraToolbox functions [139].

24. Checking metabolic functions of a model.

    If you are not using a human model then use the testPathway function to achieve similar effects with other species, you will need to apply appropriate uptake exchange constraints and obtain theoretically correct production fluxes from the literature. "[Flux, FBAsolution, model] = testPathway(model, MetIn, MetOut, AdditionalMetsInorOut, ObjectiveOption)" [140].

25. FastCore [9].

    Fastcore is a model pruning algorithm of the MBA-like type [7], building models based on a consistent generic model and containing a core set of reactions, known to be expressed, linked by other necessary reactions to create a consistent model. It is implemented within the CobraToolbox, as part of the createContextSpecificModel function. One of the attractions of this method is its speed, this may allow for the testing of multiple cutoffs for inclusion in the core reaction set [9]. Fastcore also claims to balance model compactness with the inclusion of redundant pathways, provided that there are core reactions present in both pathways.

26. GIMME [5].

    Gene Inactivity Moderated by Metabolism and Expression or GIMME algorithm is another model pruning method implemented in the CobraToolbox within the createTissue-SpecificModel function. According to Opdam et al. it lies within its own group [7]. It functions by removing reactions from a base model and then adding reactions back in until required metabolic functions, the biomass objective function, are feasible [5]. It is possible that the selection of the required metabolic functions may overly influence which reactions are included [4, 7], reactions that are lowly expressed but responsible for production of biomass will be included, this assumption may be less appropriate in mammalian cells than in bacteria.

27. iMAT [24].

    iMAT is a method for pruning metabolic models based on the closely related INIT method (see below), it is one of those implemented in the CobraToolbox. iMAT operates in a similar manner to INIT [7, 141], described below, but, in the CobraToolbox implementations, iMAT allows for more definition of thresholds for exclusion of low expression genes and inclusion of high expression genes and a user defined core set of reactions, which may not have high expression but can be defined based on user knowledge of the system.

28. INIT [25].

    The INIT algorithm or integrative network inference for tissues is also implemented in the createTissueSpecificModel function of the CobraToolbox. INIT requires the input of gene expression data and a base model, it then optimizes the exclusion of the lowly expressed reactions versus inclusion of highly expressed genes [25]. In the analysis by Opdam et al. INIT was one of the most accurate techniques with respect to prediction of essential genes; however, it also showed high variability depending on the thresholds selected for gene inclusion and exclusion [7].

29. MBA [26].

    The Model Building Algorithm or MBA may be implemented via the createTissueSpecificModel function in the CobraToolbox. It requires a list of high and a list of moderately expressed reactions and a generic model. The aim is then to find the smallest consistent model that includes all of the highly expressed reactions and the necessary moderately expressed reactions [26]. According to Opdam et al. the MBA algorithm is one of the most accurate with respect to gene essentiality prediction [7].

30. mCADRE [8].

    Metabolic Context-specificity Assessed by Deterministic Reaction Evaluation or mCADRE is the sixth and final metabolic model pruning algorithm available via the CobraToolbox. mCADRE uses a core set of highly expressed reactions with other reactions ranked according to their expression and connectivity, reactions are then removed according to their ranking until a model is obtained that is the smallest model consistent with the required metabolic functions [8]. mCADRE has been identified by Opdam et al. as one of the most accurate at predicting gene essentiality but with a tendency to remove genes with a moderately high expression if linked to reactions of lower expression [7].

5 Script

```
%%%%%%%%%%i. Install software
%%%%%%%%TO BE PERFORMED IN GITBASH NOT MATLAB
%%% $ git clone --depth=1 https://github.com/opencobra/cobratoolbox.git↙
cobratoolbox
%%%%%%%%Beginning of Matlab Code
initCobraToolbox %% Intitiate the cobratoolbox, this will display the avaliable↙
solvers

%%%%%%%%%%%%%%%%%%%%%%%%%%%%%%%%%%%%%%%%%%%%%%%%%%%%%%%%%%%%%%%%%%%%%%%%%%%%
%%%%%%%%%%ii. Select and load genome scale metabolic reconstruction
%%Usage of the read SBML function to load Recon2.2
%%%%%model = readSBML(fileName, defaultBound) the read SBML function
Recon2 = readSBML('RECON2m2.xml',1000);
Recon2 = readSBML('C:/Documents/RECON2m2.xml',1000);
%%Usage of the read cobra model function to load Recon2.2

Recon2 = readCBModel('RECON2m2.mat');

%%%%%%%%%%%%%%%%%%%%%%%%%%%%%%%%%%%%%%%%%%%%%%%%%%%%%%%%%%%%%%%%%%%%%%%%%%%%
%%%%%%%%%%iii. Ensure reconstruction feasibility
%%%%%%%%%%iii.i Checking the biomass function
%%Usage of the check objective function to check that the model has the
%%biomass function as the objective and check it's compostion
%%%%%objectiveAbbr = checkObjective(model)
objectiveR2 = checkObjective(Recon2)%For Recon2.2 this is 'biomass_reaction' and↙
they could be used interchangeably below

printRxnFormula(Recon2,objectiveR2)

[R2newBiomass,~] = addReaction(Recon2,'new_biomass','20.6508 h2o[c] + 20.7045 atp↙
[c] + 0.385872 glu_L[c] + 0.352607 asp_L[c] + 0.036117 gtp[c] + 0.505626 ala_L[c] +↙
0.279425 asn_L[c] + 0.046571 cys_L[c] + 0.325996 gln_L[c] + 0.538891 gly[c] + 0.392525↙
ser_L[c] + 0.31269 thr_L[c] + 0.592114 lys_L[c] + 0.35926 arg_L[c] + 0.153018 met_L[c]↙
+ 0.023315 pail_hs[c] + 0.039036 ctp[c] + 0.154463 pchol_hs[c] + 0.055374 pe_hs[c] +↙
0.020401 chsterol[c] + 0.002914 pglyc_hs[c] + 0.011658 clpn_hs[c] + 0.009898 dgtp[n] +↙
0.009442 dctp[n] + 0.013183 datp[n] + 0.053446 utp[c] + 0.013091 dttp[n] + 0.275194↙
g6p[c] + 0.126406 his_L[c] + 0.159671 tyr_L[c] + 0.286078 ile_L[c] + 0.545544 leu_L[c]↙
+ 0.013306 trp_L[c] + 0.259466 phe_L[c] + 0.412484 pro_L[c] + 0.005829 ps_hs[c] +↙
```

```
0.017486 sphmyln_hs[c] + 0.352607 val_L[c] -> 20.6508 h[c] + 20.6508 adp[c] + 20.6508 pi[c] ');%biomass reaction from Recon3D
R2newBiomass = changeObjective(R2newBiomass,'new_biomass');
R2newBiomass = removeRxns(R2newBiomass,'biomass_reaction');
surfNet(Recon2,'biomass_reaction')%Click on a blue highlighted component in the output to see more detail about that component
trprxns = findRxnsFromMets(Recon2,'trp_L[c]');
trprxnsnotBM = setdiff(DNA,objectiveAbbr);%%%Ignores the biomass reaction
printRxnFormula(RECON2_econ2,trprxnsnotBM)
%%%%%%%%%%iii.ii Checking reaction activity and metabolite usage
outputMets = detectDeadEnds(Recon2);
DeadEnds = model.mets(outputMets);
[rxnList, rxnFormulaList] = findRxnsFromMets(Recon2, DeadEnds);
Recon2.lb(find(ismember(Recon2.rxns,rxnList)));
Recon2.ub(find(ismember(Recon2.rxns,rxnList)));
[allGaps, rootGaps, downstreamGaps] = gapFind(Recon2, 'true');%%%Find gaps that are not at the end of a pathway, downstream gaps
BlockedReactions = findBlockedReaction(Recon2);%%%Find reactions that are not able to carry flux under the given conditions
%%%%%%%%%%iii.iii Checking the chemical feasibility of the model
%%%Identify exchanges
if ~isfield(Recon2,'SIntRxnBool')
Recon2 = findSExRxnInd(Recon2,size(Recon2.S,1),1);
end
%%%Close exchanges
Recon2close = Recon2;
Recon2close.lb(~Recon2close.SIntRxnBool)=0;
Recon2close.ub(~Recon2close.SIntRxnBool)=0;
%%%Set up optional variables for flux balance analysis so that it optimizes
%%%for ATPsynthase by sparse FBA
Recon2atp = changeObjective(Recon2close,'ATPS4m');%change objective to ATP-Synthase
osenseStr='max';%Maximise
minNorm='zero';
allowLoops=1;
zeroNormApprox='all';%test all approximations available and return the best one
sparseFBAsolutionBounded = optimizeCbModel(Recon2atp, osenseStr, minNorm, allowLoops, zeroNormApprox);
sparseFBAsolutionBounded.obj
```

```
%%%Find the reactions with flux above a cutoff value
cutoff= 1e-6;
for n=1:nRxn
if abs(sparseFBAsolutionBounded.v(n))>cutoff
formula=printRxnFormula(Recon2atp, Recon2atp.rxns{n}, 0);
end
end
%%%Alter the lower bound of some reactions to 0 making them irreversible
Recon2irrev1 = changeRxnBounds(Recon2,selectedRxns,0,'l');
%%%Set up the above model with closed exchanges so that it has open
%%%reversible internal reactions
Recon2atpopen = Recon2atp;
Recon2atpopen.lb(Recon2atpopen.SIntRxnBool)=-1000;%%open bounds in negative direction
Recon2atpopen.ub(Recon2atpopen.SIntRxnBool)=1000;%%open bounds in positive direction
sparseFBAsolutionUnBounded = optimizeCbModel(Recon2atpopen, osenseStr, minNorm, allowLoops, zeroNormApprox);
sparseFBAsolutionBounded.obj %Possibly non-zero
%%%Set up the above model with closed exchanges and all open transport
%%%reactions
allTransportRxnBool=transportReactionBool(RECON2m2atp);
Recon2atptr = RECON2m2atp;
Recon2atptr.lb(allTransportRxnBool)=-1000;
Recon2atptr.ub(allTransportRxnBool)=1000;
sparseFBAsolutionBounded = optimizeCbModel(Recon2atptr, osenseStr, minNorm, allowLoops, zeroNormApprox);
sparseFBAsolutionBounded.obj%Is likely to be non-zero
%%%Set up the above model with closed exchanges and sets of open transport
%%%reactions between two compartments
originCompartment='c';% origin cytoplasm or e for extracellular to check plasma membrane
destinationCompartment='m';% destination m (mitochondria) but also check other compartments x,g,r,l or c for plasma membrane
unidirectionalBool=0;
cmTransportRxnBool=transportReactionBool(Recon2atp,originCompartment, destinationCompartment,unidirectionalBool);
%%%%Return to the version with closed exchanges and internal reactions as
%%%%established RECON2m2atp
Recon2atptrcm = RECON2m2atp;
Recon2atptrcm.lb(cmTransportRxnBool)=-1000;%%%Open all selected transport reactions (c to m) in negative direction
Recon2atptrcm.ub(cmTransportRxnBool)=1000;%%%Open all selected transport reactions (c to m) in positive direction
sparseFBAsolutionBounded = optimizeCbModel(Recon2atptrcm, osenseStr, minNorm, allowLoops, zeroNormApprox);
sparseFBAsolutionBounded.obj
```

```
%%%%Repeat for other pairs of compartments

%%%%%%%%%%iii.iv Other metabolic features to be checked
%%%Set up a closed version of the model
modelClosed = Recon2irrev1;%Use the final version of the model produced after the↙
end of section iii.iii
modelexchanges1 = strmatch('Ex_',modelClosed.rxns);
modelexchanges2 = strmatch('DM_',modelClosed.rxns);
modelexchanges3 = strmatch('sink_',modelClosed.rxns);
modelexchanges4 = strmatch('EX_',modelClosed.rxns);
 selExc = (find( full((sum(abs(modelClosed.S)==1,1) ==1) & (sum(modelClosed.S~=0)↙
== 1))))';
modelexchanges = unique([modelexchanges1;modelexchanges2;modelexchanges3;modelexchanges4;selExc]);↙
 modelClosed.lb(find(ismember(modelClosed.rxns,modelClosed.rxns(modelex-changes))))↙
=0;
 modelClosed.ub(find(ismember(modelClosed.rxns,modelClosed.rxns(modelex-changes))))↙
=1000;
modelClosedOri = modelClosed;
%%%%Check if you get ATP from water (you shouldn't)
modelClosed = modelClosedOri;
modelClosedATP = changeObjective(modelClosed,'DM_atp_c_');
modelClosedATP = changeRxnBounds(modelClosedATP,'DM_atp_c_',0,'l');
modelClosedATP = changeRxnBounds(modelClosedATP,'EX_h2o[e]',-1,'l');
FBAatph2o=optimizeCbModel(modelClosedATP);
%%%%Check if you get ATP from water and oxygen (you shouldn't)
modelClosed = modelClosedOri;
modelClosedATP = changeObjective(modelClosed,'DM_atp_c_');
modelClosedATP = changeRxnBounds(modelClosedATP,'DM_atp_c_',0,'l');
modelClosedATP = changeRxnBounds(modelClosedATP,'EX_h2o[e]',-1,'l');
modelClosedATP = changeRxnBounds(modelClosedATP,'EX_o2[e]',-1,'l');
FBAatph2oo2=optimizeCbModel(modelClosedATP);
%%%Check if you make protons (m and c will need to be done separately for
%%%cytoplasm and mitochondria) (you don't want it to)
modelClosed = modelClosedOri;
modelClosed = addDemandReaction(modelClosed,'h[m]');%%switch m for c
modelClosed = changeObjective(modelClosed,'DM_h[m]');%%switch m for c
modelClosed.ub(find(ismember(modelClosed.rxns,'DM_h[m]'))) = 1000;%%switch m for c
FBAhm = optimizeCbModel(modelClosed,'max');%%switch m for c
%%%Check important metabolic features in a human model
[TestSolution,TestSolutionName,~,~] = test4HumanFctExt(Recon2,'Recon1');

%%%%%%%%%%iv. Apply constraints based on medium composition
```

```
% Exchanges associated with medium read from an excel file
medium_composition = xlsread('medium_file.xlsx',1,'A1:A30');%Figure 1a
% Medium concentrations associated with above exchanges read from an excel file
met_Conc_mM = xlsread('medium_file.xlsx',1,'B1:B30');%Figure 1a
mediumCompounds = {'EX_co2(e)';'EX_h(e)';'EX_h2o(e)';'EX_hco3(e)';'EX_nh4(e)';'EX_pi(e)';'EX_so4(e)'};
mediumCompounds_lb = -100;
%%%% Any other constraints needed
customizedConstraints = {'EX_o2(e)';'EX_strch1(e)';'EX_acetone(e)';'EX_glc(e)';'EX_his_L(e)';'EX_val_L(e)';'EX_met_L(e)'};
customizedConstraints_lb = [-2.3460;0;0;-500;-100;-100;-100];
customizedConstraints_ub = [500;0;0;500;500;500;500];
%%%Basic experimental information such as cell number, volume and weight
%%%and frequency of medium changes
cellConc = 2.17 * 1e6;
cellWeight = 3.645e-12;
t = 48;
current_inf = 1000;
set_inf = 1000;
[modelMedium, ~] = setMediumConstraints(RECON2m2, set_inf, current_inf, medium_composition, met_Conc_mM, cellConc,t, cellWeight, mediumCompounds, mediumCompounds_lb, customizedConstraints, customizedConstraints_ub, customizedConstraints_lb);

%%%%%%%%%%v.   Apply constraints based on expression data (transcript/proteomic)
%%%% Import the data set list of IDs and the conversion list
[~,bgentoent,cgentoent] = xlsread('gentoent.xlsx');%Figure 1b
[~,bgentoaffy,~] = xlsread('ArrayIDs.xlsx');%Figure 1c
%%%% Match the IDs in the data set to their converted equivalent
k = 1;
for i = 1:length(bgentoaffy)
    if isempty(bgentoaffy(i,2)) == 1
        k = k + 0;
    else
        a = find(ismember(bgentoent,bgentoaffy(i,2)));
          if length(a) == 0
            k = k + 0;
        else
               for j = 1:length(a)
                    IDs(k,:) = [bgentoaffy(i,:),cell2mat(cgentoent(a(j),2))];
                   k = k + 1;
               end
        end
    end
end
%%%%Read in the data set
[~,bsampleTr_A,csampleTr_A] = xlsread('sampleTr_A.xlsx');%Figure 1d
%%%%Match the converted IDs
```

```
k = 1;
for i = 1:length(IDs)
      a = find(ismember(bsampleTr_A,IDs(i,1)));
      if length(a) == 0
            k = k + 0;
      else
            for j = 1:length(a)
                  Aid(k,:) = [csampleTr_A(a(j),:),IDs(i,4)];
                  k = k + 1;
          end
      end
end
%%%%Define expressionData.gene for use later
expressionDataA.gene = modelMedium.genes;
% %%%%Load in a gene list from excel in which the format exactly matches the
% %%%%DAVID output
% [gmatch,~,~] = xlsread('mediumgenes.xlsx');
gmatch = modelMedium.genes;
%%%%Match the IDs to the model genes and deal with duplicates
for i = 1:length(modelMedium.genes)
a = gmatch(i);
bsampleTr_A = find(cell2mat(Aid(:,3)) == a);
if length(bsampleTr_A) == 0
      expressionDataA.value(i) = 0;
elseif length(bsampleTr_A) == 1
expressionDataA.value(i) = cell2mat(Aid(bsampleTr_A,2));
else
      expressionDataA.value(i) = mean(cell2mat(Aid(bsampleTr_A,2)));
end
end
expressionDataA.value = transpose(expressionDataA.value);
%%%%Map genes to reactions
[expressionRxnsA,~] = mapExpressionToReactions(modelMedium,expressionDataA);
%%%%Define core reactions from expression
cutoff = 0.1,
coreA = modelMedium.rxns(find(expressionRxnsA >= cutoff));
coreA = vertcat(coreA,'PFK');
%%%%Setup options and run create tissue specific model
optionsA.solver = 'GIMME';
optionsA.expressionRxnsM4
optionsA.threshold = 0.1;
%%%%%optionsM4.obj_frac%optional option to allow a model to be created with
%%%%%less flux through the biomass reaction, the default is 0.9
tissueModelA = createTissueSpecificModel(model_Medium, optionsA);
%%%%Check medium metabolites retained
missing_med = setdiff(medium_composition,tissueModelA.rxns);
length(missing_med)
missing_cust = setdiff(customizedConstraints,tissueModelA.rxns);
```

```
length(missing_cust)

%%%%%%%%%%vi. Apply reaction flux constraints from specific metabolomics data
%%%%%%vi.i Semi-quantitative
%%%%Read in all of the data
modelstart = tissueModelA;% or modelMedium
ex_RXNS = xlsread('massspec_info.xlsx',1,'A1:A30');%Figure 1a
lod_mM = xlsread('massspec_info.xlsx',1,'B1:B30');%Figure 1a
%[lod_mM] = calculateLODs(theo_mass, lod_ngmL);
exclude_upt = {};
exclude_secr = {};
add_upt = {};
add_secr = {};
essAA_excl = {'EX_his_L(e)'; 'EX_ile_L(e)'; 'EX_leu_L(e)'; 'EX_lys_L(e)';↙
'EX_met_L(e)';'EX_phe_L(e)'; 'EX_thr_L(e)'; 'EX_trp_L(e)'; 'EX_val_L(e)'};
data_RXNS = xlsread('massspec_info.xlsx',2,'A1:A25');%Figure 1e
input_A = xlsread('massspec_info.xlsx',2,'B1:E25');% control TP 1 control TP 2↙
Cond TP 1 Cond TP 2%Figure 1e
input_B = input_A;
tol = 0.05;
[cond1_uptake, ~, cond1_secretion, ~, ~] = defineUptakeSecretionProfiles(inpu-
t_A,↙
input_B, data_RXNS, tol, essAA_excl, exclude_upt, exclude_secr, add_secr, add_upt);
cond1_uptake_LODs(:,1) = lod_mM(find(ismember(ex_RXNS, cond1_uptake(:))));
cond1_secretion_LODs(:,1) = lod_mM(find(ismember(ex_RXNS, cond1_secretion(:))));
ambiguous_metabolites = {};
basisMedium = {'EX_o2(e)'; 'EX_strch1(e)'; 'EX_acetone(e)'; 'EX_glc(e)'; 'EX_-
his_L↙
(e)'; 'EX_ca2(e)'; 'EX_cl(e)'; 'EX_co(e)';'EX_fe2(e)'; 'EX_fe3(e)'; 'EX_k(e)';
'EX_na1↙
(e)'; 'EX_i(e)'; 'EX_sel(e)'; 'EX_co2(e)'; 'EX_h(e)'; 'EX_h2o(e)'; 'EX_hco3↙
(e)';'EX_nh4(e)'; 'EX_o2(e)'; 'EX_pi(e)'; 'EX_so4(e)'};
[model_A] = setQualitativeConstraints(modelatart, cond1_uptake, cond1_uptake_-
LODs,↙
cond1_secretion, cond1_secretion_LODs,cellConc, t, cellWeight, ambiguous_metabo-
lites,↙
basisMedium);%celConc,t and cellWeight defined with medium
%%%%%Check that the bounds make sense
switched = model_A.rxns(find(model_A.lb > model_A.ub));%%%%Bounds set the wrong↙
way round
switcheduptk = intersect(cond1_uptake,switched);
switchedsecr = intersect(cond1_secretion,switched);
switchedidx = find(model_A.lb > model_A.ub);
%%%%%Put them the right way (depending on uptake or secretion
for i = 1:length(switched)

      if ismember(switcheduptk,switched(i)) == 1
            model_A.lb(switchedidx(i)) = mean(model_A.lb(find(ismember(model_A.
```

```
rxns,↙
cond1_uptake))));
        model_A.ub(switchedidx(i)) = mean(model_A.ub(find(ismember(model_A.rxns,↙
cond1_uptake))));
        elseif ismember(switchedsecr,switched(i)) == 1
        model_A.lb(switchedidx(i)) = mean(model_A.lb(find(ismember(model_A.rxns,↙
cond1_secretion))));
        model_A.ub(switchedidx(i)) = mean(model_A.ub(find(ismember(model_A.rxns,↙
cond1_secretion))));
        else
        model_A.lb(switchedidx(i)) = mean(model_A.lb);
        model_A.ub(switchedidx(i)) = mean(model_A.ub);
end
end
GrowthRxn = 'biomass_reaction';
tolerance = 20;
doublingTimeA = 19.6; %MOLT4 cells
[model_A_BM] = setConstraintsOnBiomassReaction(modelA, GrowthRxn, doublingTi-
meA,↙
tolerance);
% dataGenes = expressionData.gene(find(expressionData.value < 0)); % set of
genes↙
absent
% [model_A_GE] = integrateGeneExpressionData(model_A_BM, dataGenes);
% theshold = 1e-6;
% model = model_A_GE;
% [model_Afinal] = extractConditionSpecificModel(model, theshold);
%%%%%vi.i Quantitative
%%%Perform regression to find data
% % Convert concentration data from MS to flux data
% % Use regression to fit a line through all data points and use slope as change↙
in concentration and convert it to flux.
% % Unit of flux: micromol/gmDryWt/hr
[NUM1,TXT1,RAW1] = xlsread('massspec_infoALTERNATE.xlsx',1,'A1:AT13'); %New data
%↙
Figure 1f
metNames = TXT1(1,2:end);
Vol = 0.001; %L New data
tp1 = 4;hrs
tp2 = 8;hrs
tp3 = 24;hrs
%%%Cell Weight and concentration as above
%% HUVEC
fluxdata = [{'Metabolites'},{'Rate'},{'LB'},{'UB'}];
for i = 2:length(TXT1)
```

```
        time0 = NUM1((find(NUM1(:,1) == 0)),i);
        time1 = NUM1((find(NUM1(:,1) == tp1)),i);
        time2 = NUM1((find(NUM1(:,1) == tp2)),i);
        time3 = NUM1((find(NUM1(:,1) == tp3)),i);

        y= [transpose(time0);transpose(time1);transpose(time2);transpose(time3)];
        x = [0;tp1;tp2;tp3];
        xreg = cat(2,[x;x;x],ones(3*length(x),1));
        [b,bint] = regress([y(:,1);y(:,2);y(:,3)],xreg);

        flux = (Vol*((b(1)/(cellConc*cellWeight)))*1000); % micromol/gmDryWT/hr
        lb = (Vol*((bint(1,1)/(cellConc*cellWeight)))*1000); % micromol/gmDryWT/
hr
        ub = (Vol*((bint(1,2)/(cellConc*cellWeight)))*1000); % micromol/gmDryWT/
hr
        fluxdata = cat(1,fluxdata,[RAW1(1,i),flux,lb,ub]);

end
fluxdata = fluxdata(2:end,:);
samples = HUVEC;
exchanges = fluxdata(:,1);
metData = fluxdata(:,2);
modelstart = tissueModelA;% or modelMedium
test_max = 1000;
test_min = 1e-6;
outputPath = pwd;
prepIntegrationQuant(modelstart, metData, exchanges, samples, test_max, test_-
min,↙
outputPath);

uptL = length(find(metData < 0));
secrL = length(find(metData > 0));
nmets = max(vertcat(uptL,secrL));
[mapped_exchanges, minMax, mapped_uptake, mapped_secretion] =↙
checkExchangeProfiles(samples, outputPath, nmets);

solverQuant = 'ibm_cplex';
changeCobraSolver(solverQuant, 'LP');
minGrowth = 0.001;% lower bound to the biomass reaction obj
obj = 'biomass_reaction';
no_secretion = {'EX_o2(e)'};
no_uptake = {'EX_o2s(e)', 'EX_h2o2(e)'};
addExtraExch = {};% metabolite exchanges that are added to the upper and lower↙
bounds
addExtraExch_value = 1;% flux values that are added to the upper and lower bounds
medium = {};% reactions that should be excluded from minimization of exchanges
tol = 1e-6;
model = tissueModelA;% or modelMedium
```

```
epsilon = 1e-6;
[ResultsAllCellLines, OverViewResults] = setQuantConstraints(model, samples, tol, minGrowth, obj, no_secretion,no_uptake, medium, addExtraExch, addExtraExch_value, outputPath);

%%%%%%%%%%vii. Check and gene essentiality
startmodel = RECON2m2;
tol = 1e-4;
GrowthR = zeros(length(startmodel.rxns),1);
for i = 1:length(startmodel.rxns)
        modelt = changeRxnBounds(startmodel,startmodel.rxns(i),0,'b');
        FBAt = optimizeCbModel(modelt);
        if length(FBAt.obj) == 0
                GrowthR(i) = 0;
        else GrowthR(i) = FBAt.obj;
        end
end
LethalR = startmodel.rxns(find(GrowthR == 0));
SubLethalR = setdiff(startmodel.rxns(find(GrowthR < tol)),LethalR);

startmodel = RECON2m2;
tol = 1e-4;
GrowthG = zeros(length(startmodel.genes),1);
for i = 1:length(startmodel.genes)
        [~,rxnsfromgenes] = findRxnsFromGenes(startmodel,startmodel.genes(i),0,1);
        modelt = changeRxnBounds(startmodel,rxnsfromgenes(:,1),0,'b');
        FBAt = optimizeCbModel(modelt);
        if length(FBAt.obj) == 0
                GrowthG(i) = 0;
        else GrowthG(i) = FBAt.obj;
        end
end
LethalG = RECON2m2.genes(find(GrowthG == 0));
SubLethalG = setdiff(RECON2m2.genes(find(GrowthG < tol)),LethalG);

%%%%%%%%%%viii. If necessary create intersection models
base = RECON2m2;
model_A = read_CbModel('model_A.mat')%A model derived from base
model_B = read_CbModel('model_B.mat')%A model derived from base
Rxns = Unique(vertcat(modelA.rxns,modelB.rxns));
notRxns = setdiff(base.rxns,Rxns);
baseDEL = deleteRxns(base,notRxns);
notA = setdiff(baseDEL.rxns,modelA.rxns);
notB = setdiff(baseDEL.rxns,modelB.rxns);

modelALL_A = changeRxnBounds(base,notA,0,'b');
```

```
for i = 1:length(model_A.rxns)
        modelALL_A = changeRxnBounds(modelALL_A,model_A.rxns(i),model_A.lb(i),'l');
        modelALL_A = changeRxnBounds(modelALL_A,model_A.rxns(i),model_A.ub(i),'u');
end

modelALL_B = changeRxnBounds(base,notB,0,'b');
for i = 1:length(model_B.rxns)
        modelALL_B = changeRxnBounds(modelALL_B,model_B.rxns(i),model_B.lb(i),'l');
        modelALL_B = changeRxnBounds(modelALL_A,model_B.rxns(i),model_B.ub(i),'u');
end
%%%%%%%%%%ix Run Random sampling
warmupn = 2000;
stepsPerPoint = 500;
maxTime = 3600000;

nFiles = 10;
pointsPerFile = 1000;
fileBaseNo = 0;
outputPath = pwd;

fileName = 'pointsA';
performSampling(modelALL_A, warmupn, fileName, nFiles, pointsPerFile, stepsPerPoint, fileBaseNo, maxTime, outputPath);

fileName = 'pointsB';
performSampling(modelALL_B, warmupn, fileName, nFiles, pointsPerFile, stepsPerPoint, fileBaseNo, maxTime, outputPath);

%%%%%%%%%%x Compare sampling outputs
%%%Reload and join together all of the points from the above files
load('pointsA_1.mat')
pointsA = points;
load('pointsA_2.mat')
pointsA = horzcat(pointsA,points);
load('pointsA_3.mat')
pointsA = horzcat(pointsA,points);
load('pointsA_4.mat')
pointsA = horzcat(pointsA,points);
load('pointsA_5.mat')
pointsA = horzcat(pointsA,points);
load('pointsA_6.mat')
pointsA = horzcat(pointsA,points);
load('pointsA_7.mat')
pointsA = horzcat(pointsA,points);
```

```
load('pointsA_8.mat')
pointsA = horzcat(pointsA,points);
load('pointsA_9.mat')
pointsA = horzcat(pointsA,points);
load('pointsA_10.mat')
pointsA = horzcat(pointsA,points);
load('pointsB_1.mat')
pointsB = points;
load('pointsB_2.mat')
pointsB = horzcat(pointsB,points);
load('pointsB_3.mat')
pointsB = horzcat(pointsB,points);
load('pointsB_4.mat')
pointsB = horzcat(pointsB,points);
load('pointsB_5.mat')
pointsB = horzcat(pointsB,points);
load('pointsB_6.mat')
pointsB = horzcat(pointsB,points);
load('pointsB_7.mat')
pointsB = horzcat(pointsB,points);
load('pointsB_8.mat')
pointsB = horzcat(pointsB,points);
load('pointsB_9.mat')
pointsB = horzcat(pointsB,points);
load('pointsB_10.mat')
pointsB = horzcat(pointsB,points);

sampleStats = calcSampleStats({pointsA,pointsB});
meanA = sampleStats.mean(:,1);
meanB = sampleStats.mean(:,2);
ratio = zeros length(meanA);
for i = 1:length(meanA)
      ratio(i) = meanA(i)/meanB(i);
end
[~, pVals] = compareTwoSamplesStat(pointsA,pointsB,'ks');
ratioSIGidx = find(pVals < 0.01);
ratioAupidx = find(ratio > 1);
ratioBupidx = find(ratio < 1);

SIGUPA = intersect(model_A.rxns(ratioSIGidx),model_A.rxns(ratioAupidx));
SIGUPB = intersect(model_B.rxns(ratioSIGidx),model_B.rxns(ratioBupidx));
SIGUPAidx = find(ismember(model_A.rxns,SIGUPA));
SIGUPBidx = find(ismember(model_B.rxns,SIGUPB));

SIG = model_A.rxns(ratioSIGidx);

enrichedA = FEA(model_A,SIGUPAidx, 'subSystems');
enrichedB = FEA(model_B,SIGUPBidx, 'subSystems');
```

```

plotSampleHist(SIG,{pointsA,pointsB}, {model_A,model_B},100,9)

%%%%%%%%%%xi MOMA
% Perform MOMA:
[solutionB, solutionA, totalFluxDiff, ~] = MOMA(modelALL_A, modelALL_B,'max');

% Identify the subsystems that need most adjustment for model A to take on
% a phenotype closer to B:
highest_rxns_adj = zeros(length(modelALL_A.rxns),1);
for i=1:length(modelA.rxns)
       highest_rxns_adj(i) = abs((solutionB.x(i)-solutionA.full(i))/solutionB.x(i));
end

% Take care of NaN or Inf values:
for i=1:length(highest_rxns_adj)
       if isnan(highest_rxns_adj(i)) || isinf(highest_rxns_adj(i))
              highest_rxns_adj(i) = abs(solutionA.full(i));
       end
end
% Sort the reactions based on amount of adjustment:
[a_sorted, a_order] = sort(highest_rxns_adj,'descend');
ranked_reactions = modelALL_A.rxns(a_order,:);

% Perform FEA analysis:
id_A_to_B = find(ismember(modelALL_A.rxns,ranked_reactions(1:100)));
subsystems_A_to_B = FEA(modelALL_A,id_A_to_B, 'subSystems');
```

References

1. Resendis-Antonio O (2013) Constraint-based modeling. In: Encyclopedia of systems biology. Springer, New York, NY, pp 494–498
2. Thiele I, Palsson B (2010) A protocol for generating a high-quality genome-scale metabolic reconstruction. Nat Protoc 5:93–121
3. Becker S, Feist A, Mo M et al (2007) Quantitative prediction of cellular metabolism with constraint-based models: the COBRA Toolbox. Nat Protoc 2:727–738
4. Machado D, Herrgård M (2014) Systematic evaluation of methods for integration of transcriptomic data into constraint-based models of metabolism. PLoS Comput Biol 10: e1003580
5. Becker SA, Palsson BO (2008) Context-specific metabolic networks are consistent with experiments. PLoS Comput Biol 4: e1000082
6. Orth J, Thiele I, Palsson B (2010) What is flux balance analysis? Nat Biotechnol 28:245–248
7. Opdam S, Richelle A, Kellman B et al (2017) A systematic evaluation of methods for tailoring genome-scale metabolic models. Cell Syst 4:318–329.e6
8. Wang Y, Eddy JA, Price ND (2012) Reconstruction of genome-scale metabolic models for 126 human tissues using mCADRE. BMC Syst Biol 6:153
9. Vlassis NN, Pacheco MPM, Sauter TTT et al (2014) Fast reconstruction of compact

context-specific metabolic network models. PLoS Comput Biol 10:e1003424
10. Schultz A, Qutub AA (2016) Reconstruction of tissue-specific metabolic networks using CORDA. PLoS Comput Biol 12:e1004808
11. Aurich MK, Fleming RMT, Thiele I (2016) MetaboTools: a comprehensive toolbox for analysis of genome-scale metabolic models. Front Physiol 7:327
12. García Martín H, Kumar VS, Weaver D et al (2015) A method to constrain genome-scale models with 13C labeling data. PLoS Comput Biol 11:e1004363
13. Patella F, Schug Z, Persi E et al (2015) Proteomics-based metabolic modeling reveals that fatty acid oxidation (FAO) controls endothelial cell (EC) permeability. Mol Cell Proteomics 14:621–634
14. Ryu JY, Kim HU, Lee SY (2017) Framework and resource for more than 11,000 gene-transcript-protein-reaction associations in human metabolism. Proc Natl Acad Sci 114: E9740–E9749
15. Haraldsdóttir HS, Preciat Gonzalez GA, Systems Biochemistry Group U of L Atomically resolve a metabolic reconstruction—the COBRA Toolbox. https://opencobra.github.io/cobratoolbox/stable/tutorials/tutorialAtomicallyResolveReconstruction.html
16. Fleming R, Thiele I Proton shuttle testing with sparse flux balance analysis—the COBRA Toolbox. https://opencobra.github.io/cobratoolbox/stable/tutorials/tutorialSparseFBA_protonShuttle.html
17. Thiele I, Fleming R FastGapFill tutorial—the COBRA Toolbox. https://opencobra.github.io/cobratoolbox/stable/tutorials/tutorialFastGapFill.html
18. Thiele I Example use of functions listed in the Standard operating procedure for metabolic reconstruction—the COBRA Toolbox. https://opencobra.github.io/cobratoolbox/stable/tutorials/tutorialReconstructionSOP.html
19. Thiele I, Fleming R Test physiologically relevant ATP yields from different carbon sources for a metabolic model—the COBRA Toolbox. https://opencobra.github.io/cobratoolbox/stable/tutorials/tutorialModelATPYield.html
20. Fleming R, Thiele I Testing chemical and biochemical fidelity—the COBRA Toolbox. https://opencobra.github.io/cobratoolbox/stable/tutorials/tutorialFidelityTesting.html
21. Thiele I, Fleming R Testing basic properties of a metabolic model (aka sanity checks)—the COBRA Toolbox. https://opencobra.github.io/cobratoolbox/stable/tutorials/tutorialModelSanityChecks.html
22. Achour B, Dantonio A, Niosi M et al (2017) Quantitative characterization of major hepatic UDP-glucuronosyltransferase enzymes in human liver microsomes: comparison of two proteomic methods and correlation with catalytic activity. Drug Metab Dispos 45:1102–1112
23. Edfors F, Danielsson F, Hallström BM et al (2016) Gene-specific correlation of RNA and protein levels in human cells and tissues. Mol Syst Biol 12:883
24. Zur H, Ruppin E, Shlomi T (2010) iMAT: an integrative metabolic analysis tool. Bioinformatics 26:3140–3142
25. Agren R, Bordel S, Mardinoglu A et al (2012) Reconstruction of genome-scale active metabolic networks for 69 human cell types and 16 cancer types using INIT. PLoS Comput Biol 8:e1002518
26. Jerby L, Shlomi T, Ruppin E (2010) Computational reconstruction of tissue-specific metabolic models: application to human liver metabolism. Mol Syst Biol 6:401
27. Pfau T, Richelle A Extraction of context-specific models—the COBRA Toolbox. https://opencobra.github.io/cobratoolbox/stable/tutorials/tutorialExtractionTranscriptomic.html
28. Aurich MK, Arreckx S Metabotools tutorial I—the COBRA Toolbox. https://opencobra.github.io/cobratoolbox/stable/tutorials/tutorialMetabotoolsI.html
29. Aurich MK, Arreckx S Metabotools tutorial II – integration of quantitative metabolomic data—the COBRA Toolbox. https://opencobra.github.io/cobratoolbox/stable/tutorials/tutorialMetabotoolsII.html
30. Feist AM, Palsson BO (2010) The biomass objective function. Curr Opin Microbiol 13:344–349
31. Yuan H, Cheung CYM, Hilbers PAJ et al (2016) Flux balance analysis of plant metabolism: the effect of biomass composition and model structure on model predictions. Front Plant Sci 7:537
32. Müller AC, Bockmayr A (2013) Fast thermodynamically constrained flux variability analysis. 29:903–909
33. Gudmundsson S, Thiele I (2010) Computationally efficient flux variability analysis. BMC Bioinformatics 11:489
34. Megchelenbrink W, Huynen M, Marchiori E (2014) optGpSampler: an improved tool for uniformly sampling the solution-space of

genome-scale metabolic networks. PLoS One 9:e86587
35. Schellenberger J, Palsson BØ (2009) Use of randomized sampling for analysis of metabolic networks. J Biol Chem 284:5457–5461
36. De MD, Mori M, Parisi V (2015) Uniform sampling of steady states in metabolic networks: heterogeneous scales and rounding. PLoS One 10:e0122670
37. Haraldsdóttir HS, Cousins B, Thiele I et al (2017) CHRR: coordinate hit-and-run with rounding for uniform sampling of constraint-based models. Bioinformatics 33:1741–1743
38. Haraldsdóttir HS, Preciat Gonzalez GA Uniform sampling—the COBRA Toolbox. https://opencobra.github.io/cobratoolbox/stable/tutorials/tutorialUniformSampling.html
39. Segre D, Vitkup D, Church GM (2002) Analysis of optimality in natural and perturbed metabolic networks. Proc Natl Acad Sci 99:15112–15117
40. MOMA—the COBRA Toolbox. https://opencobra.github.io/cobratoolbox/stable/modules/analysis/MOMA/index.html?highlight=moma
41. Heirendt L, Arreckx S, Pfau T et al (2019) Creation and analysis of biochemical constraint-based models using the COBRA Toolbox v.3.0. Nat Protoc 14:639–702
42. Schellenberger J, Que R, Fleming R et al (2011) Quantitative prediction of cellular metabolism with constraint-based models: the COBRA Toolbox v2.0. Nat Protoc 6:1290–1307
43. ArrayExpress; EMBL-EBI. https://www.ebi.ac.uk/arrayexpress/
44. Kolesnikov NN, Hastings EE, Keays MM et al (2015) ArrayExpress update—simplifying data submissions. Nucleic Acids Res 43: D1113–D1116
45. Kauffmann A, Rayner TF, Parkinson H et al (2009) Importing ArrayExpress datasets into R/bioconductor. 25:2092–2094
46. Home - GEO - NCBI. https://www.ncbi.nlm.nih.gov/geo/
47. Barrett T, Troup DB, Wilhite SE et al (2011) NCBI GEO: archive for functional genomics data sets—10 years on. Nucleic Acids Res 39: D1005–D1010
48. PRIDE Archive. https://www.ebi.ac.uk/pride/archive/
49. Vizcaino JA, Cote R, Csordas A et al (2013) The Proteomics Identifications (PRIDE) database and associated tools: status in 2013. Nucleic Acids Res 41:1063–1069
50. PeptideAtlas. http://www.peptideatlas.org/. O112.026617
51. Deutsch EW, Csordas A, Sun Z et al (2017) The ProteomeXchange consortium in 2017: supporting the cultural change in proteomics public data deposition. Nucleic Acids Res 45: D1100–D1106
52. Ruffier M, Kähäri A, Komorowska M et al (2017) Ensembl core software resources: storage and programmatic access for DNA sequence and genome annotation. Database (Oxford) 2017
53. Welcome to MassIVE. https://massive.ucsd.edu/ProteoSAFe/static/massive.jsp?redirect=auth
54. iProX – integrated proteome resources. https://www.iprox.org/
55. Ma J, Chen T, Wu S et al (2019) iProX: an integrated proteome resource. Nucleic Acids Res 47:D1211–D1217
56. jPOST. Japan proteome standard repository/database. https://jpostdb.org/
57. Okuda S, Watanabe Y, Moriya Y et al (2017) jPOSTrepo: an international standard data repository for proteomes. Nucleic Acids Res 45:D1107–D1111
58. The Human Protein Atlas. https://www.proteinatlas.org/
59. MetaboLights – metabolomics experiments and derived information. https://www.ebi.ac.uk/metabolights/
60. Haug K, Salek RM, Conesa P et al (2013) MetaboLights—an open-access general-purpose repository for metabolomics studies and associated meta-data. Nucleic Acids Res 41: D781–D786
61. GenomeRNAi – a database for RNAi phenotypes and reagents. http://www.genomernai.org/
62. Schmidt EE, Pelz O, Buhlmann S et al (2013) GenomeRNAi: a database for cell-based and in vivo RNAi phenotypes, 2013 update. Nucleic Acids Res 41:D1021–D1026
63. Japanese Genotype-Phenotype Archive – home. https://www.ddbj.nig.ac.jp/jga/index-e.html
64. Home. European Genome-Phenome Archive. https://www.ebi.ac.uk/ega/home
65. Home – dbGaP – NCBI. https://www.ncbi.nlm.nih.gov/gap/
66. Tryka KA, Hao L, Sturcke A et al (2014) NCBI's database of genotypes and phenotypes: dbGaP. Nucleic Acids Res 42: D975–D979
67. Lappalainen I, Almeida-King J, Kumanduri V et al (2015) The European Genome-

Phenome Archive of human data consented for biomedical research. Nat Genet 47:692–695

68. Thiele I, Swainston N, Fleming R et al (2013) A community-driven global reconstruction of human metabolism. Nat Biotechnol 31:419–425
69. Brunk E, Sahoo S, Zielinski DC et al (2018) Recon3D enables a three-dimensional view of gene variation in human metabolism. Nat Biotechnol 36:272–281
70. Swainston N, Smallbone K, Hefzi H et al (2016) Recon 2.2: from reconstruction to model of human metabolism. Metabolomics 12:109
71. Duarte N, Becker S, Jamshidi N et al (2007) Global reconstruction of the human metabolic network based on genomic and bibliomic data. Proc Natl Acad Sci 104:1777–1782
72. Björnson E, Mukhopadhyay B, Asplund A et al (2015) Stratification of hepatocellular carcinoma patients based on acetate utilization. Cell Rep 13:2014–2026
73. Mardinoglu A, Agren R, Kampf C et al (2013) Integration of clinical data with a genome-scale metabolic model of the human adipocyte. Mol Syst Biol 9:649
74. BiGG models: a platform for integrating, standardizing and sharing genome-scale models. http://bigg.ucsd.edu/
75. Virtual Metabolic Human. https://www.vmh.life/
76. Noronha A, Modamio J, Jarosz Y et al (2018) The Virtual Metabolic Human database: integrating human and gut microbiome metabolism with nutrition and disease. Nucleic Acids Res 47:D614–D624
77. Metabolics Atlas. https://metabolicatlas.org/
78. Sigurdsson MI, Jamshidi N, Steingrimsson E et al (2010) A detailed genome-wide reconstruction of mouse metabolism based on human Recon 1. BMC Syst Biol 4:140
79. Mardinoglu A, Shoaie S, Bergentall M et al (2015) The gut microbiota modulates host amino acid and glutathione metabolism in mice. Mol Syst Biol 11:834–834
80. Küken A, Nikoloski Z (2019) Computational approaches to design and test plant synthetic metabolic pathways. Plant Physiol 179:894–906
81. Scheunemann M, Brady SM, Nikoloski Z (2018) Integration of large-scale data for extraction of integrated Arabidopsis root cell-type specific models. Sci Rep 8:7919
82. de OD'MCG, Quek L-E, Palfreyman RW et al (2010) AraGEM, a genome-scale reconstruction of the primary metabolic network in Arabidopsis. Plant Physiol 152:579–589
83. Lakshmanan M, Lim S-H, Mohanty B et al (2015) Unraveling the light-specific metabolic and regulatory signatures of rice through combined in silico modeling and multiomics analysis. Plant Physiol 169:3002–3020
84. Yuan H, Cheung CYM, Poolman MG et al (2016) A genome-scale metabolic network reconstruction of tomato (Solanum lycopersicum L.) and its application to photorespiratory metabolism. Plant J 85:289–304
85. Seaver SMD, Bradbury LMT, Frelin O et al (2015) Improved evidence-based genome-scale metabolic models for maize leaf, embryo, and endosperm. Front Plant Sci 6:142
86. Saha R, Suthers PF, Maranas CD (2011) Zea mays iRS1563: a comprehensive genome-scale metabolic reconstruction of maize metabolism. PLoS One 6:e21784
87. Bogart E, Myers CR (2016) Multiscale metabolic modeling of C4 plants: connecting nonlinear genome-scale models to leaf-scale metabolism in developing maize leaves. PLoS One 11:e0151722
88. Hastings J, Mains A, Artal-Sanz M et al (2017) WormJam: a consensus C. elegans metabolic reconstruction and metabolomics community and workshop series. Worm 6: e1373939
89. Ma L, Chan AHC, Hattwell J et al (2017) Systems biology analysis using a genome-scale metabolic model shows that phosphine triggers global metabolic suppression in a resistant strain of *C elegans*. biorxiv 2017:144386
90. Yilmaz LS, Walhout AJM (2016) A Caenorhabditis elegans genome-scale metabolic network model. Cell Syst 2:297–311
91. Gebauer J, Gentsch C, Mansfeld J et al (2016) A genome-scale database and reconstruction of Caenorhabditis elegans metabolism. Cell Syst 2:312–322
92. Witting M, Hastings J, Rodriguez N et al (2018) Modeling meets metabolomics—the WormJam consensus model as basis for metabolic studies in the model organism Caenorhabditis elegans. Front Mol Biosci 5:96
93. Massaiu I, Pasotti L, Sonnenschein N et al (2019) Integration of enzymatic data in Bacillus subtilis genome-scale metabolic model improves phenotype predictions and enables in silico design of poly-γ-glutamic acid production strains. Microb Cell Factories 18:3
94. Zou W, Xiong X, Zhang J et al (2018) Reconstruction and analysis of a genome-scale metabolic model of Methylovorus sp. MP688, a

high-level pyrroloquinolone quinone producer. Biosystems 172:37–42
95. McDermott JE, Yoon H, Nakayasu ES et al (2011) Technologies and approaches to elucidate and model the virulence program of salmonella. Front Microbiol 2:121
96. Simeonidis E, Price ND (2015) Genome-scale modeling for metabolic engineering. J Ind Microbiol Biotechnol 42:327–338
97. Wang H, Marcišauskas S, Sánchez BJ et al (2018) RAVEN 2.0: a versatile toolbox for metabolic network reconstruction and a case study on Streptomyces coelicolor. PLoS Comput Biol 14:e1006541
98. Bordbar A, Lewis NE, Schellenberger J et al (2010) Insight into human alveolar macrophage and M. tuberculosis interactions via metabolic reconstructions. Mol Syst Biol 6:422
99. Magnúsdóttir S, Heinken A, Kutt L et al (2016) Generation of genome-scale metabolic reconstructions for 773 members of the human gut microbiota. Nat Biotechnol 35:81–89
100. Rosario D, Benfeitas R, Bidkhori G et al (2018) Understanding the representative gut microbiota dysbiosis in metformin-treated type 2 diabetes patients using genome-scale metabolic modeling. Front Physiol 9:775
101. Machado D, Andrejev S, Tramontano M et al (2018) Fast automated reconstruction of genome-scale metabolic models for microbial species and communities. Nucleic Acids Res 46:7542–7553
102. KEGG: Kyoto Encyclopedia of Genes and Genomes. https://academic.oup.com/nar/article/47/D1/D590/5128935
103. MetaCyc Metabolic Pathway Database. https://metacyc.org/
104. Enzyme Database – BRENDA. https://www.brenda-enzymes.org/
105. Barrett AJ (1995) Enzyme nomenclature. Recommendations 1992. Supplement 2: corrections and additions (1994). Eur J Biochem 232:1–1
106. Jeske L, Placzek S, Schomburg I et al (2019) BRENDA in 2019: a European ELIXIR core data resource. Nucleic Acids Res 47: D542–D549
107. Home – Reactome Pathway Database. https://reactome.org/
108. BioModels. http://www.ebi.ac.uk/biomodels/
109. Chelliah V, Juty N, Ajmera I et al (2015) BioModels: ten-year anniversary. Nucleic Acids Res 43:D542–D548
110. Hucka M, Bergmann FT, Dräger A et al (2018) The Systems Biology Markup Language (SBML): language specification for level 3 version 2 core. J Integr Bioinform 15
111. Büchel F, Rodriguez N, Swainston N et al (2013) Path2Models: large-scale generation of computational models from biochemical pathway maps. BMC Syst Biol 7:116
112. Heirendt L, Arreckx S, Pfau T, et al. The COnstraint-based reconstruction and analysis toolbox. https://opencobra.github.io/cobratoolbox/stable/
113. MATLAB – MathWorks – MATLAB; Simulink. https://uk.mathworks.com/products/matlab.html
114. CPLEX Optimizer. IBM. https://www.ibm.com/analytics/cplex-optimizer
115. Gurobi optimization – the state-of-the-art mathematical programming solver. http://www.gurobi.com/
116. Git – downloading package. https://git-scm.com/download/win
117. Installation—the COBRA Toolbox. https://opencobra.github.io/cobratoolbox/stable/installation.html
118. Fleming R, Thiele I. Sparse flux balance analysis test for a minimial stoichiometrically balanced cycle involving ATP hydrolysis—the COBRA Toolbox. https://opencobra.github.io/cobratoolbox/stable/tutorials/tutorialSparseFBA_freeATPtest.html
119. Schellenberger J, Lewis NE, Palsson BØ (2011) Elimination of thermodynamically infeasible loops in steady-state metabolic models. Biophys J 100:544–553
120. Fleming RMT, Maes CM, Saunders MA et al (2012) A variational principle for computing nonequilibrium fluxes and potentials in genome-scale biochemical networks. J Theor Biol 292:71–77
121. Le THA, Pham Dinh T, Le HM et al (2015) DC approximation approaches for sparse optimization. Eur J Oper Res 244:26–46
122. Bianconi E, Piovesan A, Facchin F et al (2013) An estimation of the number of cells in the human body. Ann Hum Biol 40:463–471
123. Huang DW, Sherman BT, Lempicki RA (2009) Systematic and integrative analysis of large gene lists using DAVID bioinformatics resources. Nat Protoc 4:44–57
124. Agarwala R, Barrett T, Beck J et al (2018) Database resources of the National Center for Biotechnology Information. Nucleic Acids Res 46:D8–D13

125. Dennis G, Sherman BT, Hosack DA et al (2003) DAVID: database for annotation, visualization, and integrated discovery. Genome Biol 4:P3
126. King ZA, Dräger A, Ebrahim A et al (2015) Escher: a web application for building, sharing, and embedding data-rich visualizations of biological pathways. PLoS Comput Biol 11:e1004321
127. escher. https://escher.github.io/#/
128. Ebrahim A, Lerman JA, Palsson BO et al (2013) COBRApy: COnstraints-based reconstruction and analysis for python. BMC Syst Biol 7:74
129. Agren R, Liu L, Shoaie S et al (2013) The RAVEN toolbox and its use for generating a genome-scale metabolic model for Penicillium chrysogenum. PLoS Comput Biol 9: e1002980
130. Brazma A, Hingamp P, Quackenbush J et al (2001) Minimum information about a microarray experiment (MIAME)-toward standards for microarray data. Nat Genet 29:365–371
131. MINSEQE: Minimum Information about a high—throughput Nucleotide SeQuencing Experiment—a proposal for standards in functional genomic data reporting (2012)
132. Smith AC, Eyassu F, Mazat J-P et al (2017) MitoCore: a curated constraint-based model for simulating human central metabolism. BMC Syst Biol 11:114
133. Thiele I, Vlassis N, Fleming RMT (2014) fastGapFill: efficient gap filling in metabolic networks. Bioinformatics 30:2529–2531
134. Salvy P, Fengos G, Ataman M et al (2018) pyTFA and matTFA: a Python package and a Matlab toolbox for thermodynamics-based flux analysis, vol 35, pp 167–169
135. matTFA. https://github.com/EPFL-LCSB/matTFA/tree/master/ext
136. Noor E, Haraldsdóttir HS, Milo R et al (2013) Consistent estimation of gibbs energy using component contributions. PLoS Comput Biol 9:e1003098
137. Haraldsdóttir HS, Thiele I, Fleming RMT (2012) Quantitative assignment of reaction directionality in a multicompartmental human metabolic reconstruction. Biophys J 102:1703–1711
138. Fleming RMT, Thiele I (2011) von Bertalanffy 1.0: a COBRA toolbox extension to thermodynamically constrain metabolic models. Bioinformatics 27:142–143
139. Thermodynamically constrain a metabolic model—the COBRA Toolbox. https://opencobra.github.io/cobratoolbox/stable/tutorials/tutorialVonBertalanffy.html
140. Reconstruction—the COBRA Toolbox. https://opencobra.github.io/cobratoolbox/latest/modules/reconstruction/index.html
141. Kim M, Yi JS, Lakshmanan M et al (2015) Transcriptomics-based strain optimization tool for designing secondary metabolite overproducing strains of Streptomyces coelicolor. Biotechnol Bioeng 113(3):651–660
142. Jóhannsson F, Guðmundsson S, Paglia G et al (2018) Systems analysis of metabolism in platelet concentrates during storage in platelet additive solution. Biochem J 475:2225–2240
143. Mcgarrity S, Anuforo Ó, Halldórsson H et al (2018) Metabolic systems analysis of LPS induced endothelial dysfunction applied to sepsis patient stratification. Sci Rep 8:6811
144. Halldorsson S, Rohatgi N, Magnusdottir M et al (2017) Metabolic re-wiring of isogenic breast epithelial cell lines following epithelial to mesenchymal transition. Cancer Lett 396:117–129

Chapter 12

Software Supporting a Workflow of Quantitative Dynamic Flux Maps Estimation in Central Metabolism from SIRM Experimental Data

Vitaly A. Selivanov, Silvia Marin, Josep Tarragó-Celada, Andrew N. Lane, Richard M. Higashi, Teresa W.-M. Fan, Pedro de Atauri, and Marta Cascante

Abstract

Stable isotope-resolved metabolomics (SIRM), based on the analysis of biological samples from living cells incubated with artificial isotope enriched substrates, enables mapping the rates of biochemical reactions (metabolic fluxes). We developed software supporting a workflow of analysis of SIRM data obtained with mass spectrometry (MS). The evaluation of fluxes starting from raw MS recordings requires at least three steps of computer support: first, extraction of mass spectra of metabolites of interest, then correction of the spectra for natural isotope abundance, and finally, evaluation of fluxes by simulation of the corrected spectra using a corresponding mathematical model. A kinetic model based on ordinary differential equations (ODEs) for isotopomers of metabolites of the corresponding biochemical network supports the final part of the analysis, which provides a dynamic flux map.

Key words Mass spectrometry, Stable isotope tracing, Isotopolog distribution, Central energy metabolism, Metabolic fluxes, Computational analysis, Kinetic models of metabolism, Stable isotope-resolved metabolomics

Abbreviations

ala	Alanine
cit	Citrate
e4p	Erythrose-4-phosphate
f6p	Fructose-6-phosphate
FBA	Flux balance analysis
fbp	Fructose bisphosphate
fum	Fumarate
g6p	Glucose-6-phosphate

Vitaly A. Selivanov and Silvia Marin contributed equally to this work.

Deepak Nagrath (ed.), *Metabolic Flux Analysis in Eukaryotic Cells: Methods and Protocols*, Methods in Molecular Biology, vol. 2088, https://doi.org/10.1007/978-1-0716-0159-4_12,

GC-MS	Gas chromatography–mass spectrometry
glc	Glucose
gln	Glutamine
glu	Glutamate
gly	Glycine
kg	Alpha-ketoglutarate
mal	Malate
MS	Mass spectrometry
NMR	Nuclear magnetic resonance
oaa	Oxaloacetate
ODEs	Ordinary differential equations
ppp	Pentose phosphate pathways
pro	Proline
pyr	Pyruvate
r5p	Ribose-5-phosphate
s7p	Sedoheptulose-7- phosphate
ser	Serine
SIM	Selected ion monitoring
SIRM	Stable isotope resolved metabolomics
t3p	Triose-3-phosphate
UHR-FTMS	Ultrahigh resolution Fourier transformed mass spectrometry

1 Introduction

Metabolism is the set of enzyme-catalyzed reactions that support all cellular functions via two coupled processes. Catabolism comprises those reactions resulting in the conversion of more complex, reduced organic compounds to simpler, more oxidized compounds, usually with the generation of metabolically usable energy in the form of ATP. Anabolism is the set of reactions that assembles complex molecules such as proteins from the simple subunits with the expenditure of metabolic energy. The rate of energy conversion in the metabolism determines a wide range of physiological processes, such as fast muscle contraction or cell proliferation [1]. Therefore, a full description of cell biochemistry also requires an analysis of metabolic rates, or fluxes.

Intracellular metabolic fluxes are often assessed by using stable isotope tracers (e.g., ^{13}C), which become incorporated into intermediates and products of metabolic networks [2]. By following the fate of precursor atoms through metabolic transformations, it is possible to reconstruct the active metabolic system and infer the relative fluxes through different pathways. This approach has many potential applications in biotechnology [3–5]. Moreover, changes in metabolic fluxes are associated with both normal physiological function (e.g., muscle contraction, heart rate) and pathological states. Although metabolism may be affected in a wide range of diseases [6], in some of them it is likely to be a driver of the disease progression, such as cancer, metabolic syndrome and diabetes [7–9]. Indeed, altered metabolism is known as a hallmark of cancer [10]. Quantitative changes in metabolism entail changes in metabolic fluxes manifested as alterations in the rates of production and utilization of metabolic intermediates. Stable isotope tracing can help discriminate between cases where use, supply or both are affected. Therefore a stable isotope-based flux analysis of primary cells isolated from patients or in ex vivo tissue slices can evaluate the difference in the metabolism of control and patient cells and their response to drugs. Such knowledge of the biochemical bases of diseases (e.g., cancer [11–18]) should lead to practical applications in biomedicine.

Stable isotopes can be traced in both the intracellular and extracellular metabolites using mass spectrometry (MS) or NMR [19, 20]. MS data, in turn, can be obtained either by direct fusion or after chromatographic separation. We present here our computational tools supporting a workflow of chromatography-based MS data. An assessment of metabolic fluxes based on stable isotope-resolved metabolomics (SIRM) data requires several sequential steps: (1) identification of the spectra of interest and extraction of the desired isotopolog distributions from raw recordings, (2) correction of the extracted data for natural isotope distribution, and (3) simulation of the corrected data using a mathematical model that provides a set of values of metabolic fluxes consistent with the data. Here we present some examples of MS data curation supported by computational tools, which we have developed. The analysis of data obtained using different MS techniques might require some modifications of the tools presented here. Figure 1 illustrates the steps of this workflow.

There are several computer programs available that, in addition to the applications already implemented in mass spectrometers, help to extract mass spectra of interest from raw MS recordings [21, 22] using a library of functions "XCMS" [23, 24], correct them for natural isotopes abundance [25–31], and simulate the artificial labeling of metabolites [32–36]. However, the existence of many similar programs indicates that no universal application can be used for all practical needs of MS data analysis. Specifically, we are not aware of computer programs adopted for the whole process of fluxomic analysis based on SIRM data. Our suite of applications supports such a study. The use of our programs allows avoiding manual transformation of the format of data between the subsequent steps of data analysis and thus makes the complete analysis automatic. In this way, this suite of programs can be applied for rapid analysis of large amounts of MS datasets to evaluate metabolic flux maps automatically. In the same time, the programs facilitate a visual check of data and intermediate results of the study.

The programs were checked for MS coupled with gas chromatography (GC-MS) and ultrahigh resolution Fourier transform MS (UHR-FTMS) data saved in NetCDF format, but they can be easily adapted to the specificity of other MS methods and formats of MS data.

2 Software

The three open source software tools support the three steps indicated in Fig. 1 of workflow of MS data analysis based on stable isotope tracing and aimed at mapping the dynamical intracellular metabolic fluxes.

1. *Ramid* is an "R" program that extracts mass spectra from NetCDF files containing the raw time course of m/z recording for metabolites of interest, specified by a short description in a text file. It is freely available at https://github.com/seliv55/ramidcor.

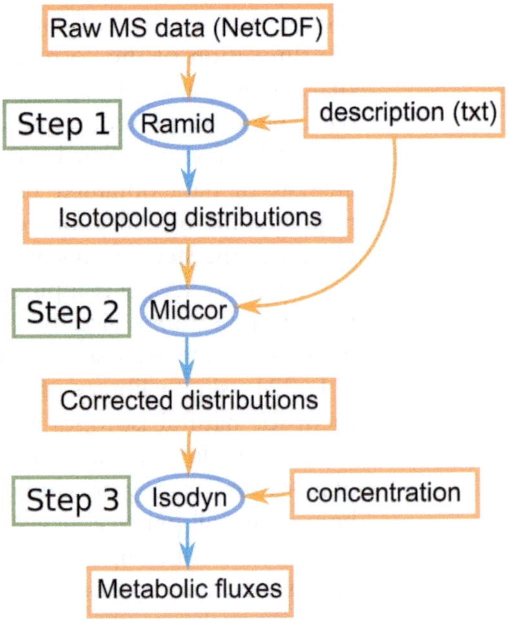

Fig. 1 Fluxomic workflow based on SIRM MS data. Three computational programs support three steps of the workflow of SIRM data analysis. Ramid extracts isotopolog distribution of the metabolites of interest from raw MS recording; Midcor corrects these distributions for natural isotope occurrence; Isodyn simulate these corrected data using a kinetic model of the corresponding reaction network that provides the metabolic fluxes in their dynamics

2. *Midcor* is an "R" program designed to further process isotopolog distribution, extracted from raw mass spectrometer recordings, by correcting it for naturally occurring isotopes and peaks overlapping [31]. The same above indicated repository includes it.

3. *Isodyn* is one of the very few software tools designed for simulation of dynamics of SIRM data under non–steady state conditions based on the numerical solution of ODEs. Isodyn is a "C++" program that implements a kinetic model to simulate dynamics of concentrations and corrected isotopolog distribution for the metabolites of interest for evaluating corresponding sets of metabolic fluxes. It is freely available at https://github.com/seliv55/isodyn.

4. *Workflow for dynamical flux map evaluation.* Midcor can use the output of Ramid directly for the isotopolog distribution correction, and Isodyn can simulate the distributions corrected by Midcor. Therefore all the three programs can be connected in the same workflow of data analysis, thus eliminating the time-consuming manual procedures of the data analysis. However, if necessary, the visual check and editing of the output of each program is also possible before starting the next step of the analysis. Such editing is often necessary, for example, to eliminate some outliers, which sometimes appear in the mass spectrometer recordings.

3 Methods

3.1 Using Ramid for Extraction of Isotopolog Distributions from Raw MS Data Saved in NetCDF files

1. Create a folder and put there the NetCDF files designed for the analysis (let us call it <cdfdir>). Each separate folder should contain NetCDF files of the same type (here either SIM, or scan, or UHR-FTMS, *see* **Notes 1** and **2**).

2. Create a text file containing general information (<info>) referred to the metabolites of interest present in each NetCDF file collected in the same directory for Ramid to process the MS data.

   (a) Table 1 shows an example of such a file for GC-MS data. An applied ionization technique can split the derivatized metabolite molecules into various ions, which can contain the whole molecule of the analyzed metabolite or a fragment of it. The space-separated table specifies retention time (RT) of the analyzed fragment, which is the same for all ions originated from a given derivatized metabolite, the m/z value (mz0) of the lightest isotopolog of the considered ion. The column "Fragment" specifies the positions of the first and last carbons of the considered fragment in the carbon skeleton of the entire molecule of the

Table 1
An example of the format of additional information needed for the extraction of isotopolog distribution for the substances registered in each of the NetCDF files provided by GCMS

#MAX number of detected ions, an indication of NOTRACER samples:					
MAX: 8,000,000					
NOTRACER: Cold					
Name	*RT*	*mz0*	*Fragment*	*Formula*	*Control*
Pyruvate	7.9	174.1	C_1C_3	$C_6H_{12}N_1O_3Si_1$	216.1
Lactate	11.1	261.1	C_1C_3	$C_{11}H_{25}O_3Si_2$	303.1
Alanine_C1C3	11.86	260.1	C_1C_3	$C_{11}H_{26}N_1O_2Si_2$	232.1
Alanine_C2C3	11.86	232.1	C_2C_3	$C_{10}H_{26}N_1OSi_2$	260.1

Columns: RT is retention time, mz0 is the lightest isotopolog, fragment is a fragment formed from the analyzed molecules due to electronic impact, formula is the complete formula of the derivatized fragment, control (*see* **Note 3**) is the mz0 value for some other fragment of the same metabolite if it exists (if not, the same value as in the column mz0 can be put here)

metabolite. The RT and mz0 values are used by Ramid to localize in the NetCDF files the distribution of isotopologs for the selected ion of a given metabolite, specified in the first column, and based on the number of carbons indicated in the column "Fragment" it calculates the desired number of isotopologs. The column "Formula" gives the numbers of all atoms that by naturally occurring isotopes affect the isotopolog distribution in the analyzed ion. Ramid does not use it, but it is necessary for the next step, correction for the natural isotope occurrence by Midcor. To verify that the peak identified for a metabolite is the correct one, the m/z value for the lightest isotopolog of a different "control" ion of the same derivatized metabolite, which always accompanies the main ion, is specified. The existence of control MS peaks with the equal RT value proves that the peak of the main ion is localized correctly (*see* **Note 3**). In the case we do not have information about a control ion, the same value mz0 should be put in the control column.

This table is preceded by a line of comments, followed by a line indicating the maximal number of ions that can be detected by the used mass spectrometer. If this number is not known, "0" should be put there, then a line indicating a word/character/text that only appears in the names of CDF files for the samples not containing tracers. In the case of absence of such files "NA" should be indicated.

(b) Table 2 shows an example of a file with the information to process UHR-FTMS data. Here the lower (mz0lowbd)

Table 2
An example of the description used for an analysis of UHR-FTMS data

Name	RT	mz0lowbd	mz0upbd	Fragment	Formula
13-bpg	35.21	264.951	264.953	C_1C_3	$C_3H_8O_{10}P_2$
2hg	15.42	147.029	147.031	C_1C_5	$C_5H_8O_5$
3pg2pg	25.78	184.985	184.987	C_1C_3	$C_3H_7O_7P$
a-kg	18.15	145.014	145.015	C_1C_5	$C_5H_6O_5$
aspartate	15.08	132.03	132.031	C_1C_4	$C_4H_7NO_4$
Cis-aconitate	29.5	173.008	173.01	C_1C_6	$C_6H_6O_6$
Citrate	27.25	191.019	191.021	C_1C_6	$C_6H_8O_7$

and upper (mz0upbd) bounds for the mz0 interval are indicated. Ramid calculates mz0 by integrating the intensities corresponding to the m/z located inside this interval. The difference in mass between ^{13}C and ^{12}C is 1.003355 atomic mass units. Therefore Ramid sets the m/z intervals for any isotopolog containing n ^{13}C atoms by adding n×1.003355 to the upper and lower bound indicated for mz0.

3. Check that names of NetCDF files contain the information, necessary to process each sample in subsequent steps of this workflow. For instance, the statistics on samples (applied after the peaks extraction and correction) requires a reference to a biological sample and injection. The subsequent simulation of the data requires grouping the samples by conditions, which will be compared, like cell line or treatment conditions. Also, the data simulation requires the characteristics of labeling of substrates used and the duration of the incubation of a given sample. The names of NetCDF files should contain this information. For instance name SW620_6h_12Glc_R3_SIM_02.CDF would indicate that the name of the cell line is SW620; 6 h is the time of the incubation with the labeled substrate; 12Glc is the labeled substrate 1,2-$^{13}C_2$-Glucose; and the data are from replicate 3, injection number 02, and registered in SIM mode. Not all the information is always necessary. For instance, if the same labeled substrate is used in all samples, this information can be omitted, the program will not select separate groups by tracers used and the information for the tracer can be introduced separately (*see* **Note 11**).
4. Check the availability of the necessary software. The computer should have installed "R" to run Ramid, the library "ncdf4" installed to read NetCDF files and the repository https://

github.com/seliv55/ramidcor downloaded as a directory (let us call it <ramidcor>).

5. To run Ramid enter the directory <ramidcor>, enter the "R" shell, read the source code, and run Ramid with <info> and <cdfdir> as parameters (*see* **Note 4**).

   Ramid reads one by one all the NetCDF files contained in the <cdfdir>, and for each file checks the availability of spectra for metabolites listed in the description file <info> (*see* Table 1). It finds the desired peak using the mz0 and RT values indicated in the table, checks for the presence of control peak, evaluates the peaks and subtracts the baseline for the desired isotopologs, and integrates the peak area. Ramid stores the absolute values of intensity after subtracting the baseline for the integrated peaks (main and control) and then finds the relative intensities to the maximal peak in the spectrum. Ramid can detect and report some issues concerning the data (such as saturation of signal in MS detector) (*see* **Notes 5–7**). There are some specificities of processing of GC-MS data recorded in SIM, or scan mode, or UHR-FTMS data (*see* **Note 4**).

6. Check the output of Ramid. The program saves several text files, one for each metabolite listed in the corresponding <info> (Tables 1 and 2), if the NetCDF files contain its mass spectra. The names of the saved files are the same as the names of metabolites indicated in the <info>. The saved files contain three parts. The first part of such files includes complete information for a visual check, exemplified for the entire molecule of alanine ("Alanine_C1C3") in the upper part of Table 3. The subsequent records of the same file show absolute values of peaks height in the spectra of the given metabolite presented in the analyzed samples (shown in the middle part of Table 3) and the values relative to the maximal peak (shown in the bottom part of Table 3).

   Midcor uses only the middle part (absolute values of intensities) of the Ramid output to correct the spectra for natural isotope occurrence. This part is terminated by a line containing the word "END." The other parts serve for visual checking of the quality of samples.

   Presumably, automated extraction of isotopolog distribution for metabolites of interest can fail in some individual experiments because of significant baseline change or spectra overlapping. Isotopolog distribution can be extracted manually, locating the peak area and background visually. Manual checking identifies such errors of an automated extraction. Comparison of the spectra obtained from the same samples (*see* **Note 8**) obtained in SIM and scan modes also is a way of checking the correctness of automated evaluation of mass spectra. We tested the performance of Ramid comparing the results of its application for SIM and scan modes with the manual mass

Table 3
Output of Ramid. File "Alanine_C1C3"

SW620_24h_12Glc_R1_PIM_SIM_01											
peak_index	**29**	**c**	**29**								
max_peak	4,203,008	c	5,021,184								
mz	259.1	260.1	261.1	262.1	263.1	264.1	265.1	c	232.1	233.1	234.1
5_timepoints	76,367	1.5E+07	3,684,352	5,034,048	993,072	364,880	52,679	c	1.8E+07	4,421,824	6,189,056
base	1191	1790	18,243	5606	7520	1931	4238	c	12,060	28,809	6688
...	...	...	...	...	...	...	...		...	...	...
absolute_values,CDF_file				Max	259.1	260.1	261.1	262.1	263.1	264.1	265.1
SW620_24h_12Glc_R1_PIM_SIM_01				4,203,008	75,176	1.5E+07	3,666,109	5,028,442	985,552	362,949	48,441
SW620_24h_12Glc_R1_PIM_SIM_02				2,819,072	53,201	1E+07	2,497,960	3,416,950	670,804	245,546	33,113
SW620_24h_12Glc_R2_PIM_SIM_01				7,221,760	121,906	2.5E+07	6,346,095	8,731,034	1,713,392	626,079	82,196
SW620_24h_12Glc_R2_PIM_SIM_02				3,654,144	67,283	1.3E+07	3,240,394	4,424,070	870,334	316,950	42,856
...				...	...	...	...	...	...	...	...
END											
relative_values,CDF_file				Max	259.1	260.1	261.1	262.1	263.1	264.1	265.1
SW620_24h_12Glc_R1_PIM_SIM_01				4,203,008	0.0051	1	0.2508	0.3441	0.0674	0.0248	0.0033
SW620_24h_12Glc_R1_PIM_SIM_02				2,819,072	0.0053	1	0.2494	0.3411	0.067	0.0245	0.0033
SW620_24h_12Glc_R2_PIM_SIM_01				7,221,760	0.0048	1	0.2495	0.3432	0.0674	0.0246	0.0032
SW620_24h_12Glc_R2_PIM_SIM_02				3,654,144	0.0052	1	0.2494	0.3405	0.067	0.0244	0.0033
...				...	...	...	...	...	...	...	...

Upper part contains the most complete information for visual check. For each processed NetCDF file (first row) it is indicated for both principal and control (values after "c") peaks: the index of time-point after selecting the time interval for the peak, the peak intensity, the m/z values taken into consideration, the integrated intensity for 5 time-points around the maximum and correspondingly integrated baseline for each considered m/z value. Middle and bottom parts contain information for further processing in the next steps of the workflow: sample (NetCDF file name), maximal peak value, integrated peaks for all isotopologs, respectively, absolute and relative to the maximal values

Table 4
Mass spectra of the same samples processed in SIM mode evaluated manually or with Ramid or processed in scan mode using Ramid

	M–1	*M*0	*M*1	*M*2	*M*3	*M*4	*M*5
SW620_24h_12Glc_R1_PIM_SIM_01	0.0049	1.0000	0.2517	0.3470	0.0683	0.0254	0.0034
SW620_24h_12Glc_R1_PIM_SIM_01	0.0051	1.0000	0.2510	0.3441	0.0675	0.0248	0.0033
SW620_24h_12Glc_R1_PIM_SCAN_01	0.0001	1.0000	0.2452	0.3364	0.0645	0.0236	0.0030
SW620_24h_12Glc_R2_PIM_SIM_01	0.0046	1.0000	0.2515	0.3465	0.0679	0.0247	0.0034
SW620_24h_12Glc_R2_PIM_SIM_01	0.0049	1.0000	0.2290	0.0944	0.0141	0.0024	0.0003
SW620_24h_12Glc_R2_PIM_SCAN_01	0.0003	1.0000	0.2240	0.0907	0.0139	0.0024	0.0002
SW620_24h_12Glc_R3_PIM_SIM_01	0.0049	1.0000	0.2516	0.3465	0.0678	0.0249	0.0034
SW620_24h_12Glc_R3_PIM_SIM_01	0.0051	1.0000	0.2510	0.3454	0.0679	0.0249	0.0033
SW620_24h_12Glc_R3_PIM_SCAN_01	0.0001	1.0000	0.2482	0.3377	0.0678	0.0248	0.0033
SW620_24h_cold_R2_PIM_SIM_01	0.0047	1.0000	0.2293	0.0947	0.0141	0.0025	0.0003
SW620_24h_cold_R2_PIM_SIM_01	0.0049	1.0000	0.2291	0.0943	0.0141	0.0024	0.0003
SW620_24h_cold_R2_PIM_SCAN_01	0.0011	1.0000	0.2266	0.0927	0.0136	0.0024	0.0003

SW620_6h_12Glc_R1_PIM_SI							
M_01	0.0068	1.0000	0.2398	0.2548	0.0467	0.0166	0.0024
SW620_6h_12Glc_R1_PIM_SI							
M_01	0.0072	1.0000	0.2413	0.2543	0.0471	0.0162	0.0025
SW620_6h_12Glc_R1_PIM_SC							
AN_01	0.0117	1.0000	0.2344	0.2507	0.0462	0.0173	0.0022
SW620_6h_12Glc_R2_PIM_SI							
M_01	0.0084	1.0000	0.2420	0.2557	0.0487	0.0165	0.0024
SW620_6h_12Glc_R2_PIM_SI							
M_01	0.0079	1.0000	0.2404	0.2526	0.0471	0.0160	0.0025
SW620_6h_12Glc_R2_PIM_SC	0.0053	1.0000	0.2406	0.2574	0.0475	0.0168	0.0029
AN_01							
SW620_6h_12Glc_R3_PIM_SI							
M_01	0.0069	1.0000	0.2401	0.2542	0.0488	0.0161	0.0020
SW620_6h_12Glc_R3_PIM_SI							
M_01	0.0073	1.0000	0.2413	0.2539	0.0477	0.0163	0.0024
SW620_6h_12Glc_R3_PIM_SC							
AN_01	0.0089	1.0000	0.2458	0.2569	0.0502	0.0178	0.0016

Isotopolog distribution of derivatized alanine ($C_{11}H_{26}N_1O_2Si_{26}$) obtained from the same injections processed in SIM mode manually (green) or with Ramid (grey), or another injection of the same sample processed in scan mode (orange)

spectra extraction in SIM mode. The isotopolog distribution obtained manually from SIM recording should be very similar to the other two distributions.

Table 4 exemplifies the isotopolog distributions obtained for Alanine_C1C3 using manual integration of spectra recorded in SIM mode, and using Ramid in spectra recorded using SIM and scan modes. This comparison proves that the program gives practically the same isotopolog distributions as manual peaks integration. The distributions from other injections of the same sample processed in scan mode and processed by the program is also the same.

To check the reliability of the resulted isotopolog distributions obtained from UHR-FTMS data using Ramid, a file

Table 5
Isotopolog distribution of the unlabeled metabolites extracted by Ramid from a NetCDF file registered by UHR-FTMS

Metabolite	# carbons	*M*	*M*1	*M*2
13-bpg	3	0.9753	0.0246	0.0001
2hg	5	0.9519	0.0478	0.0001
a-kg	5	0.9457	0.0541	0.0001
Aspartate	4	0.9672	0.0328	0
Cis-aconitate	6	0.9367	0.063	0.0002
Citrate	6	0.9408	0.0591	0.0001
Ribose-5-p	5	0.9432	0.0566	0
Sedoheptulose-7-p	7	0.9236	0.0763	0.0001
F6P	6	0.9352	0.0647	0.0001
Fructose-1-6-bp	6	0.9456	0.0534	0.0009
Fumarate	4	0.9595	0.0404	0.0001
Glucose-1-p	6	0.9396	0.0604	0
Glycerol-3-p	3	0.9657	0.0342	0.0001
Isocitrate	6	0.9325	0.0672	0.0002
Malate	4	0.9548	0.0451	0.0001
Malonate	3	0.9692	0.0308	0
Pyruvate	3	0.9602	0.034	0.0001
Succinate	4	0.955	0.0449	0.0001

corresponding to a sample with cells incubated without any artificially labeled metabolites was tested. Table 5 shows the actual isotopolog distribution extracted from such a file. All isotopolog distributions, extracted by Ramid from this file, contain practically only the lightest isotopolog (M0) and that containing just one ^{13}C (M1) (the content of heavier isotopologs is much lower that makes them practically indeterminable). This result is expected when there are no artificially labeled substrates and no derivatization is used. As the natural abundance of the ^{13}C is ~1.1%, then for molecules containing n carbon atoms the probability of finding one ^{13}C is approximately $0.011 \times n$. The fractions of the isotopolog M1 correspond to this expectation for various metabolites listed in Table 5.

The performance of Ramid was checked for limited formats of MS data. However, it can be easily adapted for a wide range of mass spectrometry data. The benefits of using Ramid reside not only in saving time compared to manual integration to extract isotopolog distributions, but also in providing a link to the subsequent steps of data analysis: correction for natural isotopes occurrence and subsequent simulation of the corrected data aimed at evaluation of consistent metabolic fluxes.

3.2 Correction of Ramid Output for Natural Isotope Abundance Using Midcor

1. Get the necessary programs. The "R" code of Midcor ("midcor.R") is located in the subdirectory "R" of <ramidcor>, the same where Ramid is located (downloaded from https://github.com/seliv55/ramidcor). This tool corrects the raw mass spectra for natural isotope abundance, which interferes with the artificial isotope enrichment, and peaks overlapping. Whereas there are several software tools that perform the former function [25–30], we are not aware of other programs performing the latter function. The principles of its functioning are described in a previous publication of our group [31].
2. Check the availability of input data. Midcor takes the necessary information from the same input file <info> prepared for Ramid (Table 1) and a directory containing files with the uncorrected isotopolog distributions saved by Ramid (<ramid_outdir>) (Table 3). Midcor reads all the files generated by Ramid, one for each metabolite.
3. To run Midcor enter the directory <ramidcor>, enter in "R" shell, read the source code, and run Midcor with <info> and <ramid_outdir> as parameters (*see* **Note 9**).
4. Midcor generates a file for each metabolite listed in the <info> in the same directory of input files. The name of generated file has the extension ".txt" added to the name of respective input file (*see* **Note 10**). Each file includes the corrected isotopolog distributions, that is, containing only the distribution of ^{13}C propagated from artificially labeled substrates, for the given metabolite calculated for each of available NetCDF files (Table 6, first part). A significant deviation from the natural isotope abundance in unlabeled samples would indicate the existence of peaks overlapping. The difference between the actual isotopolog distribution and the one calculated for unlabeled sample based on natural isotopes occurrence and chemical formula of the derivatized metabolite, named here "correction factor," is used for the additional correction of the labeled samples. The fully corrected isotopolog distributions are presented in the lower part of the files saved by Midcor.
5. Midcor saves the corrected data into several separate files. Each of these files contains the isotopolog distributions in one metabolite evaluated for various samples (NetCDF files found in the <ramid_outdir>) that may refer to various conditions (e.g., different labeling of substrates used), which the model

Table 6
Output of Midcor

Samples corrected for natural isotope occurrence only[a]					
Sample_file	*Max_int*	*m0*	*m1*	*m2*	*m3*
SW620_24h_12Glc_R1_PIM_SIM_01	4,203,008	0.7916	0.0169	0.1898	0.0023
SW620_24h_12Glc_R1_PIM_SIM_02	2,819,072	0.7939	0.0158	0.1884	0.0024
SW620_24h_12Glc_R2_PIM_SIM_01	7,221,760	0.7927	0.016	0.1896	0.0024
SW620_24h_12Glc_R2_PIM_SIM_02	3,654,144	0.7943	0.0158	0.188	0.0025
...	...	...	...	...	...
Additional correction for overlapping with other components of medium[b]					
		Correction factor			
		CF_m0	CF_m0	CF_m0	CF_m0
		0.0034	−0.0016	−0.0021	−0.0007
Sample_file	Max_int	m0	m1	m2	m3
SW620_24h_12Glc_R1_PIM_SIM_01	4,203,008	0.795	0.0153	0.1877	0.0016
SW620_24h_12Glc_R1_PIM_SIM_02	2,819,072	0.7973	0.0142	0.1863	0.0017
SW620_24h_12Glc_R2_PIM_SIM_01	7,221,760	0.7961	0.0144	0.1875	0.0017
SW620_24h_12Glc_R2_PIM_SIM_02	3,654,144	0.7977	0.0142	0.1859	0.0018
...	...	...	...	...	...

The distributions for each sample are preceded by the absolute value of intensity of the maximal peak in the corresponding spectrum

[a]The first part is isotopolog distribution for each sample corrected only for natural occurrence of isotopes of elements presented in derivatized substance according to the chemical formula presented in Table 1

[b]The second part is (1) the correction factor, calculated by deviation of actual isotopolog distribution from the one calculated for unlabeled sample based on natural isotopes occurrence, and (2) additionally corrected distributions using the correction factor

should simulate separately. On the other hand, mapping of metabolic fluxes occurring in particular conditions requires data of labeling of various metabolites coming from all these different files. Thus, the function "isoform()", which is included in the source code for Midcor, reads all the files with corrected isotopolog distribution for each metabolite and groups in one file all the available data that should be fit simultaneously (*see* **Note 11**). Table 7 shows an example of its content. This file is readable by Isodyn.

3.3 Generation of a Flux Map Simulating the Isotopolog Distribution with Isodyn

1. Get the necessary software. The complete code of Isodyn can be downloaded from https://github.com/seliv55/isodyn. It is necessary to compile all the code using a C++ compiler. We used the GNU g++ compiler. *See* **Note 12** for the details of compiling.

Table 7
An example of the content of the Isoform output file used to adjust the kinetic model for accessing the metabolic fluxes

Time (h)[a]: 0 6 24 – 1
Name[b]: Alanine_C1C3,0,2
t^c = 24 mean: 0.795 0.017 0.187 0.002 sd: 0.012 0.011 0.011 0
t = 6 mean: 0.859 0.01 0.13 0.001 sd: 0.011 0.011 0.011 0.011
Name: Glycine_C2,1,1
t = 24 mean: 0.99 0.01 sd: 0.001 0.001
t = 6 mean: 0.998 0.002 sd: 0.001 0.001
Name: Lactate,0,2
t = 24 mean: 0.795 0.015 0.189 0.001 sd: 0.002 0.001 0.002 0
t = 6 mean: 0.851 0.009 0.139 0.001 sd: 0.012 0.001 0.011 0
Name: Malate,0,3
t = 24 mean: 0.927 0.02 0.049 0.003 0.001 sd: 0.004 0.002 0.002 0 0
t = 6 mean: 0.972 0.007 0.02 0 0 sd: 0.001 0.001 0 0 0

[a]Times of cell incubation when the samples were taken

[b]The analyzed fragment name, first, and last carbon atoms referred to the whole molecule

[c]The time of incubation, mean content and standard deviation of isotopologs relative to the whole amount of a metabolite

2. Provide input data. The script "aiso.sh" situated in the working directory specifies the input information (*see* **Note 13**).
3. Run the program. Execution of the script "iso.sh" runs the program. As specified in the provided example, it reads the parameters from a file <par>, and both the measured metabolite concentrations in the incubation media and the isotopolog distributions from the files <conc> and <forisodyn>, respectively. Then it settles the output of calculated data to compare with the corresponding measured concentrations of metabolites and isotopologs in the relevant time points.
4. Get the output of the program. The subsequent calculation simulates the incubation of cells in the presence of an artificially labeled substrate until the final time point. If there are isotopolog distribution data measured in the intermediate time points, Isodyn accounts for the corresponding calculations in the χ_r^2. In addition to the screen output shown in Table 8, Isodyn provides graphs that visualize the differences between the measured and calculated labeling (Fig. 2) and concentration (Fig. 3) of metabolites. The lines in Figs. 2 and 3 are done using the set of best fit parameters.

Table 8
Screen output of Isodyn after a single simulation

Metabolite*\time:	0	360	1440:	exper ↔ chisq
Alanine_C1C3_m0:	1	0.879	0.77:	0.795 ⟶ 21.3
Aspartate_C1C4_m0:	1	0.983	0.93:	0.928 ⟶ 0.466
Glycine_C1C2_m0:	1	0.999	0.988:	0.988 ⟶ 1.21
Lactate_m0:	1	0.875	0.807:	0.795 ⟶ 3.69
Malate_m0:	1	0.97	0.923:	0.927 ⟶ 0.872
Proline_C1C5_m0:	1	0.996	0.981:	0.932 ⟶ 3.56
Serine_C1C3_m0:	1	0.995	0.983:	0.967 ⟶ 10.3
Gluc_c:	8.64	7.82	5.35**	5.28 ⟶ 0.019
Glutamin_c:	2.92	2.83	2.57**	2.64 ⟶ 0.0601
Lac_c:	0.807	2.57	7.59**	7.85 ⟶ 0.113
Glutamate2-5_c:	0.101	0.0687	0.0312**	0.03 ⟶ 0.155

File saved: 250,002: xi = 71.2; xm = 12.7; time = 3.23
* unlabeled fraction (_m0)
** total concentration (_c), calculated concentration corresponding to the end of incubation

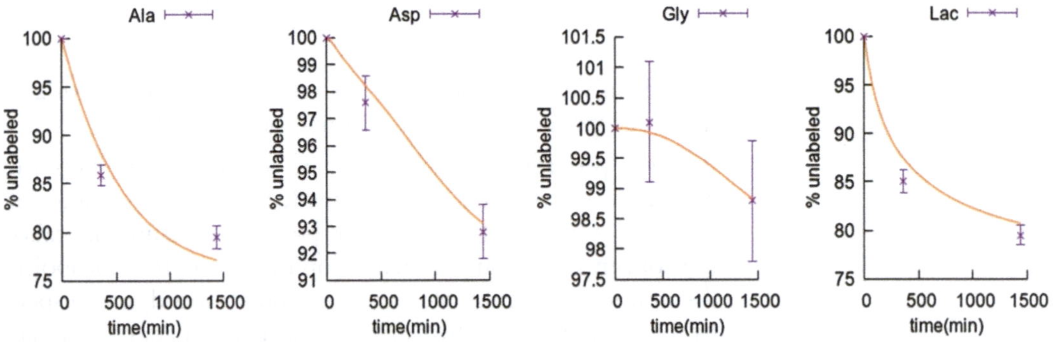

Fig. 2 Dynamics of labeling of some selected metabolites (decrease of unlabeled fraction m0). Points are MS measurements M0 isotopolog of intracellular alanine (Ala), aspartate (Asp), glycine (Gly), and extracellular lactate (Lac) at 0, 6, and 24 h after starting the incubation of SW620 cells with 50% [1,2-^{13}C]-Glucose and 50% unlabeled glucose. Lines are the respective calculations using a kinetic model that simulates the reactions shown in Fig. 4

5. Isodyn applies Simulated Annealing algorithm for optimization of the model parameters, corresponding to the maximal reaction rates, by changing them to minimize the difference between the measured isotopolog distribution, in its dynamics, and corresponding computed data. It calculates the difference as the reduced chi-squared function (χ_r^2) for each time point of labeled and total metabolite concentrations and estimates the

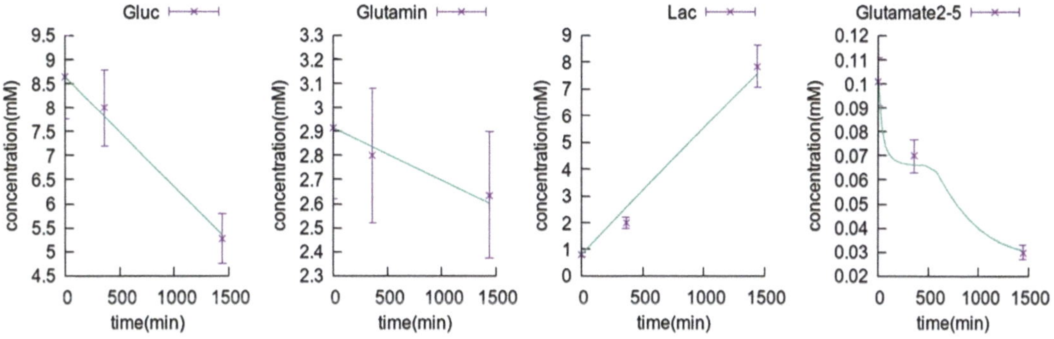

Fig. 3 Change of total metabolite concentration during the time of cells incubation. Points are measured extracellular concentrations of glucose, glutamine, lactate, and glutamate under conditions described in the legend to Fig. 2. Lines obtained in the same simulation as that shown in Fig. 2

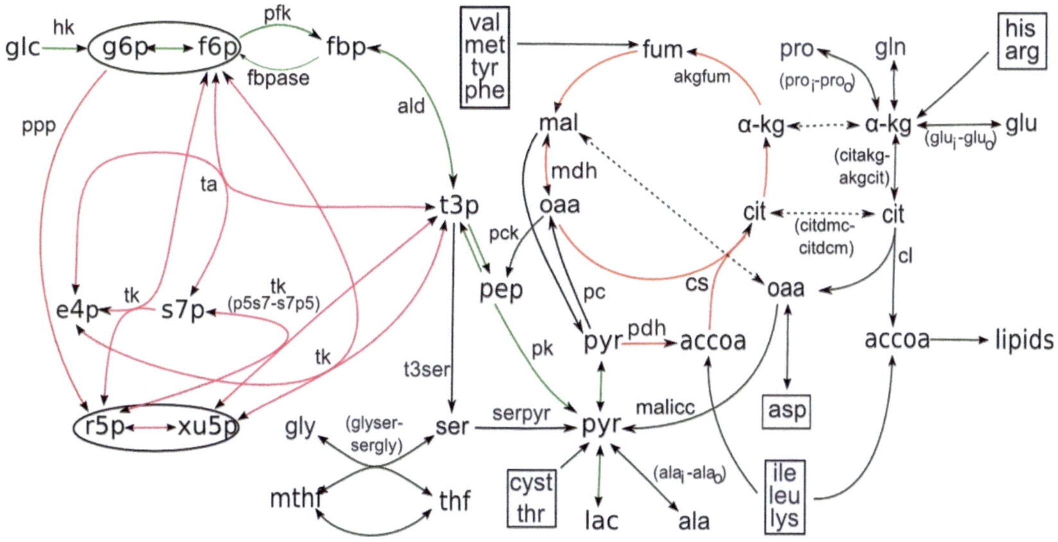

Fig. 4 Scheme of the biochemical reactions simulated using the kinetic model implemented in Isodyn in the considered here example. It simulates the reactions of central metabolism: glycolysis, Krebs cycle, and pentose phosphate pathways

confidence intervals for best-fit fluxes by Monte Carlo simulations as described [37]. The best fit, if it satisfies to the criteria of goodness of fit, gives the values of fluxes in the reaction network implemented in the model (here the reactions shown in Fig. 4). An example of such fluxes is shown in Table 9. There could be some issues with data fitting that are described in **Note 14**. The tools are continuously updated (**Note 15**).

Table 9
A set of metabolic fluxes with 0.99 confidence intervals bounds (CI bounds), found as described in [38]

Name	CI bounds		Name	CI bounds		Name	CI bounds	
hk	1.85981	2.4382	akgdmc	0.13332	0.37441	akgdcm	0.28627	0.53881
pfk	2.87347	3.77688	akgdcm	0.28627	0.53881	coaout	9.6E−05	0.00021
fbpase	0.92956	1.57897	coaout	9.6E−05	0.00021	citakgl	0.0008	0.00172
t3pep	3.69917	4.84934	citakgl	0.0008	0.00172	atpase	2.8238	3.72059
pept3	1.9E−07	3.3E−07	gln_in	0.11841	0.1542	resp	0.01081	0.02248
pk	3.69917	4.84934	gluin	0.37	0.66465	aldfl	2.66906	3.52675
pyrlac	4.51879	5.99338	gluout	0.36392	0.64596	aldrev	0.63194	1.40822
lacpyr	0.00309	0.00415	t3ser	0.00056	0.00143	f6s7a	9E−05	0.00012
pyrdcm	0.12316	0.24845	serpyr	0.00082	0.00112	s7f6a	4.7E−05	6.3E−05
pyrdmc	0.15161	0.36549	asp_o	0.63787	1.21935	sergly	0.00023	0.00928
pdh	0.06384	0.11063	asp_i	1.00614	1.7305	glyser	0.00127	0.00897
citakg	0.06296	0.10008	ala_o	2.07525	2.69381	cs0	0.06385	0.10147
akgsuc	0.18425	0.27173	ala_i	2.4472	3.20778	p5s7	4.7E−08	6.8E−08
sucmal	0.18428	0.27176	r5_o	0.00997	0.01307	s7p5	6.4E−07	9.5E−07
maloa	0.00315	0.01306	glycogin	0.00026	0.00034	f6p5	1.2E−05	1.6E−05
oamal	0.00014	0.06989	glycogou	0.00022	0.00029	gln_pr	0.01444	0.18648
pc	0.0669	0.16523	cystin	0.05319	0.06972	ser_pr	0.00026	0.00041
malicm	0.07706	0.31254	proin	0.01249	0.03575	trpala	0.00024	0.00032
malicc	0.34931	0.64728	proout	0.01672	0.02715	mthf	0.00375	0.00491
ppp	0.00985	0.01291	citdmc	0.00114	0.00247	thf	1.2E−06	1.6E−06
oacd	0.01116	0.15693	citdcm	0.00025	0.00054	kgin	0.00016	0.00021
mald	0.01377	0.17873	akgdmc	0.13332	0.37441			

4 Notes

1. *General structure of MS raw data*. The raw data comprise many mass spectra records (scans) obtained at sequential time points (acquisition time) counted starting from the injection of the sample into a chromatographic column. Each mass spectrum is recorded at approximately equal time intervals. Each scan comprises a vector of detected numbers of ions corresponding to some discrete m/z values. The amount of metabolite ions

detected depends on the time passed from starting chromatography (retention time, RT), ionization efficiency and ion suppression effects, as well as on the MS detection mode. Generally, the RT value for a concrete chromatographic method as well as *m/z* values of the peak is characteristic of the molecular type. The isotopolog distribution of metabolites of interest is extracted as peak intensities, detected by number of ions in the known *m/z* regions and appeared at acquisition times close to the known retention times. The examples of processing of GC-MS data saved in SIM (selected ion monitoring) and scan modes, and UHR-FTMS (ultrahigh resolution Fourier transformed MS) data are described here.

The GC-MS data used in the presented here example were obtained by a 7890A mass spectrometer coupled with a 5675C gas chromatograph (Agilent Technologies). Two modes of MS data recording were used: SIM mode, registering only the intensity of specific *m/z* values; or scan mode, registering all mass spectra inside the whole *m/z* interval indicated to the mass spectrometer.

2. *The UHR-FTMS data*. This data, used in the presented here example, was obtained using Tribrid Fusion Orbitrap (Thermo Scientific, San Jose, CA, USA), interfaced with an Advion Triversa Nanomate (Advion Biosciences, Ithaca, NY, USA) (provided by Dr. Higashi, University of Kentucky). The precision of *ultrahigh resolution* mass spectrometry data, presented here as an example, was 10^{-5}, which is sufficient for distinguishing ions that contain isotopes of different elements such as 1 ^{13}C versus 1 ^{15}N which have the same nominal mass increment [39, 40]. Such precision requires a narrow and precise definition of *m/z* intervals for each isotopolog of the metabolites of interest. Therefore, the fractions of the registered ^{13}C isotopologs should correspond to the natural abundance of ^{13}C and not of the other atoms.

3. *A meaning of "control" peak*. There is a probability that the program erroneously takes wrong peak because, first, the retention time can vary slightly between samples and, second, the sample may contain other compounds that elute at close retention times and with MS peaks in a close *m/z* region to the one of the ion of interest. If the peak is found correctly, not only the main ion (characteristic of the fragment to be analyzed) but also a control ion (i.e., other ion characteristic of the ionization of the derivatized molecule, and specified in the column "control") would appear simultaneously. If not, then the found peak is probably not referred to the searched metabolite. Checking the presence of a control ion at the same retention time ensures finding the right peak for the desired metabolite. Such a check permits to avoid using incorrect data in the subsequent steps of

the analysis: correction for natural isotope occurrence and the data simulation.

4. *Running Ramid.* There are several ways to run Ramid (e.g., using R-studio, or creating a library containing all the necessary functions). Maybe the simplest way to run it in Linux terminal is using the following commands:

```
$ cd <ramidcor>
$ R
>source ('R/lib.R')
>source ('R/ramid.R')
>ramid (infile='path to<info>', cdfdir='path
to<cdfdir>', fiout=<output>, md='<mode>')
```

here '<mode>' can be either 'sim' for GC-MS data in SIM mode, or 'scan' for GC-MS data in scan mode, or 'uhr' for UHR-FTMS.

In **SIM** mode at each acquisition time the mass spectrometer registers signal intensities corresponding to separated intervals of m/z specified for each metabolite of interest with the fixed distance between neighbor m/z inside the intervals, which usually is set to 1. To get the desired isotopolog distribution, the program finds the specified (by mz0, Table 1) interval in m/z vectors, determines the actual retention time (maximum of corresponding intensities, which could slightly differ from the RT specified a priori in the input file) and get the values of intensities corresponding to the selected m/z mz0 and retention time.

In **scan** mode, a vector of m/z at each moment of acquisition is just one large interval with variable distance between neighbor values. To get the desired isotopolog distribution it is necessary to integrate the intensities for an interval of m/z values and ascribe the integrated intensity to a corresponding isotopolog of interest.

In **uhr** mode Ramid accounts for the difference in mass between ^{13}C and ^{12}C is 1.003355 atomic mass units. Therefore it sets the m/z intervals for any isotopomer containing n ^{13}C atoms by adding n×1.003355 to the upper and lower bound indicated for mz0. As the NetCDF files containing UHR-FTMS data usually are very big, the algorithm is designed to putting into the memory only a variable (e.g., acquisition times, m/z values, intensities values) that is currently processed and cleaning it immediately after processing.

If the measured interval of m/z (for SIM mode) is sufficiently large, the Ramid accounts for the m/z interval from mz0–1 to mz0 + n + 5, where n is the number of carbon atoms in the fragment of the carbon skeleton.

5. *Low values of peaks in mass spectrum*. Ramid indicates issue if the ratio of maximal peak in a pattern to baseline is less than 3.
6. *High values of peaks*. Peaks for some isotopologs can be higher than the upper limit of sensitivity of the MS detector. Ramid detects such a case and prints a message.
7. *High fraction of $M-1$ peak*. Ionization in GC-MS can result in a proton loss and appearance of the peak of the compound ($M-1$) that is lighter than the isotopolog consisted of lightest isotopes ($M0$). The ratio of intensities of ($M-1$) to that of $M0$ usually is less than 2–3%. A higher fraction of ($M-1$) after subtracting the baseline indicates overlapping with another metabolite.
8. *The samples*. They represent cell extracts of SW620 obtained after 6 h of incubation in the presence of media containing glucose 50% enriched in [1,2-^{13}C]-Glucose (12Glc) or glucose unlabeled (cold). The samples were derivatized with 2% methoxamine hydrochloride in pyridine for 90 min at 37 °C and (*N*-methyl-*N*-(tert-butyldimethylsilyl) trifluoroacetamide + 1% tertbutyldimethylchlorosilate) for 60 min at 55 °C and next were processed on an Agilent 7890A GC coupled to a 7890A mass spectrometer either in SIM or scan mode.
9. *Running Midcor*. It is similar to that for Ramid (*see* **Note 4**):

```
$ cd <ramidcor>
$ R
>source ('R/lib.R')
>source ('R/midcor.R')
>rumidcor(infile='path to <info>', dadir='path to
<ramid_outdir>')
```

Midcor process the data presented in the indicated input file (Table 1) row by row. It takes metabolite names (first column), calculates the number of carbons in the fragments of interest (from column "Fragment"), and the kind and number of atoms in the derivatized molecule (from column "Formula") that can naturally contribute to forming isotopomers heavier than mz0. Then it opens a file named as the metabolite name. From this file, it reads only the middle part containing the absolute values of isotopolog distribution (Table 3). It calculates, using the provided chemical formulas, the expected contribution of naturally occurring isotopes into the isotopolog distribution for given metabolites, and corrects the measured spectra, not only for the natural isotope abundance but also for peaks overlapping, as we described elsewhere [31].

10. *The output.* Midcor corrects the isotopolog distributions for each <file> present in <ramid_outdir> and saves the results as <file.txt> in the same <ramid_outdir>.

11. *Function isoform().* It should be called after rumidcor() (**Note 9**):

```
>isoform(isofi='path to <sinfo>', dadir='path to
<ramid_outdir>')
```

Here dadir is the same path to the directory of Ramid output as in the above notes. <sinfo> is a text file with short instructions how to prepare the data needed in the same simulation by Isodyn. Here is an example of such file:

Names: Lactate,0,2 Alanine_C1C3,0,2 Alanine_C2C3,1,2 Aspartate_C1C4,0,3 Serine_C2C3,1,2

time(h): 0 6 24 –1

tracers: 12Glc

Glucose 48 0.5

The first line indicates metabolites whose isotopolog distributions should be combined in the same file, and the positions of the first and the last carbons of the registered fragment with respect to the whole molecule, starting from 0. The second line indicates the times at which samples were collected after beginning of the incubation with labeled substrates, with '−1' indicating the end of sampling. The third line is the tracer as it is indicated in the names of the NetCDF files. The last line shows the specific isotopomer used as a tracer. It represented by the binary numbers, assuming that 1 is ^{13}C and 0 is ^{12}C. In this way 1,2-^{13}C -Glucose is expressed as binary 110,000 that, being transformed into the decimal, means 48. The last number is the fraction of the tracer in the provided substrate.

Using this information the function isoform() selects the records corresponding to the metabolite fragments shown in first row, incubation times and tracer used, indicated in titles of the NetCDF files, and fraction of labeled substrates, and forms a single file readable by Isodyn, let us call it <forisodyn>, which content is exemplified in Table 7.

12. *Download and compiling Isodyn.* Create a directory for the code of Isodyn downloaded from the repository https://github.com/seliv55/isodyn, let it be <isodyn>. The repository contains the necessary makefiles, therefore to compile the whole code use the commands:

```
$ cd <isodyn>
$ make clean
$ make
```

Isodyn provides a flux map consistent with the corrected isotopolog distribution using simulations of isotope propagation in cellular biochemical networks by a mathematical model. A set of simulated fluxes that fits the corrected isotopomer distribution is a possible candidate of the actual distribution of fluxes in the corresponding real biochemical system. Commonly, steady state flux models are used for flux balance analysis (FBA) resulted in flux map evaluation [32–36]. FBA is a well-established approach for constraint-based modeling, which disregards the dynamics of intracellular metabolism thus avoiding the complications that sometimes gives solution of ordinary differential equations (ODEs) of a kinetic model. However, many biological systems may not maintain a metabolic or achieve isotopic steady state during the time of the study [41, 42], especially in common experimental cell culture designs, or in vivo. A dynamic flux analysis is more appropriate in such cases [43]. Moreover, the dynamics of cellular response to a perturbation is always more informative for studying metabolism, than just steady-state flux distributions, which is a special case of the dynamics. Dynamic flux analysis can be applied equally well to the analysis of steady-state conditions if they can be shown to be achieved. This requires measurement at more than one time point.

Isodyn was the first tool that implements a kinetic model for dynamic SIRM data analysis [44–46]. In its present implementation it can be connected with the described above two preliminary steps of SIRM data analysis, thus concluding the fluxomic analysis: fitting the implemented kinetic model to the SIRM data provides consistent flux map. Isodyn simulates the dynamics of metabolite labeling and possible dynamics of the fluxes themselves. To this end, a kinetic model, implemented in Isodyn, has a module that simulates the dynamics of isotopomers and, thereby, isotopolog distribution [46]. It requires neither metabolic nor even isotopic steady state for measuring isotopolog distribution.

A kinetic model used is a set of ODEs associating the change of metabolites concentrations with the metabolic fluxes of their production and consumption, expressed using general rate laws (mainly Michaelis-Menten). To analyze the isotopolog distribution as a result of the flux distribution in a network, Isodyn calculates the actual splitting of metabolic fluxes into the individual fractions contributing to the change of concentration of each isotopomer, accounting for atom transitions in each reaction [38, 44–46]. In this way, it splits the equations for metabolite concentrations into subsets of ODEs for isotopomers concentrations. After adjusting the model to the data, it estimates the goodness of fit, performs a statistical analysis of selected sets of fluxes and determines the confidence intervals for the fluxes that underlie the measured labeling [38].

Isodyn was used to reveal the metabolic fluxes in the central energy metabolism; therefore, the current version of the model simulates the reactions of glycolysis, Krebs cycle, pentose phosphate pathways, and its variables correspond to the concentrations of isotopomers for the metabolites of these pathways. In the example, presented here, Isodyn implements a kinetic model, corresponding to the metabolic network shown in the scheme in Fig. 4. However, this model can be easily extended or cut to simulate specific pathways according to a manual reconstruction, for example, for cancer cells [47–51] and macrophages. The model itself was created independently of the data, although the data and the model should refer to the same biochemical system. The model takes for fitting only the data that it can simulate. Good fit of the model to the data (according to the criteria described in detail in [38]) evaluates the set of metabolic fluxes in the simulated biochemical network that is compatible with the analyzed data.

13. *The script "aiso.sh"*. This script specifies (a) the path to <forisodyn>, that contains the corrected by Midcor dynamics of isotopolog distributions designed for fitting in the same simulations; (b) the path to a file with measured concentrations of metabolites in the incubation media, let it be <conc>. Here is an example of data that it contains:

time_h 0 6 24	// time of sampling
Glucose 8.636 8. 5.282	//respective concentrations (mM)

(c) the path to a file containing the values of parameters <par> that determine the calculated rates of reaction simulated by the model.

Isodyn sets the initial values and labeling of metabolites by the measured data provided in <conc> and <forisodyn>. It starts a simulation of labeled atoms propagation in metabolites by numerically solving the set of ODEs corresponding to the set of biochemical reactions, shown in Fig. 4, for the metabolites and their isotopomers. More details about the process of ODEs solving implemented in Isodyn are described in [46].

14. *Possible issues*. Fitting the model to the corrected isotopolog distributions could be related with some complications because of many parameters that contain the kinetic models of complex biochemical systems, and, respectively, many degrees of freedom in optimizing the models to obtain the flux map consistent with the data.

(a) Isodyn fails to fit the data automatically. There could be one of the two reasons for that, either (1) the implemented algorithm fails to find a point in the multidimensional space of parameters that is consistent with data, or (2) the kinetic model is not compatible with the analyzed process. In the case (1) sometimes a manual intervention can solve the problem: manually change parameters responsible for the reaction rates affecting the labeling of the metabolite, whose calculated data are far from the experiment. The case (2) indicates that the current understanding of the mechanics of the simulated process should be corrected, thus stimulating better understanding of the process.

(b) Isodyn finds different sets of fluxes consistent with the data. In this case, additional data, which discriminate between the resulted sets, should be obtained and included in the analysis.

(c) Minimizing the deviations from the data Isodyn comes to the area in the space of parameters where the time of each simulation highly increases, making practically impossible further optimization. It means that the ODE system of the model becomes stiff. Such a behavior of the model might not be consistent with the real action of the biological system, and Isodyn has particular criteria of optimization that help to avoid entering in such areas of parameters.

(d) Similar to the above mentioned, Isodyn has particular criteria of optimization that help to avoid entering in the areas of parameters where the variables reflecting concentrations of metabolites reach too high values.

15. Please check actualization of the tools and instructions in https://github.com/seliv55.

Acknowledgments

This work was supported by the European Commission (PhenoMeNal EC-654241), MINECO-European Commission FEDER funds—"Una manera de hacer Europa" (SAF2017-89673-R and SAF2015-70270-REDT), Instituto de Salud Carlos III and Centro de Investigación Biomédica en Red de Enfermedades Hepáticas y Digestivas (CIBEREHD, CB17/04/00023 and INB-Bioinformatics Platform, of the ISCIII, PT17/0009/0018), Agència de Gestió d'Ajuts Universitaris i de Recerca—Generalitat de Catalunya (2017SGR-1033), and Ministerio de Educación y Formación Profesional (FPU14-05992); and by the Redox Metabolism Shared Resource of the University of Kentucky Markey Cancer Center

(P30CA177558). M.C. also acknowledges the prize "ICREA Academia" for the excellence in research, funded by ICREA foundation—Generalitat de Catalunya.

References

1. Ipata PL, Pesi R (2018) Metabolic interaction between purine nucleotide cycle and oxypurine cycle during skeletal muscle contraction of different intensities: a biochemical reappraisal. Metabolomics 14:42
2. Crown SB, Antoniewicz MR (2013) Publishing 13C metabolic flux analysis studies: a review and future perspectives. Metab Eng 20:42–48
3. Feng X, Page L, Rubens J, Chircus L, Colletti P, Pakrasi HB, Tang YJ (2010) Bridging the gap between fluxomics and industrial biotechnology. J Biomed Biotechnol 2010:460717
4. Zamboni N, Sauer U (2009) Novel biological insights through metabolomics and 13 C-flux analysis. Curr Opin Microbiol 12:553–558
5. Ahn WS, Antoniewicz MR (2013) Parallel labeling experiments with [1,2-(13)C]glucose and [U-(13)C]glutamine provide new insights into CHO cell metabolism. Metab Eng 15:34–47
6. Grimble RF (2001) Stress proteins in disease: metabolism on a knife edge. Clin Nutr 20:469–476
7. Jin L, Zhou Y (2019) Crucial role of the pentose phosphate pathway in malignant tumors. Oncol Lett 17:4213–4221
8. King BC, Blom AM (2017) Non-traditional roles of complement in type 2 diabetes: metabolism, insulin secretion and homeostasis. Mol Immunol 84:34–42
9. Fontané L, Benaiges D, Goday A, Llauradó G, Pedro-Botet J (2018) Influence of the microbiota and probiotics in obesity. Clin Investig Arterioscler 30:271–279
10. Hanahan D, Weinberg RA (2011) Hallmarks of cancer: the next generation. Cell 144:646–674
11. Badur MG, Metallo CM (2018) Reverse engineering the cancer metabolic network using flux analysis to understand drivers of human disease. Metab Eng 45:95–108
12. Fan TW, Lorkiewicz PK, Sellers K, Moseley HN, Higashi RM, Lane AN (2012) Stable isotope-resolved metabolomics and applications for drug development. Pharmacol Ther 133:366–391
13. Jayaraman A, Kumar P, Marin S, de Atauri P, Mateo F, Thomson TM, Centelles JJ, Graham SF, Cascante M (2018) Untargeted metabolomics reveals distinct metabolic reprogramming in endothelial cells co-cultured with CSC and non-CSC prostate cancer cell subpopulations. PLoS One 13:e0192175
14. Tarrado-Castellarnau M, de Atauri P, Tarragó-Celada J, Perarnau J, Yuneva M, Thomson TM, Cascante M (2017) De novo MYC addiction as an adaptive response of cancer cells to CDK4/6 inhibition. Mol Syst Biol 13:940
15. Marín de Mas I, Marín S, Pachón G, Rodríguez-Prados JC, Vizán P, Centelles JJ, Tauler R, Azqueta A, Selivanov V, López de Ceraín A, Cascante M (2017) Unveiling the metabolic changes on muscle cell metabolism underlying p-phenylenediamine toxicity. Front Mol Biosci 4:8
16. Tarrado-Castellarnau M, de Atauri P, Cascante M (2016) Oncogenic regulation of tumor metabolic reprogramming. Oncotarget 7:62726–62753
17. Diaz-Moralli S, Aguilar E, Marin S, Coy JF, Dewerchin M, Antoniewicz MR, Meca-Cortés O, Notebaert L, Ghesquière B, Eelen G, Thomson TM, Carmeliet P, Cascante M (2016) A key role for transketolase-like 1 in tumor metabolic reprogramming. Oncotarget 7:51875–51897
18. Aguilar E, Marin de Mas I, Zodda E, Marin S, Morrish F, Selivanov V, Meca-Cortés Ó, Delowar H, Pons M, Izquierdo I, Celià-Terrassa T, de Atauri P, Centelles JJ, Hockenbery D, Thomson TM, Cascante M (2016) Metabolic reprogramming and dependencies associated with epithelial cancer stem cells independent of the epithelial-mesenchymal transition program. Stem Cells 34:1163–1176
19. Jang C, Chen L, Rabinowitz JD (2018) Metabolomics and Isotope Tracing. Cell 173:822–837
20. Lane AN, Fan TW, Bousamra M II, Higashi RM, Yan J, Miller DM (2011) Stable isotope-resolved metabolomics (SIRM) in cancer research with clinical application to nonsmall cell lung cancer. OMICS 15:173–182

21. Costa C, Maraschin M, Rocha M (2016) An R package for the integrated analysis of metabolomics and spectral data. Comput Methods Prog Biomed 129:117–124
22. Deng K, Zhang F, Tan Q, Huang Y, Song W, Rong Z, Zhu ZJ, Li Z, Li K (2019) WaveICA: a novel algorithm to remove batch effects for large-scale untargeted metabolomics data based on wavelet analysis. Anal Chim Acta 1061:60–69
23. Tautenhahn R, Patti GJ, Kalisiak E, Miyamoto T, Schmidt M, Lo FY, McBee J, Baliga NS, Siuzdak G (2011) metaXCMS: second-order analysis of untargeted metabolomics data. Anal Chem 83:696–700
24. Tautenhahn R, Patti GJ, Rinehart D, Siuzdak G (2012) XCMS online: a web-based platform to process untargeted metabolomic data. Anal Chem 84:5035–5039
25. Brauman JI (1966) Least squares analysis and simplification of multi-isotope mass spectra. Anal Chem 38:607–610
26. Katz J (1989) Studies of glycogen synthesis and the Krebs cycle by mass isotopomer analysis with [U-13C]glucose in rats. J Biol Chem 264:12994–13004
27. Fernandez CA, Des Rosiers C, Previs SF, David F, Brunengraber H (1996) Correction of 13C mass isotopomer distributions for natural stable isotope abundance. J Mass Spectrom 31:255–262
28. Lee WN, Byerley LO, Bergner EA, Edmond J (1991) Mass isotopomer analysis: theoretical and practical considerations. Biol Mass Spectrom 20:451–458
29. van Winden WA, Wittmann C, Heinzle E, Heijnen JJ (2002) Correcting mass isotopomer distributions for naturally occurring isotopes. Biotechnol Bioeng 80:477–479
30. Millard P, Letisse F, Sokol S, Portais JC (2012) IsoCor: correcting MS data in isotope labeling experiments. Bioinformatics 28:1294–1296
31. Selivanov VA, Benito A, Miranda A, Aguilar E, Polat IH, Centelles JJ et al (2017) MIDcor, an R-program for deciphering mass interferences in mass spectra of metabolites enriched in stable isotopes. BMC Bioinformatics 18:88
32. Zamboni N, Fischer E, Sauer U (2005) FiatFlux-a software for metabolic flux analysis from 13C-glucose experiments. BMC Bioinformatics 6:209
33. Feist AM, Zielinski DC, Orth JD, Schellenberger J, Herrgard MJ, Palsson BØ (2010) Model-driven evaluation of the production potential for growth-coupled products of Escherichia coli. Metab Eng 12:173–186
34. Blazeck J, Alper H (2010) Systems metabolic engineering: genome-scale models and beyond. Biotechnol J 5:647–659
35. Chiewchankaset P, Siriwat W, Suksangpanomrung M, Boonseng O, Meechai A, al TM (2019) Understanding carbon utilization routes between high and low starch-producing cultivars of cassava through flux balance analysis. Sci Rep 9:2964
36. Foguet C, Marin S, Selivanov VA, Fanchon E, Lee WN, Guinovart JJ et al (2016) HepatoDyn: a dynamic model of hepatocyte metabolism that integrates 13C isotopomer data. PLoS Comput Biol 12:e1004899
37. Press WH, Flannery BP, Teukolsky SA, Vetterling WT (2002) Numerical recipes in C: the art of scientific computing. Cambridge University Press, New York, NY
38. de Mas IM, Selivanov VA, Marin S, Roca J, Orešič M, Agius L, Cascante M (2011) Compartmentation of glycogen metabolism revealed from 13C isotopologue distributions. BMC Syst Biol 5:175
39. Higashi RM, Fan TW, Lorkiewicz PK, Moseley HN, Lane AN (2014) Stable isotope-labeled tracers for metabolic pathway elucidation by GC-MS and FT-MS. Methods Mol Biol 1198:147–167
40. Fan TW, Lane AN, Higashi RM, Yan J (2011) Stable isotope resolved metabolomics of lung cancer in a SCID mouse model. Metabolomics 7:257–269
41. Nöh K, Grönke K, Luo B, Takors R, Oldiges M, Wiechert W (2007) Metabolic flux analysis at ultra short time scale: isotopically non-stationary 13C labeling experiments. J Biotechnol 129:249–267
42. Farbehi N, Patrick R, Dorison A, Xaymardan M, Janbandhu V, Wystub-Lis K et al (2019) Single-cell expression profiling reveals dynamic flux of cardiac stromal, vascular and immune cells in health and injury. elife 8: e43882
43. Zhao X, Noack S, Wiechert W, Lieres EV (2017 Dec) Dynamic flux balance analysis with nonlinear objective function. J Math Biol 75(6–7):1487–1515
44. Selivanov VA, Marin S, Lee PW, Cascante M (2006) Software for dynamic analysis of tracer-based metabolomic data: estimation of metabolic fluxes and their statistical analysis. Bioinformatics 22:2806–2812
45. Selivanov VA, Meshalkina LE, Solovjeva ON, Kuchel PW, Ramos-Montoya A, Kochetov GA et al (2005) Rapid simulation and analysis of isotopomer distributions using constraints based on enzyme mechanisms: an example

from HT29 cancer cells. Bioinformatics 21:3558–3564

46. Selivanov VA, Puigjaner J, Sillero A, Centelles JJ, Ramos-Montoya A, Lee PW, Cascante M (2004) An optimized algorithm for flux estimation from isotopomer distribution in glucose metabolites. Bioinformatics 20:3387–3397
47. Henze AT, Mazzone M (2016) The impact of hypoxia on tumor-associated macrophages. J Clin Invest 126:3672–3679
48. Kuang DM, Zhao Q, Peng C, Xu J, Zhang JP, Wu C, Zheng L (2009) Activated monocytes in peritumoral stroma of hepatocellular carcinoma foster immune privilege and disease progression through PD-L1. J Exp Med 206:1327–1337
49. Ganeshan K, Chawla A (2014) Metabolic regulation of immune responses. Annu Rev Immunol 32:609–634
50. Qian BZ, Pollard JW (2010) Macrophage diversity enhances tumor progression and metastasis. Cell 141:39–51
51. Zhu Y, Herndon JM, Sojka DK, Kim KW, Knolhoff BL, Zuo C et al (2017) Tissue-resident macrophages in pancreatic ductal adenocarcinoma originate from embryonic hematopoiesis and promote tumor progression. Immunity 47:323–338

Chapter 13

Inferring Metabolic Flux from Time-Course Metabolomics

Scott Campit and Sriram Chandrasekaran

Abstract

The metabolic activity of a mammalian cell changes dynamically over time and is tied to the changing metabolic demands of cellular processes such as cell differentiation and proliferation. While experimental tools like time-course metabolomics and flux tracing can measure the dynamics of a few pathways, they are unable to infer fluxes at the whole network level. To address this limitation, we have developed the Dynamic Flux Activity (DFA) algorithm, a genome-scale modeling approach that uses time-course metabolomics to predict dynamic flux rewiring during transitions between metabolic states. This chapter provides a protocol for applying DFA to characterize the dynamic metabolic activity of various cancer cell lines.

Key words Dynamic flux activity, Constraint-based modeling, Flux balance analysis, Genome-scale metabolic models, Time-course metabolomics, Cancer metabolism

1 Introduction

Cancer is a complex disease associated with uncontrolled cellular growth and proliferation. Recently, several studies suggest that the metabolism of cancer cells contribute to tumorigenesis and cancer aggressiveness [1]. For example, cancer cells rewire their metabolism to synthesize DNA, proteins, and additional cellular components. Understanding metabolic rewiring in tumors can identify metabolic enzymes that are promising targets for cancer therapies. However, it is challenging to experimentally measure the activity of thousands of reactions in a cell and identify metabolic bottlenecks.

Computational approaches can bridge this gap and predict the metabolic activity of a system without resorting to time- and labor-intensive experiments. Traditional computational models implement enzyme kinetics and metabolite concentration into a system of ordinary differential equations to infer the activity of metabolic reactions [2]. However, analysis of thousands of metabolic

Electronic supplementary material The online version of this chapter (https://doi.org/10.1007/978-1-0716-0159-4_13) contains supplementary material, which is available to authorized users.

Deepak Nagrath (ed.), *Metabolic Flux Analysis in Eukaryotic Cells: Methods and Protocols*, Methods in Molecular Biology, vol. 2088, https://doi.org/10.1007/978-1-0716-0159-4_13,

reactions in a tumor is currently not possible using these methods due to the lack of key biological parameters, such as the binding constant K_M or the catalytic rate k_{cat} [3].

Flux balance analysis (FBA) is an alternative approach to infer metabolic activity without requiring kinetic parameters. This method assumes metabolic steady state and identifies all metabolic flux states by optimizing a specific cellular objective, such as biomass production [4]. FBA makes use of genome-scale metabolic reconstructions (GEMs), which contain the current knowledge of biochemical network topology of an organism [5]. FBA uses the network topology from GEMs to estimate the metabolic flux through each reaction [4]. High throughput -omics data such as transcriptomics can be used with these models to obtain more condition-relevant flux states [6].

Although FBA identifies steady state fluxes, it fails to capture dynamic metabolic processes, such as the differentiation of stem cells or malignant transformation of cancer cells. To address this, we developed the Dynamic Flux Activity (DFA) approach. This algorithm combines time-course metabolomics with a genome-scale metabolic model to infer the metabolic state that best fits the metabolomics data [7]. DFA has many advantages over unsteady-state FBA [8], a related approach for modeling time-course metabolomics. DFA can use both absolute and relative metabolomics values, while uFBA requires absolute quantified data using internal standards. Secondly, DFA can use as little as two time points while uFBA requires several time points. Further, the strength of metabolomics constraints can be easily tuned in DFA with a single simple parameter *Kappa*, which is proportional to the relative optimization strength of the growth objective.

This tutorial describes the DFA approach and outlines the steps in applying this tool to construct dynamic metabolic models. We will use time-course metabolomics data from cancer cell lines as an example to build dynamic metabolic models using DFA. The metabolic fluxes obtained from DFA reveal metabolic differences between rapid- and slow-proliferating cancer cells.

2 Materials

2.1 Dynamic Flux Activity Algorithm

The Dynamic Flux Activity (DFA) algorithm [7] was implemented in MATLAB (http://www.mathworks.com). It requires the Gurobi Mathematical Programming Solver (http://www.gurobi.com/) version 8.0.1. The latest version of the DFA MATLAB scripts and NCI-60 cancer cell line metabolomics dataset used in this protocol can be downloaded through the Chandrasekaran lab website http://www.sriramlab.org/software/.

2.2 Genome Scale Metabolic Modeling Tools

The Constraint-Based Reconstruction and Analysis (COBRA) toolbox is a collection of metabolic modeling tools intended for analyzing metabolic pathways and predicting metabolic phenotypes [9]. Several functions provided by the COBRA toolbox were used in this tutorial, including the **deleteModelGenes** function for gene knockout analysis. The MATLAB version of the COBRA toolbox can be cloned from https://github.com/opencobra/cobratoolbox.

2.3 Genome-Scale Metabolic Models

Genome scale metabolic models are available in the BiGG database (http://bigg.ucsd.edu/) and the BioModel database (https://www.ebi.ac.uk/biomodels-main/) in .mat, .sbml, and .json formats. The core cancer genome scale metabolic network [10] was used in this tutorial (*see* **Note 1**).

2.4 Visualizing Metabolic Networks

Escher (https://escher.github.io/) is a web-based tool for visualizing COBRA metabolic models [11]. The growth rates obtained from our in silico single gene knockout analysis and the metabolic flux values predicted from the dynamic models were mapped onto the human metabolic map in Escher. Alternative tools that visualize metabolic pathway activity but were not used in this analysis include the BioCyc https://biocyc.org/overviewsWeb/celOv.shtml and iPath3 from EMBL (https://pathways.embl.de/).

2.5 NCI-60 Cancer Cell Line Transcriptomics Data

There are several biorepositories that carry transcriptomics datasets. Examples include the GEO database (https://www.ncbi.nlm.nih.gov/geo/), the NCI-60 biorepository (https://discover.nci.nih.gov/cellminer/), and the TCGA Cancer Atlas (https://cancergenome.nih.gov/).

The example described in Subheading 3 uses transcriptomics data from GEO database (accession number: GSE32474). To determine differentially expressed genes in the dataset, the transcriptomics data was $\log_2$ normalized and Z-transformed. The formula for the Z-transformation is described in Eq. 1:

$$m_z = \frac{m - \mu_m}{\sigma_m} \tag{1}$$

where m is the $\log_2$ concentration, μ_m is the mean concentration across the NCI-60 cancer cell lines, and σ_m is the standard deviation across the NCI-60 cancer cell lines. The resulting value m_z tells us how the concentration for a specific cell line is different from the remaining distribution. To determine which genes were differentially expressed, an m_z value of ± 1.5 was used to determine up- or down- regulation in a cell line relative to all other cell lines, respectively.

2.6 Time-Course Metabolomics for Tissue-Specific Cancer Cells

We used the NCI-60 cancer cell line CORE time-course exometabolomics data from Jain et al. [12]. This dataset contains 141 metabolites from 59 cancer cell culture media. Each metabolite level was measured for two time points: day 0 and day 4–5. This dataset was used in our prior study to build cell-line specific models using DFA [7]. Note that DFA can use both extracellular and intracellular metabolomics data.

3 Methods

3.1 Flux Balance Analysis and Flux Variability Analysis

Flux balance analysis (FBA) is a linear optimization method to calculate the flux solutions of a metabolic network at steady state. FBA relies on two assumptions: (1) no accumulation of metabolites for any given reaction and (2) the specification of a cellular objective. The stoichiometric matrix **S** represents an **m** × **n** array with **m** metabolites and **n** reactions and the optimal flux solutions are stored in a vector **v** based on the objective function **Z**. The upper and lower flux bounds (v_{ub} and v_{lb}) are additional constraints on the solution space for **v**. FBA is mathematically summarized in Eq. 2:

$$\begin{aligned} &\text{Maximize } Z = c^T v \\ &\text{Subject to } Sv = 0 \\ &\text{and } v_{lb} \leq v_{ij} \leq v_{ub} \end{aligned} \tag{2}$$

Flux Variability Analysis (FVA) is a related method to identify the full range of possible fluxes for a given set of reactions. The optimal constraints for the upper and lower flux bound are based on the stoichiometric constraints defined in Eq. 3:

$$\begin{aligned} &v_{ub} = \max{}_v \\ &v_{lb} = \min{}_v \\ &\text{Maximize } Z = c^T v \\ &\text{Subject to } Sv = 0 \\ &\text{And } v_{lb} \leq v_{ij} \leq v_{ub} \end{aligned} \tag{3}$$

The COBRA toolbox is used to solve both FBA and FVA problems using the **optimizeCbModel()** and **fluxVariability()** functions, respectively (*see* **Note 1**). Additional documentation and tutorials on each COBRA toolbox module is available at: https://opencobra.github.io/cobratoolbox/latest/index.html.

FBA and FVA define constraints that reduce the number of possible solutions representative of the biological system. Experimental data including the substrate uptake rates and transcriptomics data can be incorporated into the metabolic model as additional constraints (*see* Subheading 3.4 for more details) [13].

While FBA and FVA reduce the number of feasible flux solutions, it is still possible to obtain several flux solutions with the same objective function. In this tutorial, parsimonious enzyme usage will be assumed to obtain a single optimal solution. This assumption specifies that a cell aims to balance the energetic cost of enzyme production and maximizing its fitness (i.e. growth rate). This is achieved by minimizing the total metabolic flux while optimizing the biomass [14].

3.2 Integrating Time-Course Metabolomics in Genome-Scale Models

To integrate time-course metabolomics data with a GEM, we will describe the **dfa** algorithm. This algorithm identifies the metabolic state that best fits the metabolomics data. The time-course metabolomics data and a genome-scale metabolic model are used as inputs. The output is a metabolic model containing reaction fluxes consistent with the time-course metabolic constraints.

3.2.1 Calculating Rates of Change in Time-Course Metabolomics

The formulation for **dfa** is similar to FBA, but includes a right-hand side (**rhs**) vector, which represents the normalized metabolite concentration rate of change. Each element in the **rhs** vector will be denoted as **epsilon** for the *i*th metabolite (ε_i). (Eq. 4). To calculate **epsilon**, the algorithm uses the MATLAB built-in **polyfit** function with one degree of freedom for linearly fitting the time-course metabolomics data for each metabolite. The function outputs the slope and the intercept corresponding to each metabolite. The slope represents the rate of change for each metabolite over time, while the intercept represents the metabolite concentration at $\boldsymbol{t_0}$. ε_i is the proportion of the slope and intercept for metabolite **i** (Eq. 6).

$$\varepsilon_i = \frac{\text{slope}}{\text{intercept}} \tag{4}$$

We normalize ε_i between $-\mathbf{1}$ and $\mathbf{1}$ to be proportional to the natural metabolite abundance while maintaining the relative magnitude and direction of metabolite changes. To get $||\varepsilon_i||$, ε_i is divided by the maximum absolute value of ε_i (Eq. 5).

$$\varepsilon_i = \frac{\varepsilon_i}{\max(\text{abs}(\varepsilon))} \tag{5}$$

If the maximum absolute value of ε_i is much larger than the average ε_i value, more advanced normalization procedures are recommended (*see* **Note 2**).

3.2.2 The Dynamic Flux Analysis Algorithm

Similar to FBA, DFA sets $\mathbf{S_{ij}}$ as the stoichiometric matrix and $\mathbf{v_j}$ as the flux state vector. For metabolites in the model not measured from the metabolomics dataset, $\varepsilon_i = 0$ (Eq. 6).

$$\varepsilon_i = \sum\nolimits_{i=1}^{M} S_{ij}v_j + \alpha_j - \beta_j, \text{where } i = 1, \ldots, M \tag{6}$$

To account for the noise within a given metabolomics dataset, we impose a soft constraint using two pseudo-reaction parameters ($\boldsymbol{\alpha}, \boldsymbol{\beta}$). α and β are positive fluxes that represent the deviation from the metabolomics data. The algorithm minimizes the sum of α and

β, affording a flux distribution that is consistent with the metabolomics-based model. The formulation for this constraint is shown below (Eq. 7):

$$\begin{aligned} &\mathrm{Min}\sum_{j=1}^{N}\left(\alpha_j+\beta_j\right),\text{where } j=1,\ldots,M \\ &\text{Subject to } \varepsilon_i=\sum_{i=1}^{M}S_{ij}v_j+\alpha_j-\beta_j,\text{where } i=1,\ldots,M \\ &\text{and } \alpha_j \text{ and } \beta_j\geq 0 \end{aligned} \tag{7}$$

To maximize the objective function and minimize the sum of α and β simultaneously, the objective function coefficients for the metabolomic constraints are set as −**kappa**. **kappa** is the strength of the metabolomic constraints relative to the objective. We also minimize the sum of absolute flux values through other reactions using the assumption of parsimonious enzyme usage. The objective function coefficients for the metabolomic constraints are set as −**kappa2** (*see* **Note 3**).

To summarize the **dfa** algorithm, there are three optimization goals that are controlled by the weights −**kappa** and −**kappa2**:

1. Maximize the objective function in the metabolic model.
2. Maximize the extent of flux distribution matches with the metabolomics data.
3. Minimize the overall flux of other reactions.

3.2.3 Matching Metabolites from the Metabolomics Dataset to the Genome-Scale Model

The first step is to match the metabolite names between the metabolomics data and the model (Table 1). As an example, L-alanine is an exometabolite that occupies the 186th position in the core model, and has two metabolomic time points. In contrast, alpha-ketoglutarate is not taken up by human cells as they lack a specific transporter. Hence this exometabolite is not present in the metabolic model and is thus excluded from the construction of the dynamic model. Data for several cell lines can be present in the same Excel file, and are specified as an argument in the **dfa** function. The metabolites can be matched manually to the metabolic model. We provide several tools below for automated matching.

MetaboAnalyst

The current MetaboAnalyst version (https://www.metaboanalyst.ca/) is available as both a web application and an R library. This tool analyzes metabolomics data [15]. The *Compound ID Conversion tool* in the *Other Utilities* page maps metabolite IDs to several metabolite databases. An approximate search can be used to select the metabolite based on some key strings in the query.

Programmatically Matching Metabolite IDs

General use languages such as Linux and Python can parse out metabolite annotations from the original model. For suggested Python libraries, refer to the documentation for **lxml**, **ElementTree**, and **BeautifulSoup** for parsing .xml and .html files.

Table 1
An example of the input file format for DFA

Metabolite	Position in model	A498 (t_0)	A498 (t_1)	...
Alpha-ketoglutarate	0	−0.00218695	0.01718404	...
Alanine	186	−0.01105939	−0.00828399	...
AMP	0	−1.66230048	−1.74863463	...
...	...	...	...	...
...	...	...	...	...
...	...	...	...	...

Metabolite Compartment Matching

Because we lose the subcellular compartment-specific information during intracellular metabolomics measurement, we assume that the intracellular metabolites represent the sum total of the three compartments (cytosol, mitochondria, and nucleus). Metabolites in the metabolic model have a suffix denoting their location in the cytosol (c), mitochondria (m), nucleus (n), or as an extracellular (e) metabolite. While matching metabolites from metabolomics data to the model, we need to specify positions (c), (m), (n), and (e) in the model. The position is used to calculate the flux coefficients (epsilon) for each species (metabolite in different compartments).

A snapshot of the MATLAB code is shown below:

```
%% dfa

% The dfa algorithm outputs the rate of change for.
% time-course metabolomics data.

(cell_linemodel, geneko_grwthrt_obj, gene-
ko_grwthrt, rxn_knockout_obj, rxnko_grwthrt_obj) =
dfa(model, timecourse_metabolomics_file, sheet-
name, celline, kappa, kappa2)

%% Inputs

% timecourse_metabolomics_file, sheetname: The time-
course metabolomics data for each time point and condi-
tion needs to be formatted in an Excel file. The rows
correspond to metabolites, while columns correspond
to different time points (Table 1). If different condi-
tions are within the same Excel file but in a different
sheet, the sheetname must be specified.

%% Optional Inputs

% genedelflag, rxndelflag: These two arguments take on values
(0,1), where 0 indicates that gene or reaction
```

```
deletions will not be performed, while 1 indicates
otherwise.
% kappa: The relative weight for the consistency of the
flux distribution compared to the weight for maximiz-
ing the biomass function. The default value = 1. If
there is no growth in the predict flux distribution or
kappa is increased with low consistency in the metabo-
lomics dataset, kappa should be reduced.

% kappa2: The relative weight for minimizing the sum of
absolute flux values of other reactions. The default
value is 1E-3. The value for kappa2 should be increased
when the magnitudes of some flux values are relatively
larger than the expected values based on known biochem-
ical knowledge.

%% Outputs
% dynamicmodel: The time-course metabolomics con-
strained model.

% grate_wt: The growth rate of the dynamic model when the
objective is to maximize the growth rate of the cell.

%% Optional Outputs

% geneko_grwthrt, rxnko_grwthrt: When these flags are set
to 1, the growth rates from gene or reaction deletion
analysis are stored in these two variables
respectively.

% slope: The rate of accumulation or depletion of metabo-
lites from time-course metabolomics data.

% solverobj, geneko_grwthrt_obj, rxnko_grwthrt_obj: The sol-
verobj is the value of the optimal objective function,
which can be used to assess the fit to the metabolomics
data. Genes or reactions deletions that lead to a reduc-
tion in the objective function suggest they are neces-
sary for satisfying the metabolic objectives of the
cell.
```

3.3 Using DFA to Infer Metabolic Flux State Based on Metabolomics Data

This tutorial will construct two cancer cell line models, one for a rapidly proliferating cancer cell line (LOXIMVI) and a slowly proliferating cell line (A498), using data from the CORE study [12]. The time-course metabolomics data from the NCI-60 cancer cell lines is available at the Chandrasekaran lab website: http://www.sriramlab.org/software/.

```
%% Fitting time-course metabolomics data to create A498 and LOXIMVI
dynamic models using the dfa algorithm
% The provided MATLAB code does the following:

  % 1) Initiate the Cobra Toolbox and set the Gurobi
    optimization solver
```

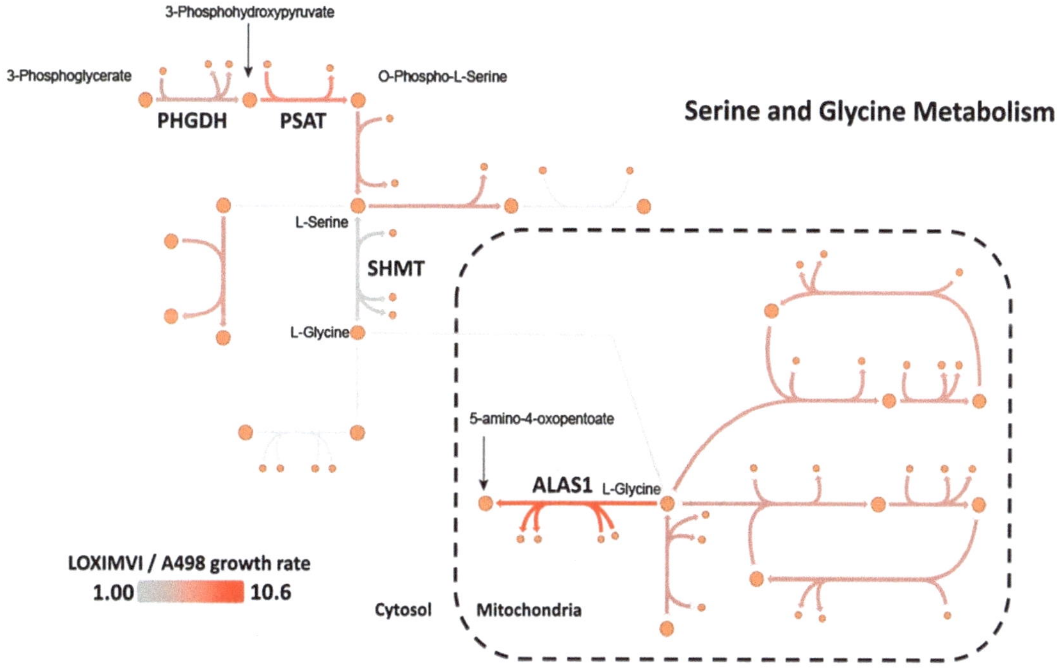

Fig. 1 Differential gene sensitivity between LOXIMVI and A498 in glycine and serine metabolism. The genes PHGDH and PSAT (red) were among the six genes that were predicted to be differentially sensitive after knockout between LOXIMVI and A498 (Z-score = 28.44, p-value = 1.73E−145, supplementary table 2, available through the Chandrasekaran lab website http:/www.sriramlab.org/software/). Note that the growth rate ratio for PSAT and ALAS1 is larger in the Escher map due to multiple gene-protein-reaction rules

```
% 2) Load the core metabolic model
% 3) Specify the biomass function for DFA
% 4) Run DFA to fit the A498 metabolomics onto the meta-
  bolic model
% 5) Repeat to create a LOXIMVI model

% Initialize the COBRA toolbox
initCobraToolbox(false);
changeCobraSolver(gurobi);

% Read in the cancer core models and make sure the bio-
mass objective function is optimized.
load('cancer_core_models_palsson.mat');
A498_core = changeObjective(core_genecomb,
'biomass_NCI60');
LOXIMVI_core = changeObjective(core_genecomb,
'biomass_NCI60');

% Run DFA for the core NCI-60 models. The input is the
Excel sheet containing the time course metabolomics
```

```
data. The sheet name is the second argument. The last
argument is the cancer cell line.
(a498_core, a498_grate_obj, a498_grate,
a498_rxn_grate_obj, A498_rxn_grate) = dfa
(A498_core, 'jainetalcore_values.xlsx',...
'Jain et al CORE Values (Cancer)', {'A498'});
(loximvi_core, loximvi_grate_obj, loximvi_grate,
loximvi_rxn_grate_obj, ...
loximvi_rxn_grate) = dfa(LOXIMVI_core, 'jainetal-
core_values.xlsx',... 'Jain et al
CORE Values (Cancer)', {'LOXIMVI'});
```

The algorithm outputs the cell-line specific metabolic model with metabolomics constraints. In addition, it provides the impact of deleting metabolic genes on the biomass growth of these cell lines. These growth rate values can be used to identify genes and reactions that one cell line relies upon more than the other (Fig. 1).

We analyzed the ratio of LOXIMVI / A498 growth rates after knocking out all 1496 metabolic genes. Six genes were identified by DFA to be differentially sensitive reactions between the two cell lines in supplementary table 1. Deletion of genes in one carbon metabolism (glycine and serine metabolism) such as PHGDH and PSAT resulted in greater growth reduction in the slow proliferating cell line compared to the fast proliferating cell line. In contrast, deleting ACLY and SLC25A1 resulted in impacting growth in fast proliferating cells.

The next section integrates gene expression data into the dynamic cancer cell model to obtain metabolic fluxes specific to both transcriptomic and metabolomics constraints.

3.4 Integrating Transcriptomics or Proteomics Within the Metabolic Network

In this section, we integrate the NCI-60 cancer cell line transcriptomic data for A498 and LOXIMVI with the dynamic model obtained from the **dfa algorithm**. This will provide metabolic fluxes consistent with both transcriptomics and metabolomics data. The **constrain_flux_regulation** algorithm [16] uses as input a list of over- or underexpressed genes or reactions, and a GEM. It outputs a flux solution consistent with the expression data. We assume that overexpressed metabolic genes are highly active; hence the corresponding reactions are likely to carry a nonzero flux. Similarly, underexpressed genes are assumed to have lower activity, but the reaction is not completely turned 'OFF' (flux = 0). Rather, we minimize metabolic flux through the underexpressed reactions. Using these two principles, the formulation for **constrain_flux_regulation** is shown below (Eqs. 8 and 9):

$$\text{Max}\sum_{i=1}^{N}(t_i + r_i), i = 1, \ldots, N$$
$$\text{Subject to } v_i - t_i(\varepsilon_i + M) \geq -M \qquad (8)$$
$$\text{and } v_i - r_i(\varepsilon_i + M) \geq M$$
$$\text{where } t_i \text{ and/or } r_i = 0 \text{ or } 1$$

The algorithm maximizes the number of active reactions that correspond to overexpressed genes. Epsilon (ε_i) is a vector of up-regulated reaction fluxes, which can be a fixed positive value or as a unique value for each reaction. $\mathbf{t_i}$ and $\mathbf{r_i}$ indicate whether the reaction is active with a value of at least ε_i with a positive or negative direction. If $\mathbf{t_i} = 1$, the reaction is active in the positive direction. Otherwise, if $\mathbf{r_i} = 1$, the reaction is active in the negative direction. The variable **M** (default value = 10,000) is a constant that is much larger than ε_i (default value = 1E−3).

constrain_flux_regulation does not set down-regulated reactions to have a flux of 0. Instead, it minimizes the flux through these reactions in a similar formulation to the PROM algorithm [17]. Like Eq. 8, the underactive reaction fluxes can be a fixed value or can be set as a unique value for each reaction. The formulation to minimize flux through down-regulated reactions is shown below:

$$\text{Max}\sum\nolimits_{i=1}^{K}\kappa(\alpha_i - \beta i), \textit{ where } i = 1, \ldots, K$$
$$\text{Subject to } v_i \geq \text{lb}_i{}' - \alpha_i \qquad (9)$$
$$\text{where } v_i \leq \text{ub}_i{}' - \beta_i$$
$$\text{and } \alpha_i \text{ and } \beta_i \geq 0$$

$\mathbf{lb_i}'$ and $\mathbf{ub_i}'$ are constraints based on transcriptional regulation (default value = 0), minimizing the flux through down-regulated metabolic enzymes. α_i and β_i are deviations from the constraints and κ is the penalty associated with the deviation. A higher value of κ (default value = 1) results in a harder constraint. This results in more constrained flux states based on gene expression.

To summarize the **constrain_flux_regulation** algorithm, there are two optimization goals that are controlled by the weights **epsilon** and **kappa**:

1. Maximize the number of reactions associated with overexpressed genes that carry flux.
2. Minimize the flux through reactions associated with underexpressed genes (*see* **Note 4**).

A snapshot of the MATLAB code is shown below:

```
%% constrain_flux_regulation

% The constrain_flux_regulation algorithm will maxi-
mize the flux in.
```

```
% reactions corresponding to upregulated genes while minimizing the flux
% corresponding to downregulated genes

(flux, growth, objective) = constrain_flux_regulation(model,... onreactions, offreactions, kappa, rho, epsilon, mode);

%% Inputs

% onreactions: a cell array containing genes that are upregulated.

% offreactions: a cell array containing genes that are downregulated.

%% Optional inputs

% rho, kappa: The relative weights for on- and off-reactions respectively.
% The default value = 1. The value should be reduced for larger models or.
% optimizations with a large number of reactions (rho = 1E-3 recommended
% for 100+ reactions).

% The weights can be set to a different value for each reaction by creating % a cell array with elements corresponding to the number of reactions that % are up or downregulated.

% epsilon: The minimum flux for on-reactions. The default value = 1E-3.

% Epsilon can be set to a different value for each reaction by creating a
% cell array with elements corresponding to the number of upregulated
% reactions. This value can be set proportional to the extent of gene
% expression.

% mode: Describes the input list. If on- and off-reactions contain gene symbols, the mode must be set to 0. Otherwise, if set to 1, BiGG reaction IDs are used.

%% Outputs

% flux: The value state of all reactions, using a formulation similar to
% parsimonious FBA (pFBA).

% growth: Predicted cellular growth rate

% objective: The optimization solver objective
```

3.5 Integrating Transcriptomics Data to Identify Differential Flux States Between Rapidly and Slowly Proliferating Cancer Cells

The following example uses the NCI-60 cancer cell line transcriptomics data. The expression datasets used in this example are available in the GEO database (GSE32474).

Jain et al. [12] observed that rapidly proliferating cancer cells like LOXIMVI are dependent on mitochondrial glycine metabolism. In contrast, slowly proliferating cancer cells such as A498 do not have this dependency. To analyze the differential metabolic activity between these two cells, we overlay transcriptomics data from these cells onto the dynamic model obtained from the **dfa** algorithm. This is achieved using the **constrain_flux_regulation** algorithm. The example datasets for this tutorial are available at the Chandrasekaran lab website: http://www.sriramlab.org/software/.

%% Using the NCI-60 transcriptomics data

to obtain A498 and LOXIMVI metabolic flux states

```
% Differential expression for the NCI-60 cancer cell lines is provided in the following files.
load("core_model.mat");

% Get genes that are upregulated and downregulated in the model
a498_up = intersect(a498_up, A498.genes);
a498_down = intersect(a498_down, A498.genes);
loximvi_up = intersect(loximvi_up, LOXIMVI.genes);
loximvi_down = intersect(loximvi_down, LOXIMVI.genes);

% Get the flux state for both cancer cell lines based on upreg/downreg gene expression. Because the model has > 100 genes, rho = 1E-3 was used.
(a498_flux, a498_grate, a498_solverobj) = constrain_flux_regulation(A498, a498_up, a498_down, 1, 1, 1E-3, 0);
(loximvi_flux, loximvi_grate, loximvi_solverobj) = constrain_flux_regulation(LOXIMVI, loximvi_up, loximvi_down, 1, 1, 1E-3, 0);
```

The loximvi_flux and a498_flux variables contain the predicted flux through all metabolic reactions in the respective cancer models.

3.6 Visualizing Metabolic Flux Predictions Through Metabolic Pathways

Metabolic flux, gene expression, and metabolomic data can be visualized on the core cancer model using Escher (https://escher.github.io/) web-based API. Figure 2 illustrates differentially active reactions between A498 and LOXIMVI in the glycine and serine metabolic pathways.

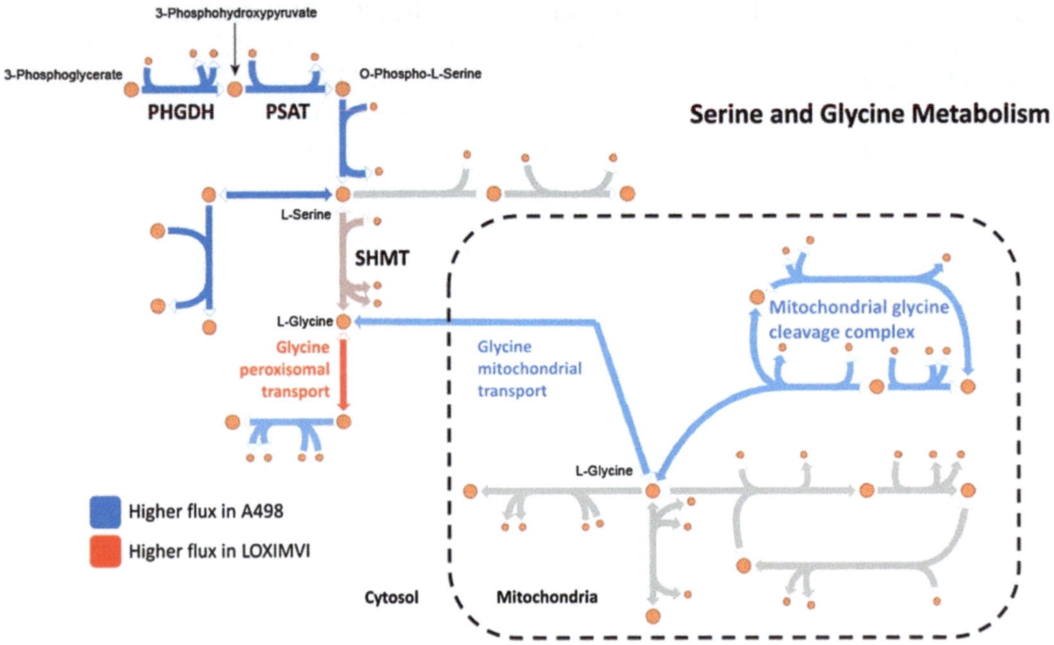

Fig. 2 Relative metabolic flux in Glycine metabolism for fast- and slow-proliferating cancer cells using both transcriptomics and metabolomics data. A498 cells shows higher efflux of glycine out of the mitochondria and out of the cell, compared to LOXIMVI (shown in blue). In LOXIMVI, glycine has high flux from the cytosol toward the peroxisome in LOXIMVI, compared to A498 (shown in red)

4 Notes

1. When using DFA, the metabolic model needs to have a specified objective function (e.g. biomass).
2. When the minimum absolute value of epsilon is very large relative to other epsilon values (~100-fold), the values in **epsilon** can be divided by percent quantile instead of the maximum.
3. In the DFA algorithm, a large **kappa2** value will result in lower flux through all reactions while a smaller **kappa2** value will lead to larger values of metabolic flux.
4. Individual constraints can be violated to maximize the global fit of the -omics data. For example, reactions that are "ON" (high expression) can be predicted to be off if this maximizes the consistency with the remaining over- or underactive reactions.

References

1. Luengo A, Gui DY, Vander Heiden MG (2017) Targeting metabolism for cancer therapy. Cell Chem Biol 24(9):1161–1180. https://doi.org/10.1016/j.chembiol.2017.08.028
2. Saa PA, Nielsen LK (2017) Formulation, construction and analysis of kinetic models of metabolism: a review of modelling frameworks. Biotechnol Adv 35(8):981–1003. https://doi.org/10.1016/J.BIOTECHADV.2017.09.005
3. Nilsson A, Nielsen J, Palsson BO (2017) Commentary metabolic models of protein allocation call for the kinetome. Cell Syst 5:538–541. https://doi.org/10.1016/j.cels.2017.11.013
4. Orth JD, Thiele I, Palsson BØ (2010) What is flux balance analysis? Nat Biotechnol 28 (3):245–248. https://doi.org/10.1038/nbt.1614
5. O'Brien EJ, Monk JM, Palsson BO (2015) Using genome-scale models to predict biological capabilities. Cell 161(5):971–987. https://doi.org/10.1016/j.cell.2015.05.019
6. Uhlén M, Hallström BM, Lindskog C, Mardinoglu A, Pontén F, Nielsen J (2016) Transcriptomics resources of human tissues and organs. Mol Syst Biol 12(4):862. https://doi.org/10.15252/msb.20155865
7. Chandrasekaran S, Zhang J, Sun Z, Zhang L, Ross CA, Huang Y-C et al (2017) Comprehensive mapping of pluripotent stem cell metabolism using dynamic genome-scale network modeling. Cell Rep 21(10):2965–2977. https://doi.org/10.1016/J.CELREP.2017.07.048
8. Bordbar A, Yurkovich JT, Paglia G, Rolfsson O, Sigurjónsson ÓE, Palsson BO (2017) Elucidating dynamic metabolic physiology through network integration of quantitative time-course metabolomics. Sci Rep 7:46249
9. Heirendt L, Arreckx S, Pfau T, Mendoza SN, Richelle A, Heinken A et al (2017) Creation and analysis of biochemical constraint-based models: the COBRA Toolbox v3.0. http://arxiv.org/abs/1710.04038
10. Zielinski DC, Jamshidi N, Corbett AJ, Bordbar A, Thomas A, Palsson BO (2017) Systems biology analysis of drivers underlying hallmarks of cancer cell metabolism. Sci Rep 7:41241
11. King ZA, Dräger A, Ebrahim A, Sonnenschein N, Lewis NE, Palsson BO (2015) Escher: a web application for building, sharing, and embedding data-rich visualizations of biological pathways. PLoS Comput Biol 11(8):e1004321. https://doi.org/10.1371/journal.pcbi.1004321
12. Jain M, Nilsson R, Sharma S, Madhusudhan N, Kitami T, Souza AL et al (2012) Metabolite profiling identifies a key role for glycine in rapid cancer cell proliferation. Science 336 (6084):1040–1044. https://doi.org/10.1126/science.1218595
13. Yizhak K, Gaude E, Le Dévédec S, Waldman YY, Stein GY, van de Water B et al (2014) Phenotype-based cell-specific metabolic modeling reveals metabolic liabilities of cancer. Elife 3. https://doi.org/10.7554/eLife.03641
14. Lewis NE, Hixson KK, Conrad TM, Lerman JA, Charusanti P, Polpitiya AD et al (2010) Omic data from evolved E. coli are consistent with computed optimal growth from genome-scale models. Mol Syst Biol 6:390. https://doi.org/10.1038/msb.2010.47
15. Chong J, Soufan O, Li C, Caraus I, Li S, Bourque G et al (2018) MetaboAnalyst 4.0: towards more transparent and integrative metabolomics analysis. Nucleic Acids Res 46(W1): W486–W494. https://doi.org/10.1093/nar/gky310
16. Shen F, Boccuto L, Pauly R, Srikanth S, Chandrasekaran S (2019) Genome-scale network model of metabolism and histone acetylation reveals metabolic dependencies of histone deacetylase inhibitors. Genome Biol 20(1):49
17. Chandrasekaran S, Price ND (2010) Probabilistic integrative modeling of genome-scale metabolic and regulatory networks in Escherichia coli and Mycobacterium tuberculosis. Proc Natl Acad Sci U S A 107 (41):17845–17850. https://doi.org/10.1073/pnas.1005139107

Chapter 14

Metabolic Network Reconstructions to Predict Drug Targets and Off-Target Effects

Kristopher Rawls, Bonnie V. Dougherty, and Jason Papin

Abstract

The drug development pipeline has stalled because of the difficulty in identifying new drug targets while minimizing off-target effects. Computational methods, such as the use of metabolic network reconstructions, may provide a cost-effective platform to test new hypotheses for drug targets and prevent off-target effects. Here, we summarize available methods to identify drug targets and off-target effects using either reaction-centric, gene-centric, or metabolite-centric approaches with genome-scale metabolic network reconstructions.

Key words Genome-scale metabolic network reconstruction (GENRE), Drug targets, Off-target effects, Constraint-based modeling, Flux balance analysis (FBA)

1 Introduction

We are beginning to capitalize on systems biology approaches to combine basic science knowledge with omics data to tackle several health challenges like more efficient and effective drug development [1]. A challenge for drug development is identifying new targets while minimizing adverse effects. Adverse, off-target effects can cause drugs to fail in preclinical trials or can result in a drug being recalled, posing a serious health concern [2, 3]. In light of this threat, there is a growing need to more accurately predict drug effects given their interaction with intended as well as unintended targets [4]. Computational models can be powerful tools used to predict drug targets as well as their off-target effects and thereby increase the yield of the drug development pipeline. GEnome-scale metabolic Network REconstructions (GENREs) are emerging as powerful computational tools to address these challenges [5–7].

Kristopher Rawls and Bonnie V. Dougherty contributed equally to this work.

Deepak Nagrath (ed.), *Metabolic Flux Analysis in Eukaryotic Cells: Methods and Protocols*, Methods in Molecular Biology, vol. 2088, https://doi.org/10.1007/978-1-0716-0159-4_14,

GENREs capture the relationships between metabolic genes, the reactions they catalyze, and associated metabolites. Constraint-based methods have been used to analyze GENREs to study the metabolism of many different organisms [8–10]. More recently, GENREs have focused on human metabolism [11–13] serving as platforms to identify drug targets, such as in cancer metabolism [14, 15] and to identify off-target drug effects [16–18].

In this book chapter, we describe examples of how to use GENREs to predict drug targets and off-target effects. There are important considerations to make using GENREs for this context such as the applicable cell type, cellular processes represented, or the kinds of data available. After the problem has been set up, there are a broad range of tools available to perform analyses to identify drug targets and off-target effects that are reaction-centric, gene-centric, or metabolite-centric. COnstraint Based Reconstruction and Analysis (COBRA) methods have been developed for these different approaches [19]. We specifically describe the steps to perform:

- Reaction-centric approaches that focus on identifying reactions in the network that when inhibited affect the cell's ability to perform key functions.
- Gene-centric approaches that focus on identifying genes that code for enzymes that are critical to maintain cellular function in either a healthy or diseased state.
- Metabolite-centric approaches that focus on identifying metabolites that are highly connected in the network and, when targeted, affect cellular function.

2 Materials

Genome-scale network reconstructions of metabolism or GENREs can serve as in silico platform for systematically identifying drug targets and drug off-target effects; for example, a GENRE was used to study cancer metabolism and identify potential drug targets [20]. A genome-scale network reconstruction is comprised of (a) reactions and associated metabolites specific to an organism or cell type (summarized into a mathematical framework known as the S matrix) and (b) the genes which code for the proteins that catalyze each reaction (represented as gene–protein–reaction or GPR rules) (Fig. 1a). Constraint-based reconstruction and analysis (COBRA) methods are used to make predictions of the functional capabilities of each reconstruction (Fig. 1b). An introduction to COBRA methods with a simplified metabolic network and associated code was recently published [21]. Flux balance analysis (FBA), a well-used COBRA method, predicts fluxes through

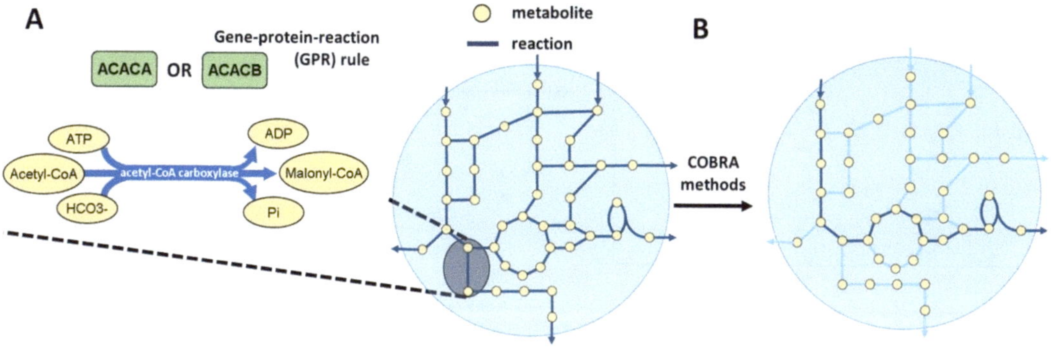

Fig. 1 Metabolic network reconstructions and their associated methods. (**a**) Genome-scale metabolic reconstructions contain the reactions that are known to be catalyzed in an organism, represented with input and output metabolites, and the genes that code for the proteins which catalyze these reactions, known as gene–protein–reaction (GPR) rules. (**b**) Since these reconstructions are represented mathematically as matrices, various optimization methods can be used to determine what reactions are used and the flux through these reactions. Constraint-based modeling and reconstruction analysis (COBRA) methods are applied to reconstructions to make predictions on fluxes through a reconstruction based on a set of constraints on inputs and outputs to the system while optimizing for flux through a specific reaction in the network

reactions in the network subject to a set of reaction constraints and given a model objective. Building on FBA as a framework, predictions of drug targets and off-target effects can be made using reaction-centric, gene-centric, or metabolite-centric approaches to determine the effect of different parts of the network on the defined objective function. In order to use these methods, your question of interest must first be framed by (1) choosing a network reconstruction, (2) choosing an applicable objective function, and (3) using data to apply constraints to the network (Fig. 2).

2.1 Choosing a Network Reconstruction or Model

The first step for predicting drug targets or off-target effects using COBRA methods is to choose a network reconstruction for your system and question of interest. There are multiple GENREs of human metabolism available [12, 13, 22–24]. From these reconstructions, various algorithms (summarized in [25, 26]) have been implemented which take in either genes or reactions, collected from various data sources, that are tissue-, cell-type-, or disease-specific to construct a tissue-, cell-type-, or disease-specific model. A commonly used data source is the human protein atlas (HPA) which contains qualitative scores for protein presence or absence in specific tissues or cancers [27]. Some algorithms have enabled the generation of compendiums of tissue or cell-type specific models of metabolism [24, 28, 29]. However, these individual models are often only in draft form and are therefore incomplete and thus do not comprehensively capture many specific tissue or cell-type functions. Individual tissue-, cell type-, or disease-specific models have been published, usually to address a specific question of interest. Curated models exist for the liver, kidney, heart, macrophage, and

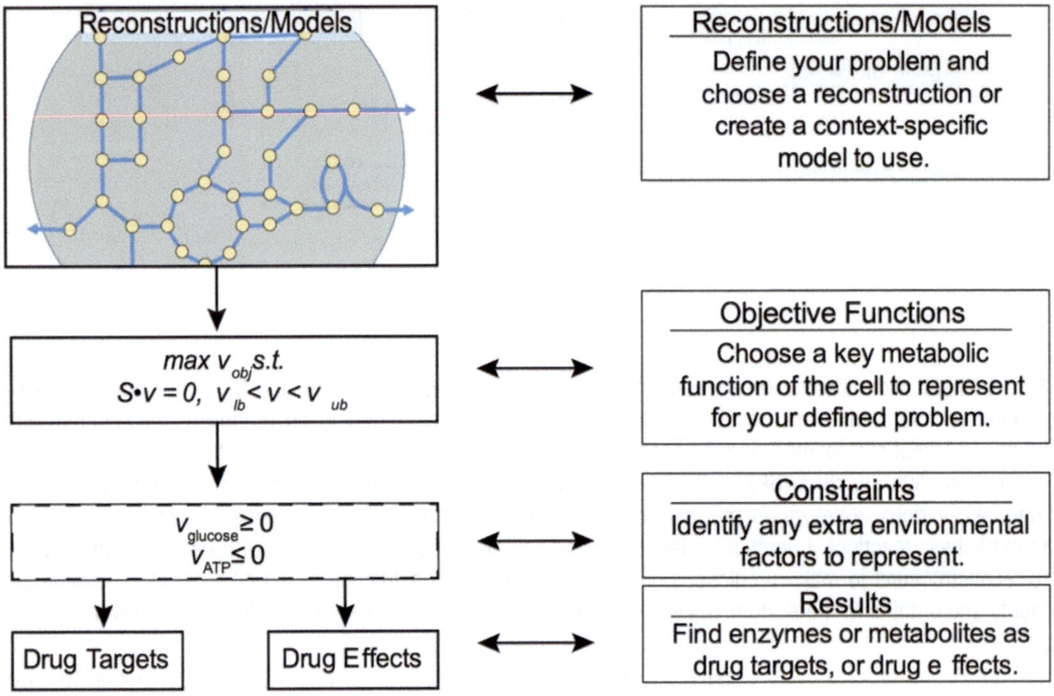

Fig. 2 Considerations to make when using GENREs to predict drug targets and off-target effects. To predict drug targets and off-target effects, you first need to choose a reconstruction that fits the organism of interest. A reconstruction or model may already exist for your organism or tissue of interest; however, you may need to adapt the reconstruction. Next, you will need to choose an objective function which describes a key function or phenotype that you are interested in studying. Finally, additional constraints, such as constraints on nutrient uptake, may be incorporated to better capture your question of interest. Once the appropriate considerations are made, constraint-based methods can be applied to your model to predict both drug targets and off-target effects

various forms of cancer [12, 16, 20, 30–35]. Finally, some authors choose to generate their model using a manual, bottom-up approach, such as through curation of a draft reconstruction. You must either choose to use a general human reconstruction, a published, curated tissue-, cell type-, or disease-specific model, or generate your own model for your system of interest. Detailed protocols for constructing metabolic network reconstructions have been published [36].

2.2 Identifying an Objective Function

After identifying the model or reconstruction that you will use, it is necessary to identify an objective function that is valid for the model you have chosen and the drug targets or drug effects you want to identify. An objective function represents a hypothesis for a metabolic functionality of the cell or tissue-type, represented mathematically as optimization of a specific reaction in the reconstruction. For example, an objective function for a cancer metabolic model might be to optimize tumor growth, where tumor growth is represented as a sum of specific metabolites in particular ratios

[20]. This objective function allows for the identification of potential therapeutic targets such as genes, reactions, or metabolites that, when removed from or perturbed in the model, inhibit in silico tumor growth. Objective functions can also be used to identify drug effects. For example, to identify off-target effects of the drug torcetrapib, an objective function was developed that captures blood pressure through secretion and absorption of different metabolites in the kidney [16]. Each metabolite was associated with either increasing or decreasing blood pressure, allowing for the identification of genes that, when perturbed in the model, either increased or decreased blood pressure [16]. In most cases, an objective function is defined for a given reconstruction or model, but the objective function is specific to the question of interest in the paper and may need to be adapted for your question of interest. Some studies do not require a defined objective function but rather rely on integrating omics data to obtain flux values for each reaction within the network [37]. Choosing an appropriate objective function allows for the prediction of flux through the network to allow for context-specific predictions for drug targets and drug effects.

2.3 Applying Constraints

Finally, after choosing a GENRE and an objective function, additional constraints can be applied to the network to add further specificity or context for identifying drug targets. Constraints can be included as either bounds on uptake reactions, representing the physiological medium that the tissue or cell is living in, or bounds on individual reactions in your model through the integration of omics data. Physiological constraints on uptake reactions representing the external media of a cell or tissue are usually published with the model. One study imposed different constraints on their model of choice (muscle) to study how different nutrient environments (i.e., starved, fed, or mixed) affected predictions of reactions necessary for ATP, TAG, and glycogen synthesis [38]. Many approaches choose to integrate gene expression data through the GPR rules in the network to provide additional constraints. For example, the metabolic transformation algorithm (MTA) uses another algorithm, iMAT, to integrate gene expression data to constrain flux through each reaction in the network to identify reactions that are differentially used in a healthy versus diseased state [37]. Another study [17] used the significance of gene expression changes for each reaction to weight the ease of using a particular reaction, so the network will increase flux through reactions with low *p*-values (significant change in expression) over reactions with high p-values (nonsignificant change in expression). Constraints placed on the network, either through bounds on uptake reactions or omics data integration, are not essential but can allow for more physiologically relevant predictions.

Each of the steps outlined above helps to frame your question of interest to identify relevant drug targets and drug effects. Once you have chosen your model system, objective function, and

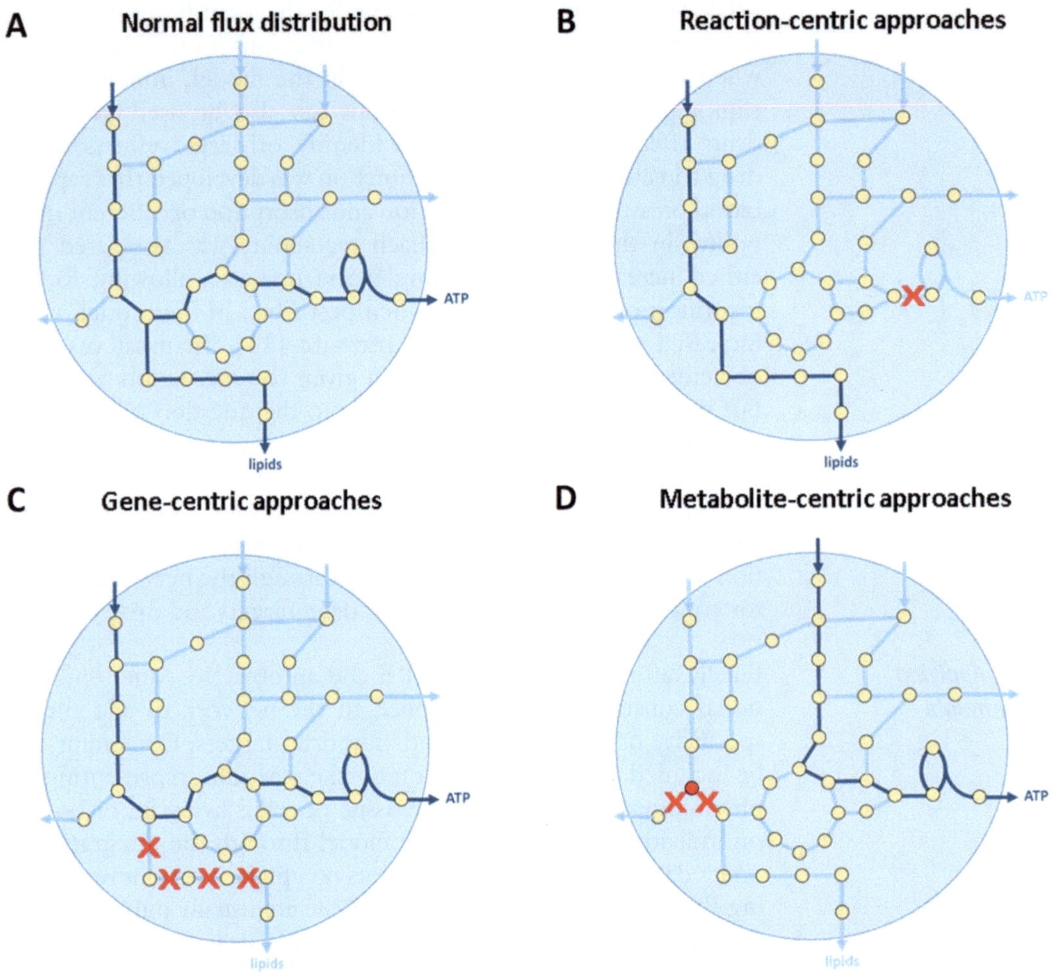

Fig. 3 Approaches to identify drug targets and off-target effects using metabolic network reconstructions. (**a**) A normal flux distribution through a metabolic network which requires the production of both ATP and lipids. (**b**) Reaction-centric approaches focus on either inhibiting or removing reactions in the network and noting the effect on the objective function, in this case the production of ATP and lipids. Reactions are usually removed one at a time. In this case, the reaction knockout represents a possible drug target because it inhibits the synthesis of ATP. (**c**) Gene-centric approaches focus on removing genes from the network by removing reactions for which that gene is essential. In most cases, this involves removing more than one reaction from the network. In some cases, this could be a linear series of reactions as shown, which inhibit the production of lipids, but this is not always the case. (**d**) Metabolite-centric approaches focus on removing metabolites in the network by removing reactions that consume that metabolite. The metabolite being removed (shown in red) is removed by disabling the two reactions that consume that metabolite. The removal of this reaction also inhibits the production of lipids which are necessary for the network. All three of these approaches yield different ways to identify drug targets, through either targeting specific reactions, genes, or metabolites

constraints, the methods below describe reaction-centric, gene-centric, or metabolite-centric approaches for identifying drug targets or drug effects for your question and system of interest (Fig. 3).

3 Methods

3.1 Reaction-Centric Approaches to Drug Target and Off-Target Identification

3.1.1 Reaction Knockouts

Reaction-centric approaches identify potential drug targets through systematically perturbing reactions in the network and are the most straightforward conceptual approach for identifying drug targets using GENREs. These approaches simulate either the removal or inhibition of reactions in the network and evaluate the downstream effect on the objective function. Reactions that are essential to enable nonzero flux through an objective function of interest can serve as potential drug targets. Once a drug target is identified, the GPR rules of the metabolic network can be used to identify a protein or enzyme that can be targeted to inhibit that reaction (*see* **Note 1**). Reactions as drug targets have been identified using a model of cancer metabolism and an objective function of cancer cell growth using the following steps [39]:

1. Determine a threshold for the objective function that determines how you will classify your reactions, where above the threshold indicates reactions that do not inhibit growth and below the threshold indicates reactions that inhibit growth.
2. Systematically constrain each reaction in the network and run FBA to determine the value of the objective function (*see* **Note 2**).
3. If the value of the objective function is below the threshold, the reaction is considered a possible drug target.

3.1.2 Metabolic Transformation Algorithm

Additional data, such as gene expression data, can be integrated with your metabolic model of choice to generate predictions for reactions as drug targets. The metabolic transformation algorithm (MTA) is an example of a reaction-centric approach which uses gene expression data and reaction knockouts with a generic human network reconstruction to predict reactions which will shift the network from one state to another, such as from diseased to healthy [37]. The reactions that shift the network from a diseased to a healthy state are potential drug targets; this approach has been used to identify drug targets for Alzheimer's disease [40] and kidney disease [31]. MTA was originally implemented with a model of yeast metabolism by integrating young (i.e., healthy) and old (i.e., diseased) gene expression data sets to predict drug targets to prevent aging using the following steps:

1. Integrate gene expression data using the iMAT algorithm to get feasible flux distributions for both the healthy and diseased states (*see* **Note 3**).
2. After integrating the expression data, classify each reaction in the network as either:

(a) Unchanged, there is no difference in flux between the healthy and diseased states.

(b) Changed, there is a difference in flux between the healthy and diseased states.

3. Systematically remove each reaction from the original reconstruction (i.e., before gene expression integration) while minimizing the change in flux through the "unchanged" reactions and maximizing the change in the "changed" reactions. This step represents keeping the network the same (minimizing the "unchanged" reactions) while pushing the network back to a healthy state (maximizing the "changed" reactions).
4. Each reaction is then assigned a score indicating the effect the reaction knockout had on the system, reflecting the ability of the knockout to maintain the original metabolic state (minimizing the "unchanged" reactions) while bringing the network back to a "healthy" state (maximizing the "changed" reactions).
5. Reactions with the lowest score, representing a minimum number of changes in unchanged reactions and a maximum change in changed reactions, serve as drug targets for shifting the network from the diseased to healthy state.

3.1.3 Using Flux Through Metabolic Reactions to Predict Drug Side Effects

Previous work leveraged metabolic modeling, data from external drug databases, and machine learning approaches to predict drug side effects using fluxes through reactions in a metabolic network [18]. DrugBank [41] and the side effect resource (SIDER) [42, 43] were mined for data to relate drugs to their gene targets and drugs with known side effects. Using these two data sources and a general human network reconstruction, predictions of side effects were made with the following steps:

1. Drugs were selected from DrugBank that (a) had only metabolic targets and (b) inhibited their targets. These drugs were incorporated into a general human network reconstruction to get flux ranges for every reaction in the network. DrugBank connects drugs with their gene targets; using the GPR rules in a general human reconstruction, reactions associated with genes that were inhibited by specific drugs were removed from the network.
2. Run flux variability analysis (FVA) to get the range of fluxes through each reaction in the network. Repeat for every drug in the DrugBank database.
3. The SIDER database connects drugs with known side effects. These side effects were incorporated with the flux data above to build a classifier to predict side effects based on flux ranges. For each side effect, select the drugs that are associated with the

side effect. Use the flux ranges for each reaction in the network with a support vector machine (machine learning approach) to predict the presence or absence of that specific side effect.

4. Once the above steps are complete, you have an array of model-based phenotype predictors (AMPP). Given a specific drug with known metabolic targets, FVA can be run to calculate the range of fluxes through each reaction in the network. Feeding these flux ranges into the AMPP will give a prediction of the presence or absence of all side effects in the AMPP.

Initially, 89 drugs with only metabolic targets that were associated with 286 side effects were used to demonstrate the predictive ability of this approach. Seventy side effects could be predicted accurately (AUC > 0.7) based on using a metabolic modeling approach with machine learning, highlighting the potential of using metabolic networks to predict drug side effects.

3.2 Gene-Centric Approaches

Proteins are widely used as drug targets because they can affect multiple cellular processes at once. Computationally, this characteristic makes them good candidates for knockouts because they can simulate potential inhibition by a drug. In GENREs, GPR rules use a logic based approach to describe the relationships between genes, proteins, and reactions. There are four types of GPR rules that exist in GENREs to describe the relationships between genes, proteins, and reactions shown in (Fig. 4) [21]. First, there are one-to-one relationships, where one reaction is connected to an enzyme. Next, there are enzymes comprised of subunits, where two or more subunits form an enzyme to catalyze a reaction. Third, there are isozymes, where multiple enzymes can catalyze a reaction independently. Finally, there are complex enzyme relationships where multiple enzymes, which can have several subunits, can either jointly or independently catalyze a reaction. Gene-centric approaches, such as single (gene essentiality) or pairwise (synthetic lethality) gene knockouts, leverage GPR rules to remove the downstream reactions in a system to identify potential drug targets.

3.2.1 Gene Essentiality

In vitro gene essentiality analysis studies the impact of individual genes on the cell's ability to grow. In silico gene essentiality analysis characterizes the importance of each gene in a metabolic network on the objective function. For example, gene essentiality was used to identify genes necessary for tumor cell proliferation [4], thus identifying potential drug targets. Depending on your model, you can have a range of objective functions to model different processes, depending on the disease in question. In models of prokaryotes, very frequently, the objective function is defined to represent biomass synthesis, so gene essentiality analysis identifies the genes that are critical for an organism to grow. For models of eukaryotes, an objective function representing biomass synthesis is often

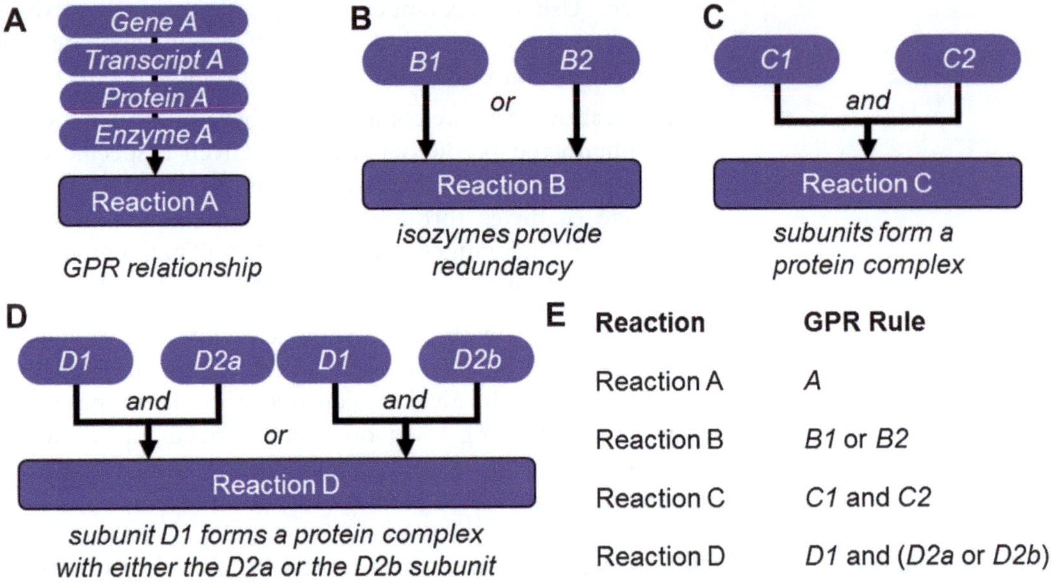

Fig. 4 Gene–protein–reaction (GPR) rules describe the relationship between phenotype and genotype. (**a**) Example of a GPR rule representing an enzymatic reaction catalyzed by the protein product of a single gene. (**b**) Example of a redundant GPR rule where either protein **b1** or protein **b2** can independently catalyze the same function. In this case, these isozymes are separated by an "OR" statement in the GPR rule. (**c**) Example of a complex GPR rule where both **c1** and **c2** are required for the catalytic reaction to occur. In this case, two nonredundant subunits that form a protein complex are separated by an "AND" statement. (**d**) Example of a complex GPR rule with redundancies where **d1** can form a protein complex with either **d2a** or **d2b**. In this case, the GPR rule can be separated by unique protein complexes or first by subunits then by redundancies as represented in (**e**). (**e**) Table summarizing genotype–phenotype relationships from **a**–**d** as Boolean GPR rules. (Reproduced from Rawls et al. (2019) [21])

explored, but other objective functions representing other complex physiological functions like the regulation of blood pressure have also been analyzed [16]. In models representing disease phenotypes, gene essentiality analysis can reveal which genes are critical to support the diseased state, such as which genes are necessary for cancer cell survival or proliferation. This analysis provides a way to find which genes should be targeted; here, we describe how to conduct gene essentiality analysis.

1. Choose an objective function to represent a cellular process.
2. Run flux balance analysis (FBA) to identify the default objective value.
3. Choose a threshold or fraction of objective value to be considered "essential." Genes which produce an objective value below this threshold are essential genes.
4. Identify genes of interest to target. You can choose a subset or use all genes captured by the model.
5. Test the essentiality of each gene on the biomass function.

(a) Find reactions catalyzed by this gene. GPR rules are used to find which reactions are affected by the loss of this one gene.
(b) Constrain the lower and upper bounds of these reactions to 0, for those reactions that are directly dependent on this gene, or reactions that form an enzyme subunit with this gene.
(c) Run FBA to calculate the value of the objective function after applying constraints.
(d) Store this output, ensuring that you connect this value to the gene that was deleted.
(e) Restore lower and upper bounds of reactions to original values.

6. Repeat **step 5** for each gene.
7. For the vector of objective values, divide by the default objective value to obtain fractional values representing the fractional amount of the objective function obtained.
8. Genes whose fractional objective values are below the threshold are labeled as essential.

Genes that produce minimal effects on healthy tissue while disrupting diseased tissue make ideal drug targets. Therefore, it is often necessary to use a healthy model and a diseased model to test whether the gene targets identified are specific to the disease, or affect both the healthy and diseased cases. When using microbial GENREs to identify microbial drug targets, it is critical to compare the identified essential genes with human homologs to ensure that there is no overlap [7, 44]. Genes that are unique to a microbe make ideal drug targets, as they are predicted to have minimal impact on human metabolism.

3.2.2 Synthetic Lethality Analysis

Synthetic lethality refers to the lethal combination of two nonlethal genes [45]. Here, the nonlethal mutation of one gene can modulate the impact of another nonlethal gene mutation to create a lethal combination. The computational simulation of synthetic lethality follows that of gene essentiality with slight modifications. Synthetic lethality explores the impact of double gene deletions of nonessential genes to ascertain if the combination of gene mutations produce an effect [46]. While the steps are consistent, we outline them here.

1. Identify all nonessential genes in the network.
2. Perform double gene deletion simulations of each pairwise combination of genes.
   (a) Select a pair of genes from the list of nonessential genes identified in **step 1**.

(b) Delete both genes by constraining the lower and upper bounds of reactions to 0 for those reactions whose GPR rules depend on the function of either gene.

(c) Run FBA to calculate a value of the objective function after applying constraints.

(d) Store this output, ensuring that you connect this value to the genes that were deleted.

(e) Restore lower and upper bounds of reactions to original values.

3. Repeat **steps 2** for each gene pair combination from the list of nonessential genes in the network.
4. Determine synthetic lethal pairs by identifying gene knockouts that produce an objective value below the threshold.

In the COBRA toolbox [19], both single gene deletion and double gene deletion analyses are built-in functions. After defining the desired parameters for your analysis, these analyses can be run in one line of code.

3.3 Metabolite-Centric Approaches

3.3.1 Identification of Antimetabolites as Targets

Antimetabolites are metabolites that have similar structure to another metabolite and can therefore competitively inhibit reactions that their corresponding metabolite would participate in. One appeal to using antimetabolites as drug targets is that metabolites participate in multiple reactions, thus making it possible to inhibit multiple reactions at once. Antimetabolites have been used as anticancer drugs [47]. For example, 5-fluorouracil (5-FU), an antimetabolite of uracil, inhibits pyrimidine biosynthesis by binding to and inhibiting thymidylate synthetase and thus inhibits DNA synthesis [48, 49]. Computationally, antimetabolites that could serve as potential drug targets were identified by simulating the inhibition of growth using a metabolic network of cancer cell metabolism [35]. We present a step-by-step process to predict antimetabolites that could serve as drugs as described by Argen et al. To use this method, the authors use a series of constraint-based approaches to simulate the effects of antimetabolites. To create healthy and cancer models, the authors developed the Task-Driven Integrative Network Inference for Tissues (tINIT) algorithm [35]. After you have developed your own models for a healthy and diseased state, you can follow the steps below to identify antimetabolites as drug targets:

1. Pick a list of metabolites that you want to test for or use the full set of metabolites in the model.
2. Define a threshold for determining if there was a significant effect on the value of the objective function. The baseline objective value is determined by running FBA.
3. Choose a metabolite to test.

4. For the metabolite chosen in **step 3**, constrain flux values through reactions that consume this metabolite to 0.
5. For reversible reactions with this metabolite as a product, constrain the lower bound to 0 so this metabolite is only produced.
6. Run FBA to determine the value of the objective function.
7. Repeat **steps 3–6** for each metabolite in list of metabolites from **step 1**.
8. Find the subset of metabolites that, when removed, produce an objective value below the threshold (determined in **step 2**).

Metabolites that are identified after this process are appropriate drug targets with a few caveats. First, metabolites that are derivatives of "pool" metabolites should be disregarded; these represent abstractions of metabolites that are used to synthesize general metabolites, such as a general phosphatidyl choline. Second, metabolites that disrupt energy and oxidation-reduction based on metabolic tasks should also be removed from the list of possible metabolites that can serve as drug targets. The authors remove these metabolites under the assumption that the disruption in energy would occur in both healthy and treatment models, making them less ideal targets to use.

3.3.2 Using MetChange to Identify Metabolites and Pathways Associated with the Pathogenesis of Side Effects

A metabolite-centric algorithm, MetChange, was used to identify metabolites and pathways that are associated with side effects, providing possible hypotheses for the pathogenesis of side effects [17]. MetChange calculates a score for each metabolite in the network from gene expression data; this score represents how easy it is, based on gene expression data, for flux to flow through the pathways in the production of that metabolite. The gene expression data used in this study was from in vitro exposure of human cells to a variety of drugs. Significantly changed pathways, termed disease-linked drug-changed (DISLoDGED) pathways, were identified using a generic human network reconstruction and provide hypotheses for the metabolic mechanisms underlying side effects. The MetChange algorithm uses the following approach:

1. For each metabolite in the network, run FBA with the objective function to maximize production of that metabolite. This value represents baseline metabolite production.
2. Summarize gene expression values as *p*-values for each gene in the network. GPR rules are used to assign an overall *p*-value for each reaction in the network, representing the ease of flux through that reaction based on the gene expression data.
3. For each metabolite in the network, run a new optimization problem with the following parameters: (a) minimize the sum of fluxes through each reaction times the calculated *p*-value of

that reaction and (b) maintain the baseline metabolite production calculated in **step 1**.

4. Calculate the MetChange score for each metabolite for each drug condition based on the matched control, representing the ease of producing that metabolite relative to a control condition. The formula for calculating the MetChange score is similar to calculating a *z*-score, indicating the difference between the drug condition and the matched control.
5. Use the SIDER database to identify side effects for each drug for which there are MetChange scores.
6. Use a genetic algorithm to predict the presence or absence of side effects for each drug with a MetChange score.
7. DISLoDGED pathways are pathways where a particular metabolite production score is a predictor of the presence of a specific side effect from the genetic algorithm.

3.4 Summary

In this chapter, we present approaches to identify drug targets and off-target effects using GENREs and their associated computational analyses. Using step-by-step descriptions of different methods, we describe reaction-centric, gene-centric, and metabolite-centric approaches to identify either drug targets or off-target effects (Fig. 3). The methods presented here represent the current state of the field. However, additional methods are rapidly developing which leverage different aspects of GENREs to predict new drug targets and predicting or minimizing potential off-target effects. In light of the stagnation in the drug development pipeline, these methods can be of tremendous value by identifying new drug targets while minimizing off-target effects to enhance the number of drugs that make it to market.

4 Notes

1. Many of the reaction-centric approaches can also be used to identify genes that could serve as drug targets. Reactions that are identified as potential drug targets or sources of drug effects can be connected back to specific genes through the GPR rules. However, genes may participate in more than one reaction so targeting a gene may not be as specific as targeting a reaction.
2. For reaction knockouts, some studies choose to completely remove reactions (i.e., set reaction bounds to 0) while some studies choose to constrain reactions to a percentage of maximal flux (i.e., set reaction bounds to 50% of maximal flux).
3. iMAT is an algorithm that integrates gene expression data to provide flux through each reaction in the network without specific consideration of an objective function.

References

1. Mougin F, Auber D, Bourqui R et al (2018) Visualizing omics and clinical data: Which challenges for dealing with their variety? Methods 132:3–18
2. Schuster D, Laggner C, Langer T (2005) Why drugs fail—a study on side effects in new chemical entities. Curr Pharm Des 11:3545–3559
3. Edwards IR, Aronson JK (2000) Adverse drug reactions: definitions, diagnosis, and management. Lancet 356:1255–1259
4. Sawada R, Iwata M, Tabei Y et al (2018) Predicting inhibitory and activatory drug targets by chemically and genetically perturbed transcriptome signatures. Sci Rep 8:156
5. Suthers PF, Zomorrodi A, Maranas CD (2009) Genome-scale gene/reaction essentiality and synthetic lethality analysis. Mol Syst Biol 5:301
6. Pey J, San José-Eneriz E, Ochoa MC et al (2017) In-silico gene essentiality analysis of polyamine biosynthesis reveals APRT as a potential target in cancer. Sci Rep 7:14358
7. Bordbar A, Lewis NE, Schellenberger J et al (2010) Insight into human alveolar macrophage and M. tuberculosis interactions via metabolic reconstructions. Mol Syst Biol 6:422
8. Orth JD, Palsson BØ, Fleming RMT (2010) Reconstruction and use of microbial metabolic networks: the core Escherichia coli metabolic model as an educational guide. EcoSal Plus 4: PMID: 26443778
9. Oberhardt MA, Puchalka J, Fryer KE et al (2008) Genome-scale metabolic network analysis of the opportunistic pathogen Pseudomonas aeruginosa PAO1. J Bacteriol 190:2790–2803
10. Förster J, Famili I, Fu P et al (2003) Genome-scale reconstruction of the Saccharomyces cerevisiae metabolic network. Genome Res 13:244–253
11. Thiele I, Swainston N, Fleming RMT et al (2013) A community-driven global reconstruction of human metabolism. Nat Biotechnol 31:419–425
12. Mardinoglu A, Agren R, Kampf C et al (2014) Genome-scale metabolic modelling of hepatocytes reveals serine deficiency in patients with non-alcoholic fatty liver disease. Nat Commun 5:3083
13. Blais EM, Rawls KD, Dougherty BV et al (2017) Reconciled rat and human metabolic networks for comparative toxicogenomics and biomarker predictions. Nat Commun 8:14250
14. Yizhak K, Chaneton B, Gottlieb E et al (2015) Modeling cancer metabolism on a genome scale. Mol Syst Biol 11:817
15. Ghaffari P, Mardinoglu A, Asplund A et al (2015) Identifying anti-growth factors for human cancer cell lines through genome-scale metabolic modeling. Sci Rep 5:8183
16. Chang RL, Xie L, Xie L et al (2010) Drug off-target effects predicted using structural analysis in the context of a metabolic network model. PLoS Comput Biol 6:e1000938
17. Zielinski DC, Filipp FV, Bordbar A et al (2015) Pharmacogenomic and clinical data link non-pharmacokinetic metabolic dysregulation to drug side effect pathogenesis. Nat Commun 6:7101
18. Shaked I, Oberhardt MA, Atias N et al (2016) Metabolic network prediction of drug side effects. Cell Syst 2:209–213
19. Heirendt L, Arreckx S, Pfau T et al (2019) Creation and analysis of biochemical constraint-based models using the COBRA Toolbox v.3.0. Nat Protoc 14:639–702
20. Folger O, Jerby L, Frezza C et al (2011) Predicting selective drug targets in cancer through metabolic networks. Mol Syst Biol 7:501
21. Rawls KD, Dougherty BV, Blais EM et al (2019) A simplified metabolic network reconstruction to promote understanding and development of flux balance analysis tools. Comput Biol Med 105:64–71
22. Ma H, Sorokin A, Mazein A et al (2007) The Edinburgh human metabolic network reconstruction and its functional analysis. Mol Syst Biol 3:135
23. Swainston N, Smallbone K, Hefzi H et al (2016) Recon 2.2: from reconstruction to model of human metabolism. Metabolomics 12:109
24. Brunk E, Sahoo S, Zielinski DC et al (2018) Recon3D enables a three-dimensional view of gene variation in human metabolism. Nat Biotechnol 36:272–281
25. Robaina Estévez S, Nikoloski Z (2014) Generalized framework for context-specific metabolic model extraction methods. Front Plant Sci 5:491
26. Blazier AS, Papin JA (2012) Integration of expression data in genome-scale metabolic network reconstructions. Front Physiol 3:299
27. Uhlen M, Hallstro m BM, Lindskog C et al (2016) Transcriptomics resources of human tissues and organs. Mol Syst Biol 12:862–862
28. Schultz A, Qutub AA (2016) Reconstruction of tissue-specific metabolic networks using CORDA. PLoS Comput Biol 12:e1004808

29. Agren R, Bordel S, Mardinoglu A et al (2012) Reconstruction of genome-scale active metabolic networks for 69 human cell types and 16 cancer types using INIT. PLoS Comput Biol 8:e1002518
30. Zhang A-D, Dai S-X, Huang J-F Reconstruction and analysis of human kidney-specific metabolic network based on omics data. https://www.hindawi.com/journals/bmri/2013/187509/
31. Sohrabi-Jahromi S, Marashi S-A, Kalantari S (2016) A kidney-specific genome-scale metabolic network model for analyzing focal segmental glomerulosclerosis. Mamm Genome 27:158–167
32. Karlstädt A, Fliegner D, Kararigas G et al (2012) CardioNet: a human metabolic network suited for the study of cardiomyocyte metabolism. BMC Syst Biol 6:114
33. Zhao Y, Huang J (2011) Reconstruction and analysis of human heart-specific metabolic network based on transcriptome and proteome data. Biochem Biophys Res Commun 415:450–454
34. Bordbar A, Mo ML, Nakayasu ES et al (2012) Model-driven multi-omic data analysis elucidates metabolic immunomodulators of macrophage activation. Mol Syst Biol 8:558
35. Agren R, Mardinoglu A, Asplund A et al (2014) Identification of anticancer drugs for hepatocellular carcinoma through personalized genome-scale metabolic modeling. Mol Syst Biol 10:721
36. Thiele I, Palsson BØ (2010) A protocol for generating a high-quality genome-scale metabolic reconstruction. Nat Protoc 5:93–121
37. Yizhak K, Gabay O, Cohen H et al (2013) Model-based identification of drug targets that revert disrupted metabolism and its application to ageing. Nat Commun 4:2632
38. Nogiec C, Burkart A, Dreyfuss JM et al (2015) Metabolic modeling of muscle metabolism identifies key reactions linked to insulin resistance phenotypes. Mol Metab 4:151–163
39. Yizhak K, Le Devedec SE, Rogkoti VM et al (2014) A computational study of the Warburg effect identifies metabolic targets inhibiting cancer migration. Mol Syst Biol 10:744–744
40. Stempler S, Yizhak K, Ruppin E (2014) Integrating transcriptomics with metabolic modeling predicts biomarkers and drug targets for Alzheimer's disease. PLoS One 9:e105383
41. Wishart DS, Feunang YD, Guo AC, et al (2018) DrugBank 5.0: a major update to the DrugBank database for 2018. Nucleic Acids Res 46:D1074–D1082
42. Kuhn M, Campillos M, Letunic I et al (2010) A side effect resource to capture phenotypic effects of drugs. Mol Syst Biol 6:343
43. Kuhn M, Letunic I, Jensen LJ et al (2016) The SIDER database of drugs and side effects. Nucleic Acids Res 44:D1075–D1079
44. Rienksma RA, Suarez-Diez M, Spina L et al (2014) Systems-level modeling of mycobacterial metabolism for the identification of new (multi-)drug targets. Semin Immunol 26:610–622
45. Guarente L (1993) Synthetic enhancement in gene interaction: a genetic tool come of age. Trends Genet 9:362–366
46. Jerby L, Shlomi T, Ruppin E (2010) Computational reconstruction of tissue-specific metabolic models: application to human liver metabolism. Mol Syst Biol 6:401
47. Raškevičius V, Mikalayeva V, Antanavičiūtė I et al (2018) Genome scale metabolic models as tools for drug design and personalized medicine. PLoS One 13:e0190636
48. Longley DB, Harkin DP, Johnston PG (2003) 5-fluorouracil: mechanisms of action and clinical strategies. Nat Rev Cancer 3:330–338
49. Mehrmohamadi M, Jeong SH, Locasale JW (2017) Molecular features that predict the response to antimetabolite chemotherapies. Cancer Metab 5:8

Chapter 15

Single-cell Digital Twins for Cancer Preclinical Investigation

Marzia Di Filippo, Chiara Damiani, Marco Vanoni, Davide Maspero, Giancarlo Mauri, Lilia Alberghina, and Dario Pescini

Abstract

Laboratory models derived from clinical samples represent a solid platform in preclinical research for drug testing and investigation of disease mechanisms. The integration of these laboratory models with their digital counterparts (i.e., predictive mathematical models) allows to set up digital twins essential to fully exploit their potential to face the enormous molecular complexity of human organisms. In particular, due to the close integration of cell metabolism with all other cellular processes, any perturbation in cellular physiology typically reflect on altered cells metabolic profiling. In this regard, changes in metabolism have been shown, also in our laboratory, to drive a causal role in the emergence of cancer disease. Nevertheless, a unique metabolic program does not describe the altered metabolic profile of all tumour cells due to many causes from genetic variability to intratumour heterogeneous dependency on nutrients consumption and metabolism by multiple co-existing subclones. Currently, fluxomics approaches just match with the necessity of characterizing the overall flux distribution of cells within given samples, by disregarding possible heterogeneous behaviors. For the purpose of stratifying cancer heterogeneous sub-populations, quantification of fluxes at the single-cell level is needed. To this aim, we here present a new computational framework called single-cell Flux Balance Analysis (scFBA) that aims to set up digital metabolic twins in the perspective of being better exploited within a framework that makes also use of laboratory patient cell models. In particular, scFBA aims at integrating single-cell RNA-seq data within computational population models in order to depict a snapshot of the corresponding single-cell metabolic phenotypes at a given moment, together with an unsupervised identification of metabolic subpopulations.

Key words Cancer heterogeneity, Constraint-based modelling, Single-cell RNA-seq

1 Introduction

Since 1951, when the first immortalized cancer cell line was derived from cervical cancer cells taken from Henrietta Lacks [1], stabilized cell lines have long been used in cancer research. They have been recently complemented by laboratory models derived from clinical samples, like patient-derived xenografts (PDX) [2], organoids (PDO) [2] and organ-on-chips (ONC) [3], which provide a renewable easily accessible, comprehensive representation of the architecture of normal and diseased tissues, an asset of major importance

Deepak Nagrath (ed.), *Metabolic Flux Analysis in Eukaryotic Cells: Methods and Protocols*, Methods in Molecular Biology, vol. 2088, https://doi.org/10.1007/978-1-0716-0159-4_15,

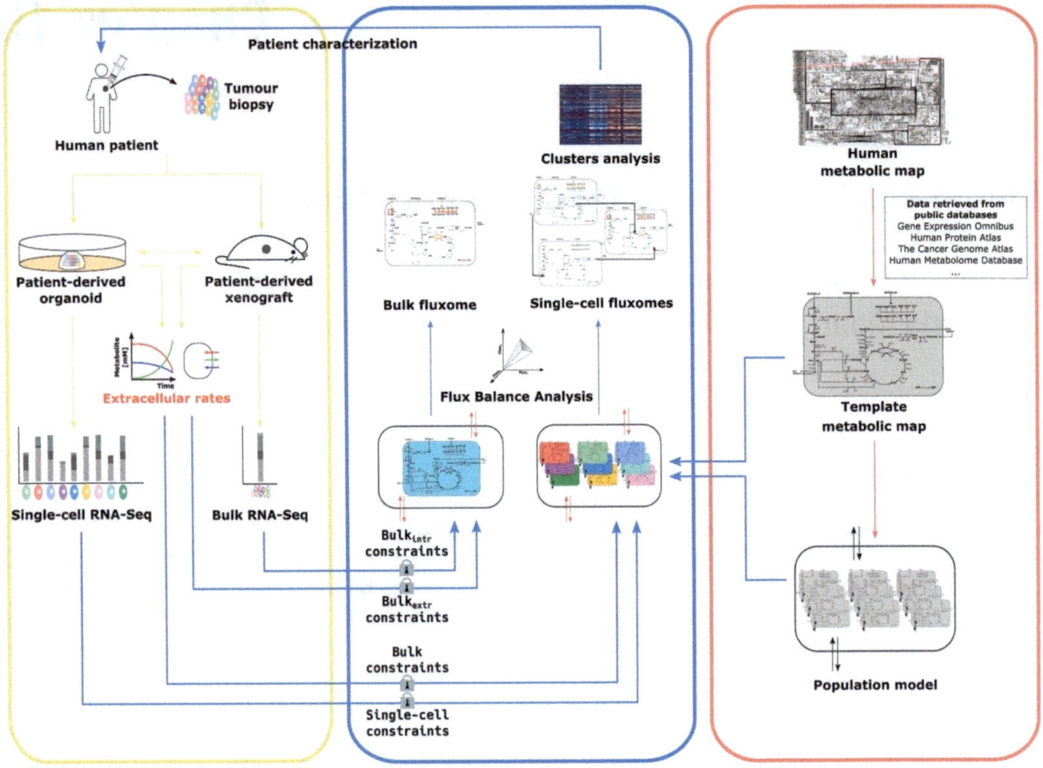

Fig. 1 Overview of scFBA approach. Stratification of cancer heterogeneous subpopulations taken from xenografts or organoids needs the quantification of fluxes at the single-cell level. scFBA integrates extracellular bulk fluxes and single-cell RNA-seq data into bulk and single-cell constraints that are imposed on a initially homogeneous population model consisting of N_{cells} replicas of the input template metabolic network. The output of this analysis is a heterogeneous set of single-cell flux patterns. The translation of single-cell transcriptomes into the corresponding single-cell fluxomes also contributes to identify metabolic clusters of cells that are characterized by different growth rates and global metabolic profiles. In case of single-cell RNA-seq data missing, extracellular fluxes ($\text{bulk}_{\text{extr}}$ constraints) and RNA-seq data depicting the average gene expression from all the population cells ($\text{bulk}_{\text{intr}}$ constraints) can be exploited to generate patient-specific net flux distributions

for drug testing and for studies aiming to reconstruct the disease mechanisms. The remarkable resemblance to the originating structure of these laboratory models both in cellular organization and genomic variability has led to a strong development towards their use in preclinical research, where they have been used for testing of patient-specific protocols derived from experimental analysis, allowing direct safe testing of experimentally-generated, patient-specific clinical protocols. To fully exploit potential of these laboratory twins, predictive mathematical models (digital twins) are needed, as shown in Fig. 1. The stumbling block of this approach is the staggering molecular complexity of an organism with trillions of cells, each containing several hundred thousands proteins, the agents which determine biological functions. It seems necessary therefore to find a correct way to reduce the complexity considering both a modular, multi-level approach in which the molecular detail (necessary for making drug action predictions) is restricted to a manageable module. Metabolomics may offer a convenient/delimited approach. Mass spectrometry (and NMR)-based methods allow to obtain high resolution "snapshots" of metabolites present in a given sample [4]. Using labelled compound metabolic fluxes originating from the labelled compound have been measured using Metabolic Flux Analysis (MFA) [5], a metabolic flux being the rate at which a metabolite is consumed or produced by a given metabolic reaction. In cell cultures, extracellular fluxes have been more easily determined from the exo-metabolome at different time points, that is bulk measurements of the concentration of metabolites in the cell culture media.

Changes in metabolism have a causal role in many complex multifactorial diseases, including cancer, mostly dependent on oncogene activation [5, 6]. Changes in metabolic fluxes directly impact on epigenetic regulation and on enzyme activities, affecting cell functions, which emerge from the network of informational and chemical fluxes taking place in each given cell [7]. Because of the close integration of metabolism with practically all cellular processes, perturbations in cellular physiology typically alter metabolic profiling of cells, tissues and biological fluids [8]. The complete human metabolic network has been reconstructed [9]. It includes 5000 metabolites and about 13,000 reactions, which have been structured in genome-wide metabolic models, whose active metabolic fluxes have been predicted by constraint-based metabolic models, knowing, from other omics analyses, which pathways are expressed in any given cell or tissue [9]. Constraint-based modelling, and in particular Flux Balance Analysis (FBA), represents the most applied computational approach. Given specific constraints on the flux of some relevant reactions, especially those regarding the intake and secretion rates of metabolites, this approach allows to compute the flux of metabolites through a given set of reactions.

In particular, we introduced a novel computational framework for data integration called Metabolic Reaction Enrichment Analysis (MaREA) [10]. MaREA aims at integrating RNA-seq data into metabolic networks by calculating for each reaction the Reaction Activity Score (RAS) exploiting the corresponding gene-protein-reaction (GPR) rule, without requiring any metabolic measurements. The usage of RAS scores instead of the only transcripts allows to characterize transcriptional deregulations of metabolic reactions under different conditions. Furthermore, this strategy offers the possibility of ranking reactions according to their activity variation, and visualizing the most critical ones directly on metabolic networks. In this way, it also provides a graphical visualization of how deregulated paths are interconnected, by enriching the map of human metabolic routes with RAS variation. Finally, MaREA, providing a new unsupervised clustering tool, allows to stratify different tumors samples according to their metabolic activity.

MaREA revealed its ability in discriminating normal and cancer samples, by reproducing well-known features of cancer deregulation and generating new hypotheses. By applying this methodology to two distinct datasets of The Cancer Genome Atlas database [11] relative to lung and breast tumours, cancer patients can be, in an unsupervised way, stratified in distinct clusters characterized by similar metabolic activity. In this way, subgroups of patients differing in terms of survival expectancy can be identified. Therefore, the reaction enrichment performed via MaREA allows to identify the metabolic patterns underlying the phenotypic and functional properties observed in different sample subgroups, as in the case of patients with distinct cancer subtypes, by visualizing the up-/down-regulated reactions directly on the metabolic networks. In this regard, MaREA enables to identify, rank and visualize critical metabolic reactions that can be targeted by using metabolic drugs, instead of targeting the expression of individual genes. Finally, the comparison of MaREA prognostic power by using a well curated core model focused on central carbon metabolism and a complete genome-wide metabolic network revealed in the first case an improvement of its prognostic power. This behavior has been observed because of the model curation in terms of GPR rules associated to metabolic reactions.

Cancer is catalogued as a heterogeneous and multi-factorial disease due to the existence of multiple tumoral types and subtypes deriving from the same or different origin primary sites that show considerable phenotypic diversity both at the intertumour and intratumour level [12–14]. In addition to genetic and epigenetic factors, differential trophic supply and variations in the tumour microenvironment (TME) plays a great role in the emergence of intratumour metabolic heterogeneity and of a complex cancer population architecture [15, 16]. According to this perspective, the outcome of the metabolic strategy adopted by a given cancer cell to survive within the overall population is both dependent on its approach used for growing together with the metabolic strategies adopted by other interacting constituents of the population, including other tumour cells, stromal cells, and the local microenvironment. Indeed, despite the reorganization of metabolic fluxes is a general feature of cancer cells, a unique metabolic program does not allow to globally describe the altered metabolic profile of all tumour cells due to a heterogeneous dependency on nutrients consumption and metabolism by multiple co-existing subclones, even within the same tumour. Intratumour metabolic heterogeneity contributes to considerably hamper cancer classification and pushes the need for exploring more in detail the set of interactions and cooperation phenomena that occur within cancer ecosystem together with the adopted metabolic programs. Their better knowledge could be used to develop more effective and personalized therapeutic strategies to hamper and potentially reverse tumour progression.

To this aim, the quantification of the flux through the metabolic pathways for each subpopulation within the entire tumor population by using constraint-based approaches would be ideal for a deep characterization of the adopted metabolic programs. Currently, fluxomics approaches just match with the necessity of characterizing the overall flux distribution of cells in bulk samples, disregarding possible heterogeneous behaviors. This is a relevant question when, in view of the above, the aim is to investigate bulk samples containing intermixed cell subpopulations. Indeed, by using classic fluxomics approaches, the retrieved flux distribution will portray, as a black box of the investigated population, the net of heterogeneous flux distributions, by hiding actual intracellular fluxes and possible cooperation phenomena.

For the purpose of stratifying cancer heterogeneous subpopulations, taken, for example, from biopsies, xenografts or organoids, quantification of fluxes at the single-cell level is needed. In this regard, we developed a new computational framework called single-cell Flux Balance Analysis (scFBA) [17], which will be better explained below in Subheading 3.

In view of investigating the role of cooperation within a population sharing a common environment, we firstly devised popFBA [18, 19], an extension of classic FBA able to cope with the presence of several subpopulations exchanging a defined set of metabolites (*see* **Note 1** for a summarized description of this methodology). The potentialities revealed by popFBA and MaREA methodologies effectively served as a basis for the set up of scFBA. scFBA aims at integrating single-cell RNA-seq data (scRNA-seq) within population models by exploiting the computation of RAS scores in order to depict a snapshot of the corresponding single-cell metabolic phenotypes at a given moment, and to identify metabolic subpopulations without a priori knowledge. In this way, scRNA-seq data act as further constraints on flux boundaries of distinct cells within the population, in order to identify the possible combinations of single-cell steady states.

Thanks to this approach, the integration of scRNA-seq data within population models efficiently reduces the space of optimal solutions as compared to the situation depicted by popFBA where no information on single cell trancriptome is available with the consequent ability of each cell to alone contribute to the 100% of the total population biomass. Indeed, after the scRNA-seq data integration, flux value of biomass synthesis for each cell correspond to a certain fraction of the total population biomass, by preventing solutions where each cell in the population can alone contribute to total biomass of the entire tumour mass. Moreover, the conversion of single-cell transcriptomes into single-cell fluxomes also allows to identify metabolic clusters of cells characterized by different growth rates, which can be exploited to better investigate intratumour heterogeneity. Finally, similarly to popFBA, possible interactions among cells within the population can be identified thanks to scFBA methodology. However, as opposed to its predecessor, scFBA is not limited to list all the theoretical subpopulations and their possible interactions network, but aims to capture the actual interactions between the subpopulation of cells within a specific population, characterized through quantitative experimental data.

2 Materials/Software

To perform a scFBA analysis three input resources are required: a template metabolic network, a scRNA-seq dataset, and the extracellular fluxes of the overall cell population corresponding to the uptake and secretion rates of main metabolites, which can be approximated from their concentration measurements in the spent cell culture medium at different time points.

The scFBA suite of MATLAB functions is available at https://github.com/BIMIB-DISCo/scFBA, whereas the "Metabolic Enrichment Analysis" tool proposed by MaREA can be found at https://orione.crs4.it/.

3 Methods

In this section, a detailed description of all the steps that are necessary to perform a scFBA analysis is provided. Three input resources need to be firstly prepared.

A **template metabolic network** A is required, suitable either as a genome-wide or a core metabolic network. The network A is defined as $A = (\mathcal{X}^A, \mathcal{R}^A, \mathcal{E}^A)$, where:

- $\mathcal{X}^A = \{X_1, \ldots, X_M\}$ corresponds to the set of metabolites;
- $\mathcal{R}^A = \{R_1, \ldots, R_N\}$ corresponds to the set of biochemical reactions taking place within the network;
- $\mathcal{E}^A = \{E_1, \ldots, E_{N_{\text{ext}}}\}$ corresponds to the set of N_{ext} exchange reactions that enable a predefined set of metabolites $\Upsilon = \{\Upsilon_1, \ldots, \Upsilon_{N_{\text{ext}}}\} \subset \mathcal{X}^A$ to be consumed or secreted from the network.

Secondly, a **scRNA-seq dataset** is required in the form of a $N_{\text{genes}} \times N_{\text{cells}}$ matrix T, where N_{genes} is the number of genes and N_{cells} is the number of single-cells under study. Each element $T_{g,c}$, $g = 1, \ldots, N_{\text{genes}}$, $c = 1, \ldots, N_{\text{cells}}$ of the dataset corresponds to the normalized read count of gene g in cell c (e.g. the Transcripts Per Kilobase Million or TPM).

Finally, the **extracellular fluxes** of the overall cell population are required. These values correspond to the uptake and secretion rates of main metabolites, and can be approximated from their corresponding concentration measurements performed in the spent cell culture medium at different time points.

Once these required input resources have been correctly prepared, the user has to proceed by following these five steps:

1. **Pre-processing of scRNA-seq dataset.** Pre-processing of scRNA-seq dataset is necessary to face the risk of having false negatives. In this regard, the information coming from bulk expression profile, when available, is exploited in the following scenarios (*see* **Note 2** for further details):
   - Genes having a zero read count both in the bulk and in each single-cell are assumed to be off in all cells. This set of genes G_{off} are deleted within the template metabolic network A. This implies that reactions for which their expression is essential are removed from the network A, by obtaining a subnetwork A^*.
   - Genes having non-zero read count in the bulk but a zero read count in each single-cell are assumed to be assigned to the bulk read count.
   - Genes having non-zero read count in the bulk and zero read count in just some of the single-cells are kept unchanged in terms of their read counts. Moreover, flux boundaries through the associated reactions is not completely prevented by setting it to a small value ϵ.
2. **Creation of the population model.** The template metabolic network map A^*, as previously described in Subheading 1, is a subnetwork of the generic model A which integrates the transcriptional information that holds for all cells in the bulk. This network corresponds to a generic single-cell and is used as a building block to automatically reconstruct the population model, which consists of identical copies of the single metabolic network, all having identical stoichiometry and capacity constraints and sharing the plasma supply of nutrients.

   The population model consists of N_{cells} replicas A^c of network A^* that can cooperate by exchanging nutrients in the tumour microenvironment. Each replica A^c corresponds to a single-cell c in the bulk, $c = 1, \ldots, N_{\text{cells}}$, whose corresponding metabolic network $A^c = (X^c, \mathscr{R}^c, C^c)$ can be reconstructed in the following way:
   - the set of metabolites X^c corresponds to the X^A set of the original template network A^*, i.e., $X^c \equiv X^A$;

- the set of internal reactions $\mathscr{R}^c$ corresponds to the $\mathscr{R}^A$ set of the original template network A^*, i.e., $\mathscr{R}^c \equiv \mathscr{R}^A$;
- a new set of cooperation reactions $C^c = \{C^c_j\}$, with $j = 1, \ldots, N_{\text{ext}}$, is introduced to allow the exchange of metabolites among single cells via a shared environment that is represented by the TME compartment. In particular, each exchange reaction $E_j \in \mathcal{E}^A$ is used to build the corresponding cooperation reaction in this way:

$$C^c_j : Y^c_j \leftrightarrow Y'_j \tag{1}$$

Due to the introduction of cooperation reactions, an additional set of metabolites $Y' = \{Y'_i\}$ with $i = 1, \ldots, N_{\text{ext}}$ must be defined in the TME compartment.

Finally, to allow a subset of metabolites $K = \{K_1, \ldots, K_{N_{\text{blood}}}\} \subset Y'$ to be exchanged with the external environment, e.g., the blood supply, a set of N_{blood} exchange reactions $\mathcal{B} = \{B_1, \ldots, B_{N_{\text{blood}}}\}$ must be defined:

$$B_j : K_j \leftrightarrow \emptyset \tag{2}$$

Overall, the population model P consists of the union set of the metabolites $X^P = \bigcup_c X^c \cap Y'$, the internal reactions $\mathscr{R}^P = \bigcup_c \mathscr{R}^c$, the cooperation reactions $C^P = \bigcup_c C^c$, and the population exchange reactions $\mathcal{B}$. A stoichiometric matrix S^P is then built for the population model, with a size of $(N_{\text{cells}} \cdot M + N_{\text{blood}}) \times (N_{\text{cells}} \cdot (N + N_{\text{ext}}) + N_{\text{blood}})$.

3. **Computation of Reaction Activity Scores.** Gene-Protein-Reaction (GPR) rules are logical expression associated to metabolic reactions, which indicate the relationships established among gene products involved in their catalysis by means of AND and OR logical operators (*see* **Note 3**). In particular, AND operator is used to join distinct genes encoding for different subunits of the same enzyme that are necessary for the reaction to occur, whereas OR operator join distinct genes encoding for isoforms of the same enzyme. Acting in this way, it is possible to associate a GPR rule to each reaction of the population model P.

   By exploiting GPR rules, a Reaction Activity Score (RAS) can be computed for each reaction $j \in \mathscr{R}$ in each single-cell $c = 1, \ldots, N_{\text{cells}}$ as a function of the expression $T_{g,c}$ of the genes encoding for the subunits or the isoforms of the associated enzymes. In particular, two scenarios are considered:

   - in case of a reaction whose GPR rule includes the AND operator, all the genes are necessary for the reaction to occur and therefore RAS is calculated as follows:

$$RAS_j^c = \min_g(T_{g,c} : g \in S_j) \quad (3)$$

where S_j corresponds to the set of genes encoding for the subunits that make up the enzyme catalyzing the reaction j.

- in case of a reaction whose GPR rule includes the OR operator, either gene is sufficient to catalyze the reaction and the RAS is calculated as follows:

$$RAS_j^c = \sum_{g \in I_j} T_{g,c} \quad (4)$$

where I_j corresponds to the set of genes encoding for the isoforms of the enzyme responsible for the catalysis of reaction j.

Composite GPR rules including both AND and OR respect the standard precedence of the two operators.

4. **Constraining of the population model.** Once the population model P is obtained and the RAS scores are computed, two kinds of constraints called bulk and single-cell constraints are imposed:
   - *bulk constraints* correspond to boundaries on the extracellular fluxes of the population model P. Accordingly, the upper and lower bound of the N_{blood} exchange reactions within set $\mathcal{B}$ are constrained according to metabolic measurements;
   - *single-cell constraints* correspond to boundaries on internal fluxes of each single-cell c, according to their RAS.

   In order to project the information of the activity score of a given reaction j in a given cell c, R_j^c, onto its flux "pipe capacity":
   - The maximal flux in both the forward (F_f) and backward direction (F_b) is estimated by just setting constraints on extracellular fluxes (bulk constraints), by leaving unbounded internal fluxes (single-cell constraints), and without setting any objective function in the system. To this aim, a Flux Variability Analysis [20] is carried out with no optimality required, and $F_j^c = \max(|F_f|, |F_b|)$ is defined.
   - The RAS of reaction R_j^c in each $c = 1, \ldots, N_{\text{cells}}$ is computed with respect to the total activity of reaction j as follows:

$$\overline{RAS}_j^c = \frac{RAS_j^c}{\sum_{c=1}^{N_{\text{cells}}} RAS_j^c}, \quad (5)$$

   - The upper bound U_j^c of reaction R_j^c is then set as portion of F_j^c and proportional to the corresponding activity score

$\overline{RAS}^c_j$. Practically, the value $\overline{RAS}^c_j, j = 1, \ldots, N; c = 1, \ldots, N_{\text{cells}}$ is remapped in the interval $[\epsilon, F^c_j]$:

$$U^c_j = \epsilon + (F^c_j - \epsilon) \cdot \overline{RAS}^c_j \tag{6}$$

In case of reactions having $\overline{RAS}^c_j = 0$, the upper bound is set to a small value ϵ rather than to 0 to mitigate the impact of false-negatives.

- According to the directionality of the reaction R^c_j, a zero lower bound $L^c_j = 0$ is assigned in case of irreversible reactions, otherwise the lower bound is set as $L^c_j = -U^c_j$ to reflect the ability of the corresponding enzyme to equally work in either direction.

5. **Simulation of the population model.** Once the *P* model is constrained (with both bulk and single-cell constraints), Linear Programming [21], as well as other standard constraint-based methods were applied. Frequently imposed objective function is the biomass maximization of the entire population.

   In addition, a single gene deletion analysis can be performed. The *in silico* deletion of a single gene implies that the reactions purely associated to that gene together with reactions associated to that gene joined in the corresponding GPR rule with other genes through the AND operator are removed from the network. After removal of these essential reactions has been performed, the population model is once more optimized for total biomass production, and the growth ratio of the new biomass flux value over the original one is computed.

   Finally, scFBA allows to extract features from scRNA-seq data for the purpose of identify metabolic clusters of cells that may be exploited to investigate other main features of cancer metabolic deregulation.

4 Notes

1. **popFBA methodology.** popFBA methodology is focused on the reconstruction and simulation of cell population metabolism, by putting emphasis on the relationships established among their individuals. By exploiting this approach, popFBA highlighted that, although the global purpose of tumour cells popolation is an enhanced proliferation rate, this objective does not constitute the common aim of all the individual components of this system. On the contrary, this approach highlighted that a cooperative behaviour within the investigated population model together with heterogeneity in terms of adopted metabolic strategies by individual subpopulations, are consistent with the achievement of the optimal tumour

biomass. Through popFBA strategy, we observed plasticity of population clones metabolism under different nutrients exchanged with plasma, or when a dishomogeneous distribution of oxygen is provided. In particular, among the explored scenarios, phenotypes of the most proliferative clones result characterized by a recurrent consumption and consequent oxidation of extracellular lactate that is secreted in the tumour microenvironment by the less proliferative clones, as experimentally demonstrated by the "reverse Warburg effect" between cancer and stromal cells in [22–24].

2. **Further details on scFBA data pre-processing.** Despite a totally safe solution to deal with the presence of false negatives in scRNA-seq data does not exist, the information coming from bulk expression profile can be exploited to manage and mitigate this risk.

   In the first presented scenario, a gene can have a zero read count both in the bulk and in each single-cell. In this case, the possibility of a false-negative in the bulk cannot be excluded. Nevertheless, false-negative can be excluded in the single-cells due to the low concentrations of scRNA-seq.

   In the second scenario, a considerable inconsistency between bulk and scRNA-seq data can emerge when a gene has non-zero read count in the bulk and a zero read count in each single-cell. This situation suggests that scRNA-seq data for this gene cannot be trusted and losing information on single-cell heterogeneity by just relying on the bulk value is preferred.

   In the third and last scenario, a gene can have a non-zero read count in the bulk and zero read count just in some of the single-cells. In this situation, a problem related to the detection of this specific gene can be excluded by hypothesizing to have a gene that is poorly expressed as compared to other cells. In this situation, the single-cell read count for this genes is retained avoiding, at the same time, to completely prevent flux passing through the associated reactions. In particular, the related flux boundaries are set to a very small value ϵ instead of leaving them to be null.

   Despite this first approximation to deal with the presence of false-negatives in scRNA-seq data, the combination with more sophisticated data pre-preprocessing techniques might help to further refine the presented methodology.

3. **Curation of gene-protein-reaction (GPR) rules.** GPR rules need to be associated to the cohort of reactions constituting the investigated metabolic model in order to connect reactions to one or multiple enzymes and every protein to the responsible genes. To ensure high quality of the reconstructed GPR rules, manual curation is required, representing the safest,

while simultaneously time consuming, way. In this regard, useful and specific information can be retrieved from reaction databases (e.g. KEGG [25, 26], Uniprot [27] and Reactome [28]), organism-specific databases, literature data as well as well-curated and up-to-date genome-wide metabolic networks in order to infer the monomeric or heteromeric state of the enzymes catalyzing given reactions [29, 30]. Curation of GPR rules represents a critical factor able to significantly affect the analyses outputs. Indeed, as shown in [10], by using the GPR rules coming from the genome-wide network of human metabolism Recon 2.2 [31], the rules for Complex I to V belonging to the human respiratory chain needed a significant manual curation from their original too strict formulation. In particular, after the manual investigation, many genes in the original rule of Complex IV resulted to be isoforms instead of subunits. The subsequent substitution of the related AND with OR operators allowed to avoid null RAS for the Complex IV reaction and the consequent empty set of optimal solutions from constraint-based simulations of the adopted core human metabolic model. Additionally, the usage of a well curated model in terms of the GPR rules associated to included metabolic reactions also allowed to improve the prognostic power of MaREA, as shown in [10].

References

1. John R Masters (2002) Hela cells 50 years on: the good, the bad and the ugly. Nat Rev Cancer 2(4):315
2. Byrne AT, Alférez DG, Amant F, Annibali D, Arribas J, Biankin AV, Bruna A, Budinská E, Caldas C, Chang DK, *et al* (2017) Interrogating open issues in cancer precision medicine with patient-derived xenografts. Nat Rev Cancer 17(4):254
3. Sontheimer-Phelps A, Hassell BA, Ingber DE (2019) Modelling cancer in microfluidic human organs-on-chips. Nat Rev Cancer 19:65–81
4. Damiani C, Colombo R, Gaglio D, Mastroianni F, Pescini D, Westerhoff HV, Mauri G, Vanoni M, Alberghina L (2017) A metabolic core model elucidates how enhanced utilization of glucose and glutamine, with enhanced glutamine-dependent lactate production, promotes cancer cell growth: the warburq effect. PLoS Comput Biol 13(9): e1005758
5. Gaglio D, Metallo CM, Gameiro PA, Hiller K, Danna LS, Balestrieri C, Alberghina L, Stephanopoulos G, Chiaradonna F (2011) Oncogenic k-ras decouples glucose and glutamine metabolism to support cancer cell growth. Mol Syst Biol 7(1):523
6. Icard P, Fournel L, Wu Z, Alifano M, Lincet H (2019) Interconnection between metabolism and cell cycle in cancer. Trends Biochem Sci 44:490–501
7. Reid MA, Dai Z, Locasale JW (2017) The impact of cellular metabolism on chromatin dynamics and epigenetics. Nat Cell Biol 19 (11):1298
8. Nielsen J (2017) Systems biology of metabolism: a driver for developing personalized and precision medicine. Cell Metab 25 (3):572–579
9. Brunk E, Sahoo S, Zielinski DC, Altunkaya A, Dräger A, Mih N, Gatto F, Nilsson A, Gonzalez GAP, Aurich MK, *et al* (2018) Recon3d enables a three-dimensional view of gene variation in human metabolism. Nat Biotechnol 36 (3):272
10. Graudenzi A, Maspero D, Di Filippo M, Gnugnoli M, Isella C, Mauri G, Medico E, Antoniotti M, Damani C (2018) Integration of transcriptomic data and metabolic networks in cancer samples reveals highly significant

prognostic power. J Biomed Inform 87:37–149
11. The cancer genome atlas (tcga). https://www.cancer.gov/tcga
12. Marjanovic ND, Weinberg RA, Chaffer CL (2013) Cell plasticity and heterogeneity in cancer. Clin Chem 59(1):168–179
13. Allison KH, Sledge GW (2014) Heterogeneity and cancer. Oncology 28(9):772–778
14. Di Filippo M, Colombo R, Damiani C, Pescini D, Gaglio D, Vanoni M, Alberghina L, Mauri G (2016) Zooming-in on cancer metabolic rewiring with tissue specific constraint-based models. Comput Biol Chem 62:60–69
15. Burrell RA, McGranahan N, Bartek J, Swanton C (2013) The causes and consequences of genetic heterogeneity in cancer evolution. Nature 501(7467):338
16. Jamal-Hanjani M, Quezada SA, Larkin J, Swanton C (2015) Translational implications of tumor heterogeneity. Clin Cancer Res 21 (6):1258–1266
17. Damiani C, Maspero D, Di Filippo M, Colombo R, Pescini D, Graudenzi A, Westerhoff HV, Alberghina L, Vanoni M, Mauri G (2019) Integration of single-cell rna-seq data into population models to characterize cancer metabolism. PLoS Comput Biol 15(2): e1006733
18. Damiani C, Di Filippo M, Pescini D, Maspero D, Colombo R, Mauri G (2017) popfba: tackling intratumour heterogeneity with flux balance analysis. Bioinformatics 33 (14):i311–i318
19. Di Filippo M, Damiani C, Colombo R, Pescini D, Mauri G (2016) Constraint-based modeling and simulation of cell populations. In: Italian workshop on artificial life and evolutionary computation. Springer, Berlin, pp 126–137
20. Mahadevan R, Schilling CH (2003) The effects of alternate optimal solutions in constraint-based genome-scale metabolic models. Metab Eng 5(4):264–276
21. Orth JD, Thiele I, Palsson BØ (2010) What is flux balance analysis? Nat Biotechnol 28 (3):245
22. Fiaschi T, Marini A, Giannoni E, Taddei ML, Gandellini P, De Donatis A, Lanciotti M, Serni S, Cirri P, Chiarugi P (2012) Reciprocal metabolic reprogramming through lactate shuttle coordinately influences tumor-stroma interplay. Cancer Res 72:5130–5140
23. Sanità P, Capulli M, Teti A, Galatioto GP, Vicentini C, Chiarugi P, Bologna M, Angelucci A (2014) Tumor-stroma metabolic relationship based on lactate shuttle can sustain prostate cancer progression. BMC Cancer 14 (1):154
24. Martinez-Outschoorn UE, Lin Z, Trimmer C, Flomenberg N, Wang C, Pavlides S, Pestell RG, Howell A, Sotgia F, Lisanti MP (2011) Cancer cells metabolically "fertilize" the tumor microenvironment with hydrogen peroxide, driving the warburg effect: implications for pet imaging of human tumors. Cell Cycle 10(15):2504–2520
25. Kanehisa M, Goto S (2000) Kegg: kyoto encyclopedia of genes and genomes. Nucleic Acids Res 28(1):27–30
26. Kanehisa M, Sato Y, Kawashima M, Furumichi M, Tanabe M (2016) KEGG as a reference resource for gene and protein annotation. Nucleic Acids Res 44(D1):D457–62
27. UniProt Consortium *et al* (2018) Uniprot: the universal protein knowledgebase. Nucleic Acids Res 46(5):2699
28. Joshi-Tope G, Gillespie M, Vastrik I, D'Eustachio P, Schmidt E, de Bono B, Jassal B, Gopinath GR, Wu GR, Matthews L, *et al* (2005) Reactome: a knowledgebase of biological pathways. Nucleic Acids Res 33 (suppl 1):D428–D432
29. Orth JD, Jeffrey D and Fleming, Ronan MT and Palsson, Bernhard O (2010) Reconstruction and use of microbial metabolic networks: the core Escherichia coli metabolic model as an educational guide. EcoSal plus 4(1). American Society for Microbiology https://www.asmscience.org/content/journal/ecosalplus/10.1128/ecosalplus.10.2.1. https://doi.org/10.1128/ecosalplus.10.2.1
30. Thiele I, Palsson B (2010) A protocol for generating a high-quality genome-scale metabolic reconstruction. Nat Protoc 5(1):93
31. Swainston N, Smallbone K, Hefzi H, Dobson PD, Brewer J, Hanscho M, Zielinski DC, Ang KS, Gardiner NJ, Gutierrez JM *et al* (2016) Recon 2.2: from reconstruction to model of human metabolism. Metabolomics 12(7):109

Chapter 16

Supply and Demand Analysis of Autophagy

André du Toit, Ben Loos, and Jan Hendrik S. Hofmeyr

Abstract

Autophagy is an intracellular protein degradation pathway that plays a vital role in cellular homeostasis. It maintains cellular function through proteostasis and the removal of unused and harmful proteins and organelles. Moreover, it also serves as an adaptive response to metabolic perturbations. Deviation in autophagy activity has been linked to the progression of several pathologies, including neurodegenerative diseases. Preclinical trials have shown that modulating autophagy holds great promise in treating neurodegenerative diseases by clearing toxic protein aggregates. The success of autophagy modulating therapies requires extensive knowledge of the molecular machinery and, importantly, an in-depth understanding of the underlying systems properties of the autophagy system. A computational approach provides a powerful platform to interrogate and analyze the regulation, control, and behavior of reaction networks. However, the complexity of interactions involved in the autophagy pathway makes it challenging to isolate and characterize individual components. By reducing the autophagy process to a supply-demand system in which autophagosome synthesis (supply) and autophagosome degradation (demand) are linked by the autophagosomes, it is possible to determine the control of the supply and demand over the steady-state autophagosome flux and autophagosome concentration. In this chapter, we describe a methodology to perform supply and demand analysis of the autophagy system, the experimental procedure to measure the autophagy variables, and the use of the supply-demand framework to determine the distribution of flux and concentration control.

Key words Autophagy, Autophagosome flux, Supply-demand analysis, Elasticity coefficient, Control coefficient, Autophagosomes, Fluorescence microscopy

1 Introduction

Macroautophagy (hereafter referred to as autophagy) is an evolutionary conserved metabolic process through which intracellular components are degraded. The autophagy pathway involves sequestration of cytoplasm in a double membrane vesicle, termed an autophagosome, which then fuses with and delivers its cargo to lysosomes where degradation occurs. The continuous removal of bulk cytoplasmic proteins, primarily long-lived proteins, and the ensuing *de novo* synthesis ensure homeostatic maintenance of functional proteins and organelles while preventing the build-up of

Deepak Nagrath (ed.), *Metabolic Flux Analysis in Eukaryotic Cells: Methods and Protocols*, Methods in Molecular Biology, vol. 2088, https://doi.org/10.1007/978-1-0716-0159-4_16,

dysfunctional and potentially harmful proteins and organelles [11]. Autophagy also serves as an adaptive response mechanism to metabolic stress; for instance, during periods of starvation autophagy is enhanced to supply amino acids for energy production. Although autophagy is vital for normal physiological development and cell function, its dysregulation can have deleterious effects. It has been implicated in the progression of several diseases: loss of autophagy function leads to the build-up of toxic compounds, such as those in neurodegenerative pathologies [18], while enhanced autophagy activity contributes to the resistance of cancer cells against chemotherapeutic drugs [15]. Pre-clinical trials aimed at modulating autophagy have shown promising results for treating neurodegenerative diseases by clearing toxic protein aggregates [17]. There is therefore an ever-increasing need to develop novel autophagy-targeting therapies to modulate and control autophagy in an effective and precise manner.

The success of autophagy-modulating therapies depends on identifying and characterizing targets, so as to allow for the safe and effective pharmacological regulation of autophagy activity; computational models have proved to be extremely useful in such endeavors. Computational modeling is a powerful technique for studying complex cellular processes such as autophagy in order to understand their regulation, control, and behaviour of the underlying reaction networks. Several computational models that model unique aspects of the autophagy process, such autophagosome vesicle dynamics [14], spatiotemporal autophagy trafficking behavior [1], autophagy cargo dynamics [5–7], and crosstalk between autophagy and apoptosis pathways [10, 19] have been described in the literature. While these models provide valuable insight into particular aspects of the autophagy system, they do not shed light on the degree to which the different steps of the autophagic system control the autophagosome flux, knowledge that is crucial for developing autophagy targeting therapies. This is in part due to the complexity of the autophagy process: autophagy involves the recruitment of numerous proteins in a highly organized manner to produce functional units within the autophagy pathway, which in turn are regulated by metabolic and stress sensing networks [12]. Creating a comprehensive computational model of autophagy would require the isolation of all enzymes and proteins involved in the pathway to study their kinetic properties separately. However, this approach would be impractical due to the complexity of interactions involved in regulating and assembling autophagosomes, and subsequent events such as their fusion with lysosomes to form autolysosomes.

The supply-demand analysis framework of Hofmeyr and Cornish-Bowden [9] does not require information on the kinetics or activity of all the enzymes and proteins involved in the pathway. All that is required is the ability to measure the response of steady-state variables to perturbations in the activities of the metabolic processes that synthesize and degrade a chosen intermediate in the pathway. By reducing the autophagy process to a supply-demand system in which autophagosome synthesis (supply) and autophagosome degradation (demand) are linked by the autophagosome pool, it is possible to determine the distribution of control of the supply and demand blocks over the steady-state autophagosome flux and concentration of autophagosomes. Why is this important given that the control distribution can only be determined for the supply and demand block (instead of a specific protein)? It allows us to experimentally determine the flux and concentration control distribution (with respect to the supply and demand) for any cell type and, importantly, under any condition such as disease states without the need to characterize individual components of the autophagy pathway. With ever-increasing drug libraries, there is no shortage of available drugs (and molecular targets); using the knowledge gained from supply and demand analysis, drug screening can be performed in a more cost and time-effective manner to identify suitable autophagy modulation drugs.

A recently developed fluorescence-based microscopy approach [4] allows the quantitative measurement of the autophagosome flux and the size of the autophagic vesicle pools that is necessary for performing a supply and demand analysis of the autophagy system. Here we will first introduce the supply and demand analysis framework, second, describe the methodology to determine the response of autophagy steady-state variables and, third, perform using hypothetical data a supply and demand analysis of the autophagy system.

2 Supply-Demand Framework

The living cell is intrinsically a molecular economy in which one metabolic network, the supply, produces intermediate(s) that are consumed by another, the demand. The supply-demand framework of Hofmeyr and Cornish-Bowden [9] is a powerful tool for quantitative analysis of how the behavior and control of the steady-state flux, J, and concentration of the intermediate that links supply and demand is determined by the properties of the so-called supply and demand rate characteristics (Fig. 1).

The supply and demand elasticity coefficients determine the behavior of the steady-state flux and concentration of the linking intermediate, and are equal to the slopes of the tangents to the supply and demand rate characteristics where they intersect at the steady state (Fig. 1b). An elasticity coefficient quantifies the sensitivity of a rate v with respect to a perturbation in a metabolic intermediate p that directly affects this rate, and is mathematically defined as:

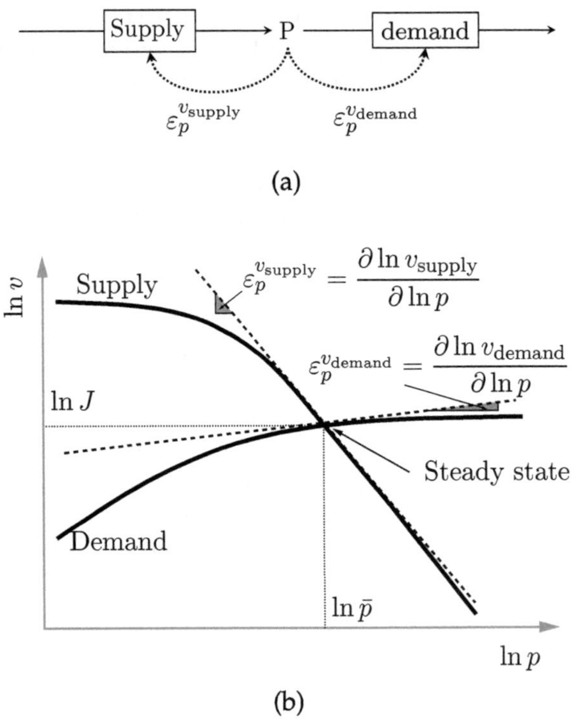

Fig. 1 Analysis of supply-demand systems. (**a**) A metabolic supply-demand system where the supply block produces an intermediate, p, which is consumed by the demand block; (**b**) Log-Log rate characteristic plots of the supply and demand rates as a function of p. The two rate characteristics intersect at the steady state, which is characterized by the flux, J, and the concentration $\bar{p}$. The behavior and control of J and $\bar{p}$ is determined by the sensitivities of the supply and demand rate to perturbations in p at steady state, respectively quantified by the elasticity coefficients $\varepsilon_p^{v_{supply}}$ and $\varepsilon_p^{v_{demand}}$. Figure adopted from Hofmeyr and Cornish-Bowden [9]

$$\varepsilon_p^v = \frac{\partial \ln v}{\partial \ln p} \tag{1}$$

The flux- and concentration-control coefficients can be calculated from the values of the supply and demand elasticities [9]. A control coefficient quantifies the sensitivity of a steady-state variable to a perturbation in the activity of the supply or demand step in the system, and is defined as

$$C_{v_i}^X = \frac{\partial \ln X}{\partial \ln v_i} \tag{2}$$

where X is a flux J or a concentration, and v_i is the supply or demand rate.

When considering the autophagy process as the formation and maturation of autophagosomes and their fusion with lysosomes to form autolysosomes (Fig. 2a), the autophagy system reduces to a

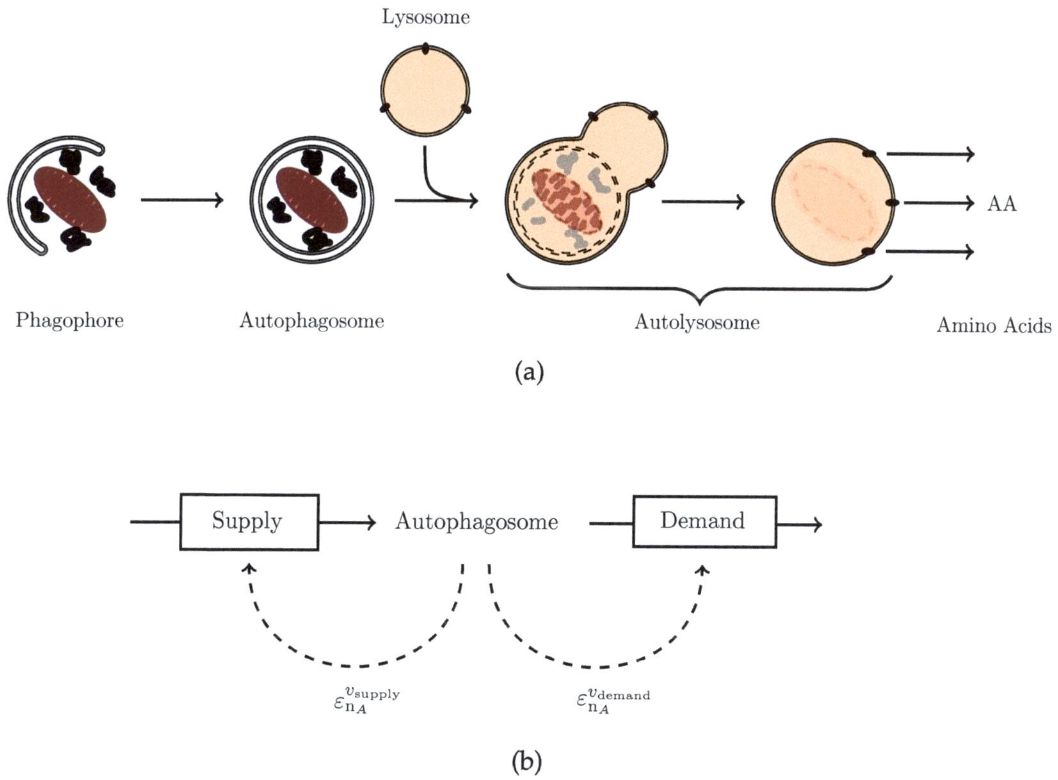

Fig. 2 The autophagy process. (**a**) Schematic representation of the autophagy process, which starts with the formation of a phagophore, that, once matured into an autophagosome, fuses with lysosomes to form autolysosomes where the autophagy cargo is degraded and recycled. The autophagosome flux is the rate of flow along the vesicular pathway which is measured as the turnover rate of autophagosomes at steady state [13]. The autophagy process can be reduced to (**b**) a supply-demand system where the supply block (autophagosome synthesis) and the demand block (autophagosome degradation) are linked by the autophagosome pool. The distribution of control over the autophagosome flux and the concentration of autophagosomes can be determined from the expressions of the control coefficients in terms of the supply and demand elasticity coefficients (see text). Figure adopted from du Toit et al. [4]

simple supply and demand system where the synthesis of autophagosomes represents the supply block and the degradation of autophagosomes represents the demand block (Fig. 2b). By measuring the response of the autophagosome flux and the steady-state concentration of autophagosomes to perturbations of the supply and demand, the flux and concentration control of autophagy can be quantified using the supply-demand framework.

3 Materials

The fluorescence-based single cell microscopy method described by du Toit et al. [4] allows the characterization of the autophagy system in a single cell in terms of the steady-state pool size of the various autophagy intermediates and the autophagosome flux (J). In short, the experimental procedure is based on quantitative measurement of the total number of autophagosomes (n_A), autolysosomes (n_{AL}), and lysosomes (n_L) over time in single cells. The transient time-dependent profiles of the autophagy intermediates (n_A, n_{AL} and n_L) are used to (1) verify whether or not the autophagy system is in steady state, (2) quantify the steady-state pool sizes of the autophagy intermediates, and (3) calculate the autophagosome flux J from the initial rate of accumulation in n_A after the complete inhibition of fusion between autophagosomes and lysosome with bafilomycin A_1 when the autophagy system is in steady state. Here we will describe the experimental setup (Subheading 3.1) for measuring the autophagy vesicles over time and characterizing the autophagy variables (Subheading 4.1), so as to perform a supply and demand analysis of autophagy (Subheading 4.2).

3.1 Materials

3.1.1 Cell Culture

Mouse Embryonic Fibroblast (MEF) cells that stably express green fluorescent protein light chain 3 (GFP-LC3) were maintained in Dulbecco's modified Eagle's medium (DMEM) (Life Technologies, #41-965-039), supplemented with 10% fetal bovine serum (Biochrom, #S-0615) and penicillin-streptomycin (Life Technologies, #15-140-122), at 37 °C in a 5% CO_2 atmosphere.

3.1.2 Chemicals

The fusion between autophagosomes and lysosomes was inhibited using 400 nM bafilomycin A_1 (LKT Laboratories Inc., #B-0025), and 25 nM rapamycin (Sigma-Aldrich, #R-0395) was used as an autophagy inducer. Culture medium was supplemented with LysoTracker red (Thermo Fisher Scientific, #L-7528) fluorescence probe according to the manufacturer's instructions to distinguish between autophagy vesicles (*See* **Note 1**).

3.1.3 Microscopy and Image Analysis

Cells were seeded onto CYTOO micro-patterned slides (*See* **Note 2**) with large fibronectin disc shapes (CYTOO, #10-003-10) and maintained for 6 h before imaging. Fluorescence microscopy was performed on an Olympus IX81 wide-field microscope, equipped with a stage incubator, using a 60 × oil immersion objective. Image stacks were acquired using an automated z-stack and stage control to measure the complete pool size of autophagy vesicles (n_A, n_{AL}, n_L) over time. An average of four images with a 0.5 μm step-width between layers were used. Images were processed and deconvolved using Cell^R, and analyzed with ImageJ/Fiji (http://rsbweb.nih.gov/ij/download.html).

4 Methods

4.1 Measuring Autophagy Variables

The experimental procedure to characterize the autophagy variables involves three steps: first, verifying that the autophagy system is at steady state (measuring the autophagy steady state intermediate concentration); second, to completely inhibit the fusion between autophagosomes and lysosomes; third, to calculate the autophagosome flux as the rate of accumulation of autophagosomes after the inhibition of fusion. Here we briefly describe each step.

1. *Verifying that the autophagy system is in steady state.* Cells are imaged 1 h apart and the n_A, n_{AL}, and n_L quantified per cell. When n_A remains constant over a period of time of at least 3 h, it signifies that the autophagy system is in a steady state where the rate of autophagosome synthesis equals the rate of autophagosome degradation. The steady-state intermediate concentrations are the averages of n_A, n_{AL}, and n_L over the steady-state period. Where there are significant changes in the n_A over time the autophagy system is in a transition state, and n_A should then be continuously monitored until a steady state has been established before proceeding to the next step.
2. *Complete inhibition of fusion between autophagosomes and lysosomes.* It is important that the fusion of autophagosomes and lysosomes is completely inhibited, as incomplete fusion inhibition will result in inaccurate measurement of flux. The concentration of fusion inhibitor required for the complete inhibition of fusion is the concentration when there is no further increase in the initial rate of n_A accumulation per cell with increasing fusion inhibitor concentration. The concentration of fusion inhibitor required should be determined for each cell and inhibitor type used.
3. *Measuring autophagosome flux.* Once the autophagy system is in a steady state, the fusion between autophagosomes and lysosomes is completely inhibited to determine autophagosome flux, J, which is calculated from the initial slope of the increase in n_A at the point of inhibition of fusion. The smaller the intervals used for calculating J, the greater the accuracy as it limits possible feedback effects of downstream metabolites.

Figure 3 shows the transient time-dependent behavior of autophagosomes, autolysosomes, and lysosomes under basal and 25 nM rapamycin induced conditions, and the quantified autophagy variables are shown in Table 1.

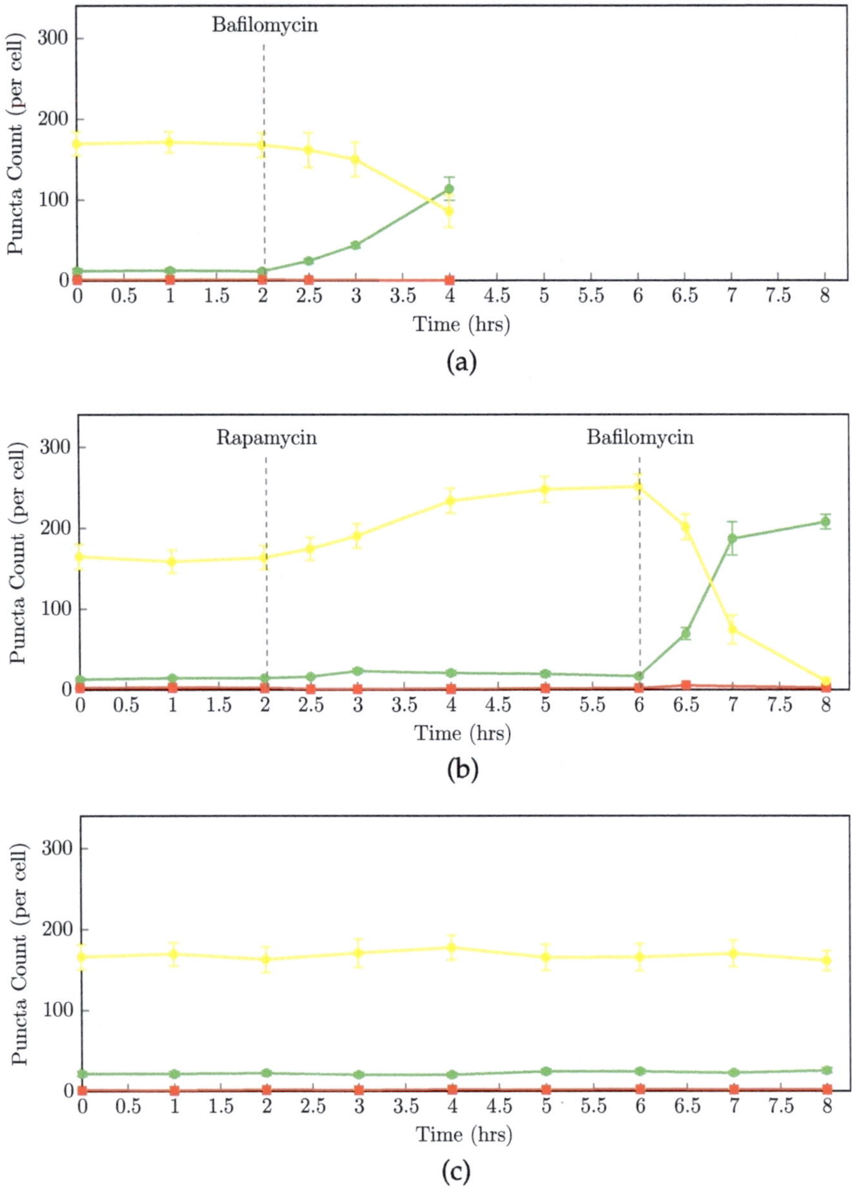

Fig. 3 The transient time-dependent profiles of autophagosomes (green filled circle), autolysosomes (yellow filled circle) and lysosomes (red filled circle). Pool sizes of the autophagy intermediates under (**a**) basal conditions (0–2 h) and after inhibition of fusion with 400 nM bafilomycin A_1 at 2 h, and (**b**) enhanced autophagy conditions after 25 nM rapamycin treatment at 2 h and after inhibition of fusion with 400 nM bafilomycin A_1 at 6 h. (**c**) Control: autophagy intermediates under basal conditions without bafilomycin A_1 treatment (0–8 h). ($n = 10$). Figure adopted from du Toit et al. [4]

4.2 Supply-Demand Analysis: Perturbing the Supply and the Demand for Autophagosomes

Above we described how to measure the steady-state response to 25 nM rapamycin induction of autophagy. In order to perform supply-demand analysis we need to determine how the autophagy steady-state variables respond to incremental perturbations of both the supply and demand blocks.

Table 1
Functional variables of autophagy for basal and rapamycin (25 nM) induced autophagy in MEF cells (A: autophagosomes; AL: autolysosomes; L: lysosomes)

Variable	Unit	Basal	Induced
Autophagosome flux, J	A/h/cell	25.4	105.4
Number of autophagosomes, n_A	A/cell	13	17
Number of autolysosomes, n_{AL}	AL/cell	165	251
Number of lysosomes, n_L	L/cell	1	1

Table adapted form du Toit et al. [4]

1. *Incrementally perturb the supply/demand blocks and measure the response of the autophagy steady-state variables.* This can be achieved by incrementally increasing the concentration of rapamycin to characterize the supply rate characteristic, while the demand rate characteristic can be determined by partially inhibiting the fusion of autophagosomes with lysosomes with increasing concentrations of bafilomycin A_1, followed by the complete inhibition of fusion. Figure 4 shows an example of how the transient time-dependent behavior of autophagosomes could look like over time in response to incremental perturbations of the supply and demand of autophagosomes. We illustrate a hypothetical situation where the control over the autophagosome flux is associated with the supply.
2. *Determine the supply and demand elasticities.* Plot the co-responses [8] of the steady-state autophagosome flux and autophagosome pool size in response to perturbation of the supply and demand of autophagosomes (Fig. 4c, d). The slopes of the tangents to the supply and demand rate characteristics at the steady-state point are equal to the supply and demand elasticity coefficients. The perturbation of the supply of autophagosomes by rapamycin allows the calculation of the demand elasticity (Fig. 4c), while the perturbation of the demand of autophagosome using bafilomycin A_1 allows the calculation of the supply elasticity (Fig. 4d).
3. *Calculate the flux- and concentration-control coefficients using the supply and demand elasticity coefficients.*

   The flux-control coefficients can be expressed in terms of supply and demand elasticity coefficients [9]:

$$C^J_{v_{\text{supply}}} = \frac{\varepsilon^{v_{\text{demand}}}_{n_A}}{\varepsilon^{v_{\text{demand}}}_{n_A} - \varepsilon^{v_{\text{supply}}}_{n_A}}$$

$$C^J_{v_{\text{demand}}} = \frac{-\varepsilon^{v_{\text{supply}}}_{n_A}}{\varepsilon^{v_{\text{demand}}}_{n_A} - \varepsilon^{v_{\text{supply}}}_{n_A}}$$

and, similarly, the concentration-control coefficients [9]:

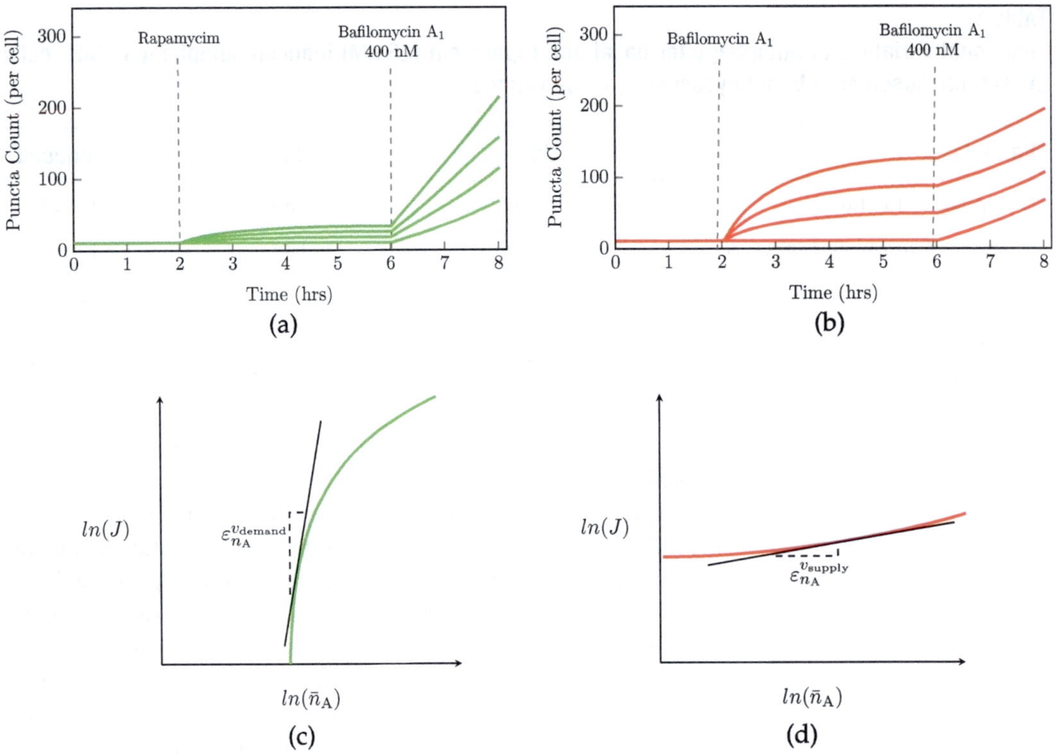

Fig. 4 Hypothetical transient time-dependent behavior in autophagosomes. (**a**) Perturbing the supply of autophagosomes by inducing autophagy with increasing concentrations of rapamycin (e.g. 1–100 nM) at 2 h and inhibiting fusion with 400 nM bafilomycin A_1 at 6 h; (**b**) Perturbing the demand of autophagosomes by partial inhibition of autophagosome/lysosome fusion with increasing concentrations of bafilomycin A_1 (e.g. 1–20 nM) at 2 h and complete inhibition of fusion with 400 nM bafilomycin A_1 at 6 h; Co-responses of J and $\bar{n}_A$ to perturbation of the (**c**) supply and (**d**) demand of autophagosomes. The supply and demand elasticity coefficients are the slopes of the tangents to the supply and demand rate characteristics at the steady-state

$$C^{n_A}_{v_{\text{supply}}} = \frac{1}{\varepsilon^{v_{\text{demand}}}_{n_A} - \varepsilon^{v_{\text{supply}}}_{n_A}}$$

$$C^{n_A}_{v_{\text{demand}}} = \frac{-1}{\varepsilon^{v_{\text{demand}}}_{n_A} - \varepsilon^{v_{\text{supply}}}_{n_A}}$$

4. *Visualizing the distribution of flux and concentration control around the steady state.* The distribution of flux and concentration control around the steady state can be visualized by plotting the supply and demand co-responses in Fig. 4c, d in logarithmic space. Comparing equal fold-changes in the supply and demand rate characteristics shows in this example that flux control resides in the supply of autophagosomes, while the regulation of the number of autophagosomes resides in the demand (Fig. 5). The steepness of the slope of the demand rate characteristic determines the degree of homeostatic maintenance of autophagosome number; the steeper the slope of

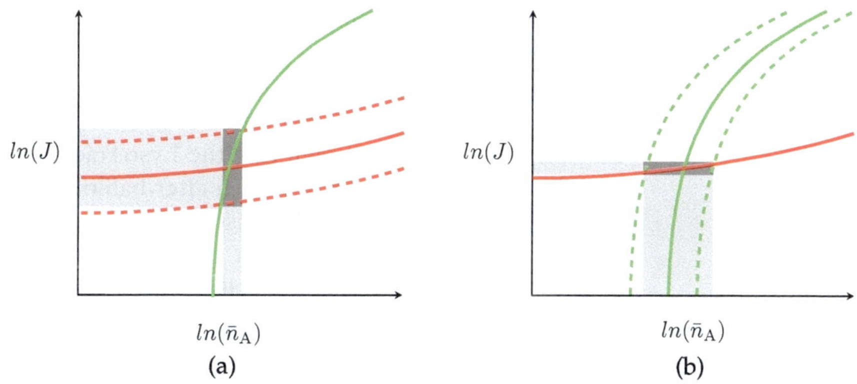

Fig. 5 The combined rate characteristics of the autophagic supply–demand system. The rate characteristics of the supply (green solid line) and demand (red solid line) plotted in log-log space, showing the intersection between the supply and demand rate characteristics; J is the steady-state autophagosome flux, and $\bar{n}_A$ is the steady-state autophagosome concentration. The dotted lines show a fixed percentage increase/decrease in supply (**a**) and demand (**b**) of autophagosomes. The shaded regions show the magnitude of the responses in J and $\bar{n}_A$

the demand rate characteristic, the narrower the magnitude of variation in autophagosome number, and the better the homeostatic maintenance of autophagosome number.

5 Notes

1. *Labeling of autophagy vesicles.* It is important to distinguish accurately between the autophagy pathway intermediates (n_A, n_{AL}, n_L) to generate precise and reliable data of the autophagy system. There are a number of alternative probes that could be used to label autophagy vesicles. One of these methods is to use tagged autophagy pathway proteins using fluorescence constructs. They can be either co-transfected such as with GFP-LC3 and red fluorescent protein tagged to lysosome-associated membrane glycoproteins (RFP-LAMP), or tandem constructs such as mRFP-GFP-LC3 [16] (this would allow for quantification of autophagosome flux, autophagosomes and autolysosomes, but not lysosomes). However, such an approach is not without its challenges; heterogeneous expression levels in the cell population as well as varying signal/noise ratio increase variability and limits the size of batch analysis. Another approach is to use fluorescent dyes with inherent binding properties, such as LysoTracker, which offers the advantage of homogeneous staining with optimal signal/noise ratio. It can also be used in combination with fluorescence constructs as presented here. Note that LysoTracker is an acidotrophic dye that accumulates and fluoresces in acidic

organelles such as lysosomes; inhibiting vacuolar-type H^+-ATPase with bafilomycin A_1 will therefore disrupt the maintenance of a low pH environment in lysosomes and consequently result in the loss of fluorescent signal of LysoTracker. However, du Toit et al. [4] have shown that the LysoTracker fluorescent signal is maintained for at least 2 h after bafilomycin A_1 treatment, allowing for a reliable measurement of autophagosome flux. It is advisable to verify that the concentration of LysoTracker used for the respective cell lines is suitable.

2. *Enchaining data quality through micro-patterning.* There are a number of factors that contribute to the observed variability in any given cell-specific population such as cell size, migration, and proliferation. While standard microscopy slides suitable for live cell imaging can be used for assessing autophagy in single cells, micro-patterned slides can be used to control cell size and position. This reduces cell variability, thereby improving data quality. It is also suitable for high-throughput analysis. Micropatterned coverslips can also be self fabricated as described by Carpi et al. [2] and mounted to chamber slides for live cell imaging [3].

Acknowledgements

We acknowledge financial support from the South African National Research Foundation (NRF), the South African Medical Research Council (SAMRC) as well as the Cancer Association of South Africa (CANSA). All microscopy work was performed at the Central Analytical Facility (CAF), Cell Imaging Unit, Stellenbosch University, South Africa.

References

1. Börlin CS, Lang V, Hamacher-Brady A, Brady NR (2014) Agent-based modeling of autophagy reveals emergent regulatory behavior of spatio-temporal autophagy dynamics. Cell Commun Signal 12, 56
2. Carpi N, Piel M, Azioune A, Fink J (2011) Micropatterning on glass with deep uv. Nat Protocol Exch 10
3. Du Toit A, De Wet S, Hofmeyr J-H, Müller-Nedebock K, Loos B (2018) The precision control of autophagic flux and vesicle dynamics–a micropattern approach. Cells 7:94
4. du Toit A, Hofmeyr J-HS, Gniadek TJ, Loos B (2018) Measuring autophagosome flux. Autophagy 14:1060–1071. PMID: 29909716
5. Han K, Kwon HW, Kang H, Kim J, Lee M-S, Choi M (2012) Dynamics of macroautophagy: modeling and oscillatory behavior. Physica A 391:686–692
6. Han K, Kim J, Choi M (2014) Computer simulations unveil the dynamics of autophagy and its implications for the cellular quality control. J Biol Syst 22:659–675
7. Han K, Kim J, Choi M (2014) Quantitative indices of autophagy activity from minimal models. Theor Biol Med Model 11:31
8. Hofmeyr J-HS, Cornish-Bowden A (1996) Co-response analysis: a new experimental strategy for metabolic control analysis. J Theor Biol 182:371–380
9. Hofmeyr J-HS, Cornish-Bowden A (2000) Regulating the cellular economy of supply and demand. FEBS Lett 476:47–51

10. Kapuy O, Vinod PK, Mandl J, Bánhegyi G (2013) A cellular stress-directed bistable switch controls the crosstalk between autophagy and apoptosis. Mol BioSyst 9:296–306
11. Kuma A, Mizushima N (2010) Physiological role of autophagy as an intracellular recycling system: with an emphasis on nutrient metabolism. Semin Cell Dev Biol 21:683–690
12. Loos B, Engelbrecht A-M, Lockshin RA, Klionsky DJ, Zakeri Z (2013) The variability of autophagy and cell death susceptibility: Unanswered questions. Autophagy 9:1270–1285
13. Loos B, du Toit A, Hofmeyr J-HS (2014) Defining and measuring autophagosome flux--concept and reality. Autophagy 10:2087–2096
14. Martin KR, Barua D, Kauffman AL, Westrate LM, Posner RG, Hlavacek WS, MacKeigan JP (2013) Computational model for autophagic vesicle dynamics in single cells. Autophagy 9:74–92
15. Mathew R, Karantza-Wadsworth V, White E (2007) Role of autophagy in cancer. Nat Rev Cancer 7:961–967
16. Mizushima N, Yoshimori T, Levine B (2010) Methods in mammalian autophagy research. Cell 140:313–326
17. Ravikumar B, Vacher C, Berger Z, Davies JE, Luo S, Oroz LG, Scaravilli F, Easton DF, Duden R, O'Kane CJ, Rubinsztein DC (2004) Inhibition of mTOR induces autophagy and reduces toxicity of polyglutamine expansions in fly and mouse models of Huntington disease. Nat Genet 36:585–595
18. Son JH, Shim JH, Kim K-H, Ha J-Y, Han JY (2012) Neuronal autophagy and neurodegenerative diseases. Exp Mol Med 44:89–98
19. Tavassoly I, Parmar J, Shajahan-Haq A, Clarke R, Baumann W, Tyson J (2015) Dynamic modeling of the interaction between autophagy and apoptosis in mammalian cells. CPT: Pharmacometrics Syst Pharmacol 4:263–272

Chapter 17

Thermodynamic Approaches in Flux Analysis

Sabine Peres and Vincent Fromion

Abstract

Networks of reactions inside the cell are constrained by the laws of mass and energy balance. Constrained-based modelling (CBM) is the most used method to describe the mass balance of metabolic network. The main key concepts in CBM are stoichiometric analysis such as elementary flux mode analysis or flux balance analysis. Some of these methods have focused on adding thermodynamics constraints to eliminate non-physical fluxes or inconsistencies in the metabolic system. Here, we review the main different approaches and how they tackle the different class of problems.

Key words Thermodynamic in constraint-based modelling, Metabolic networks, Gibbs free energy, Equilibrium constant of reactions

1 Introduction

The number of applications of thermodynamics-based network analysis methods has been increasing in the last 10 years [1]. The important use of thermodynamics-based analysis is the determination of reaction directionality, whereby the feasibility of reaction fluxes or flux distributions can be checked based on calculation of changes in Gibbs free energy (ΔG) using metabolite concentrations. Most of the proposed approaches belong to the constrained-based modelling (CBM) framework which is powerful to analyze and predict the properties of metabolic networks. The idea behind CBM approach is quite simple: it involves defining a set of constraints that the system being studied has to respect and the set of system configurations, i.e. the fluxes, that are compatible with these constraints. Since the set of configurations is generally not reduced to a singleton, it is common to consider among these configurations, the one possessing some suitable properties or achieving the value of a given optimal criterion. For the metabolic network analysis, a first constraint is associated with the mass conservation that has to be satisfied in a steady-state regime by each elementary metabolic reaction of the metabolic network. This

Deepak Nagrath (ed.), *Metabolic Flux Analysis in Eukaryotic Cells: Methods and Protocols*, Methods in Molecular Biology, vol. 2088, https://doi.org/10.1007/978-1-0716-0159-4_17,

leads, after their integration through the whole metabolic network, to the definition of the so-called stoichiometric constraints which connotes that the fluxes through the metabolic network have to satisfy a system of linear equalities. Although the stoichiometric constraints are essential, they only define a part of constraints acting on the "real metabolic network." It is, for example, well-established that the constraints related to the irreversibility of some metabolic reactions also add essential constraints. Indeed, this new set of constraints, expressed as a sign constraint on some fluxes generally reduces the set of feasible fluxes and dramatically increases the predictive power of computational tools integrating all these constraints. Behind the addition of the constraints induced by the irreversible reactions, it is more generally the question of taking into account thermodynamic constraints that are raised. Actually, the difficulty to manage stoichiometric and thermodynamic constraints together is related to the second essential issues attached to CBM approaches: how can all the defined constraints contribute to quantitative predictions about the possible behavior of the metabolic network? The relevant concept for this second question is whether or not the set of fluxes compatible with the set of constraints defines a convex set. It is clear that in the case of stoichiometric constraints, the set of possible fluxes defined a convex set (it can be empty). On the other hand, verifying if a flux configuration satisfying stoichiometric constraints is thermodynamically feasible is a problem that can be formulated as a convex optimization problem. Unfortunately, when stoichiometric and thermodynamic constraints are considered together, the set of possible flux configurations does not generally define a convex set (a simple adaptation of the example provided in [2] allows to prove this statement). In general, the fact that the set of constraints does not define a convex set makes its numerical exploration much more difficult or even impossible. This also means, for example, that it is particularly difficult to maximize biomass production that respects both stoichiometric and thermodynamic constraints. Indeed, in this specific case [2], its non-convexity implies that the problem has multiple local maxima. However, despite the lack of convexity, different methods have been proposed in the literature and the main goal of this overview is to present some of the methods dealing with the stoichiometric and thermodynamic constraints.

2 Materials

2.1 Metabolic Model

Let $S \in I\!R^{n \times r}$ be the stoichiometric matrix of a metabolic network of r reactions and n metabolites. The time course of the vector of metabolites $x = (x_1, \ldots, x_i, \ldots, x_n)^t$ can be described by differential equations $\frac{dx(t)}{dt} = S \cdot \nu(x(t))$ where $v(t) = \nu(x(t)) \in I\!R^r$ is the vector of fluxes. A steady-state regime of the metabolic network corresponds to an equilibrium of the differential system where the metabolite concentrations and the fluxes are now two constant vectors such that $S \cdot v_e = 0$ and $v_e = \nu(x_e)$. Actually, it is common to define the space of flux vectors compatible with the stoichiometric matrix constraint, i.e., $v \in \mathbb{R}^r$ such that $S \cdot v = 0$. When some reactions $\nu_i(x)$ are known to be irreversible, the space of feasible flux vectors compatible with the stoichiometric matrix and irreversible reactions is now defined by this convex polyhedral cone $C = \{v \in \mathbb{R}^r$. An elementary flux mode (EFM) $e \in \mathbb{R}^r$ is a steady-state flux ($S \cdot e = 0$) that fulfills the irreversibility constraints ($e \geq 0$), and has minimal support, which means that *supp*(e) (those reactions i such that $e_i \neq 0$) is not a proper superset of the support of any other EFM [3]. Geometrically, if the reversible reactions are split in two irreversible ones, EFMs are so-called extreme rays of a pointed polyhedral (flux) cone C. They fully characterize the solution space but their computation leads to combinatorial explosion of their number when the networks are large. In the Flux Balance Analysis (FBA) [4], an objective function of fluxes, like the biomass flux $v_{biomass}$ is maximized under the set of constraints acting on flux vectors defined by the stoichiometric constraint $S \cdot v = 0$ and bounds on the flux vectors, i.e. $l_i \leq v_i \leq u_i$.

2.2 Gibbs Free Energy

Assuming constant pressure and a closed system, according to the second law of thermodynamics, a reaction j proceeds spontaneously only in the direction of its negative Gibbs free energy $\Delta_r G_j$ [5]:

$$\Delta_r G_j < 0 \tag{1}$$

For example, for this following biochemical reaction

$$A + B \rightleftharpoons C + D$$

where A, B, C, and D are given molecules, the Gibbs free energy associated to the previous reaction is defined by

$$\Delta_r G = \Delta_r G^0 + RT\ln\frac{x_C x_D}{x_A x_B} \tag{2}$$

with x_X denotes the concentration of molecule X and where $\Delta_r G^0$ corresponds to the standard-state free energy of the reaction. ΔG^0 is related to the equilibrium constant of the reaction K_{eq}, by this simple relation: $\Delta G^0 = -RT\ln(K_{eq})$. In order that the previous reaction in a steady-state regime has a net flux of C and D production, i.e., the net flux is in the forward direction, it is necessary that $\Delta_r G < 0$. Figure 1 illustrates the influence of the metabolite concentrations on the Gibbs free energy.

Gibbs free energies of formation can be obtained from experimental values such as NIST databases [6], National Bureau of Standards Database [7] or estimated with contribution of molecular groups from experimental energies data such as eQuilibrator [8], Group Contribution Methods [9, 10].

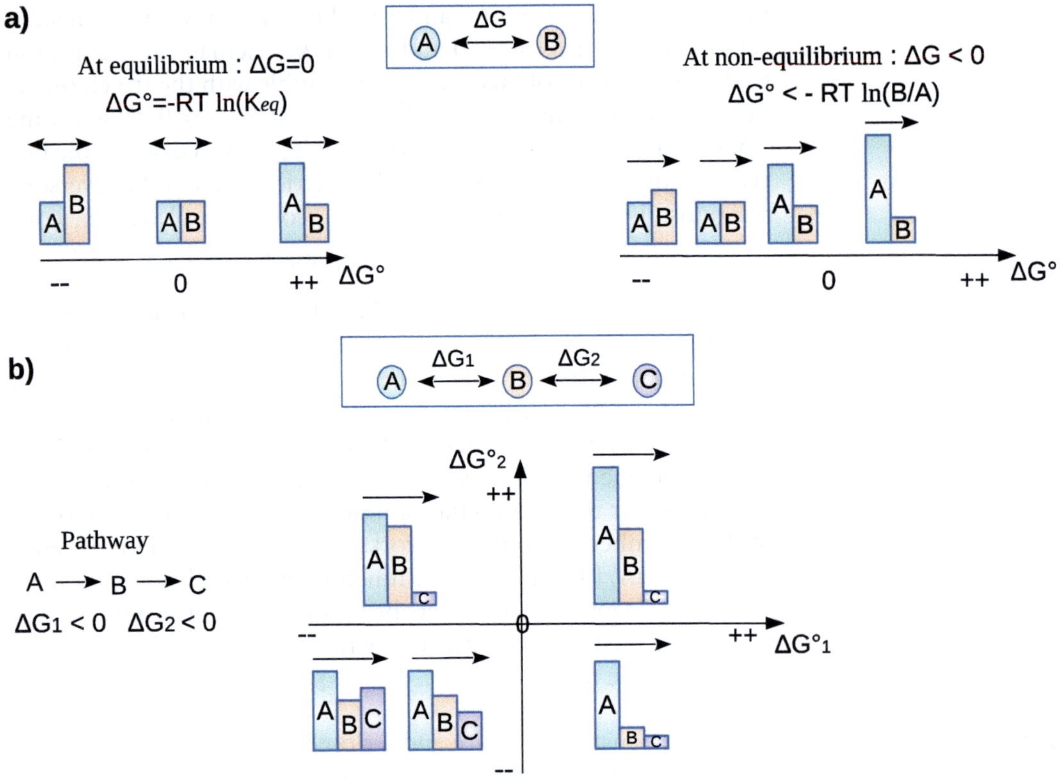

Fig. 1 Influence of concentrations and spontaneity. (**a**) Possible distribution of concentrations at equilibrium and non-equilibrium reaction. (**b**) Possible distribution of concentrations making the metabolic pathway A → B → C possible for different ΔG^0

2.3 Non-convexity of the General Problem

As indicated in the introduction, the set of fluxes and internal metabolite concentrations respecting the stoichiometric and thermodynamic constraints is generally not a convex set. This means in particular that the maximization of a linear objective function of fluxes, like the biomass flux $v_{biomass}$, respecting both constraints is generally not a convex optimization problem, and thus it is generally difficult to solve exactly. Nevertheless, despite the lack of convexity, there exist two main approaches allowing to solve exactly or approximately this non-convex optimal problem. In the first family of approaches, the problem is embedded in a given class of non-convex problems and it is solved by calling general algorithms allowing to tackle this specific class of non-convex problems, as, e.g., in [11, 12]. The second approach finds its roots in a result presented in [13, 14]. Indeed, it can be shown that if an optimal solution exists then it is reached in an EFM. In addition, in order to solve the optimal problem, it is enough to check the thermodynamic feasibility of each EFM. Then, the value of cost function for this specific EFM can be computed. The solution is then given by the best feasible EFM, see, e.g., [15–17]. Obviously and with no surprises, the exact resolution of this non-convex problem has a (high) computational cost that is paid here by the fact that the number of EFMs associated to the metabolic network could be huge (see, e.g., [3]). Such an approach has been developed in [15, 18, 19], in [20] and the efmTOOL and also Peres et al. [16] where the approach proposed in [21] based on the equilibrium constant (K_{eq}) has been revisited.

3 Methods

3.1 Energy Balance Analysis (EBA)

The energy balance analysis (EBA), defined by Beard et al. [11], is based on chemical potentials. The authors define sign patterns of fluxes and analyze which of them are thermodynamically feasible [22]. They analyze the internal cycles space and use the theory of oriented matroids. Internal cycles do not perform a net transformation of external metabolites and should, thus, be excluded from the set of relevant EFMs. Moreover, the authors show that the thermodynamic constraints (based on chemical potential) which arise from the sign patterns of the internal cycle space can be obtained directly from the stoichiometric matrix S. EBA eliminates the infeasible fluxes computed with FBA or EFMA but does not compute feasible flux distribution as the problem is not convex.

To be more formal, let us define $\mu \in I\!R^n$, the vector of chemical potential of metabolites with $\mu_i = \mu_i^0 + R.T.\ln(x_i/x_0)$ where μ_i^0 is the chemical potential of pure species i, x_i the concentration of metabolite i, $x_0 = 1$ mole/L is the unit concentration, R the molar gas constant, T the absolute temperature. The vector of potential difference is defined by $\Delta\mu_i$ and corresponds to the difference of chemical potential of the i-th reaction. A chemical potential difference satisfies a law similar to the Kirchoff's law in electrical circuit:

$$K\Delta\mu = 0 \tag{3}$$

where the row of the matrix K corresponds to a complete basis of the null-space vectors of the stoichiometric matrix S. Moreover, the second law of thermodynamics ensures that the entropy increases in each internal reaction i and hence the direction of flux v_i is from metabolites of higher chemical potential to one of lower chemical potential, i.e.

$$v_i\Delta\mu_i < 0 \tag{4}$$

where v_i and μ_i represent the i-th entries of v and $\Delta\mu$. For a flux vector v to be thermodynamically feasible, there must exist a vector $\Delta\mu$ for which we have 3 and 4 for all $i \in \{1, \ldots, r\}$. These constraints remove all cycles which are thermodynamically infeasible.

3.2 Network Embedded Thermodynamics Analysis (NET)

In [23], the authors introduced Network Embedded thermodynamics (NET) analysis where the Gibbs free energies of reactions are constrained by the mutual thermodynamic interdependencies of the reactions in a network (i.e., the reactions' simultaneous action in the network). The metabolite concentrations have to be feasible in the entire network. The NET analysis requires a predetermination of the directionality of the fluxes and the thermodynamic constraints determine if the fluxes is feasible and the feasible concentration ranges. NET analysis couples metabolite concentrations to an operating metabolic network via the second law of thermodynamics and the metabolites' Gibbs free energies of formation $\Delta_f G_i$. The following optimization problem determines the feasible range of the Gibbs free energy $\Delta_r G_j$ of a particular reaction j with its directionality d_j (+ in forward direction and − in backward direction) for metabolite concentrations x_i constrained by lower and upper bounds x_i^- and x_i^+ (concentrations are assumed to be positive and dimensionless quantities after division by the unit concentration x_0, in order to be able to consider their logarithm):

$$\min/\max \ \Delta_r G_j$$

$$\text{subjectto}$$

$$\begin{aligned} & d_j \Delta_r G_j < 0 \\ & \Delta_r G_j = \sum\nolimits_{i=1}^{m} S_{ij} \Delta_f G_i \\ & \Delta_f G_i = \Delta_f G_i^{\prime 0} + RT \ln(x_i / x_0) \\ & x_i^- \leq x_i \leq x_i^+ \end{aligned} \tag{5}$$

The NET analysis algorithm is implemented in the anNET software [24] which checks quantitative data sets for thermodynamic consistency. NET analysis has been used in [25] to characterize the flux solution space by determining EFMs that are subsequently classified as thermodynamically feasible or infeasible on the basis of experimental metabolome data. This was done by incorporating quantitative metabolite concentrations into the analysis of all stoichiometrically feasible flux distributions, then computing all EFMs and using quantitative metabolite data to test the activities of each EFM for thermodynamic feasibility.

3.3 Thermodynamics-Based Flux Analysis (TFA)

Henry et al. [12] presented the thermodynamics-based flux analysis (TFA) which uses mixed-integer linear programming formulation and computes the flux directionality based on the thermodynamically feasible concentration profiles. In this context, a reversible reaction has to be split as the difference between two non-negative terms, i.e., $v = v^+ - v^-$. By definition, the thermodynamic constraint acting on the reaction leads to have only one of these terms (v^+ or $-\ v^-$) different of 0. These alternatives for the set of reversible reactions is handled in [12] by introducing a supplementary decision variable for each reversible reaction, taking only two possible values. Following the introduction of this new decision variable, the constraint related to v_i^+ is formulated through this inequality $0 \leq v_i^+ \leq v_{i,max} z_i$ and the one on v_i^- by $0 \leq v_i^- \leq |v_{i,min}|(1 - z_i)$ where z_i is an integer taking only two possible values in $\{0, 1\}$. Furthermore, z_i satisfies two inequality constraints associated with the thermodynamic constraint acting on the i-th reaction: (1) $\Delta_r G_i - K + K z_i < 0$ and (2) $- \Delta_r G_i - K + K (1 - z_i) < 0$. Indeed, when $\Delta_r G_i > 0$ then Inequality (2) is always satisfied and Inequality (i) implies that $K(1 - z_i) > 0$ and thus that $z_i = 0$, $v_i^+ = 0$ and $0 \leq v_i^- \leq |v_{i,min}|$. In the same vein, when $\Delta_r G_i < 0$ then Inequality (i) is always satisfied and Inequality (2) implies that $K z_i > - \Delta_r G_i$ and thus that $z_i = 1$, $v_i^- = 0$ and $0 \leq v_i^+ \leq v_{i,max}$. From these preliminary elements, it is then possible to formulate the problem by combining the stoichiometric constraint given by $S \cdot v = 0$ with the set of inequalities associated to each reversible reaction. This formulation leads to define a so-called *mixed integer linear programming formulation* which is generally difficult to solve even if there exist some efficient ways to handle such a problem (see, e.g., [26] for a review). This means that the strategy used by the authors to tackle the problem is to reformulate their non-convex problem into a standard optimization problem, which is itself difficult to solve, but where many computer science researchers have developed approaches (and powerful heuristics) to solve in an exact or approximate way. TFA has been implemented in python package and MATLAB toolbox [27] which integrate quantitative metabolomics data to study the flux directionality but also estimate how far each reaction operates from their thermodynamic equilibrium.

3.4 Dealing Thermodynamics in EFMA

3.4.1 Linear Programming to Deal with Thermodynamics in EFMs

An obvious necessary condition for an EFMe to be thermodynamically feasible is that the system of inequalities $\Delta_r G_j < 0$, for $j \in supp(e)$, is consistent, i.e. has a solution in the unknown variables x_i's. Notice that this system is linear in those unknown variables $\ln(x_i)$. So a linear programming tool can easily check this condition for each EFMe. In [25], the authors showed that a flux distribution containing an infeasible EFM is always infeasible. The thermodynamic EFMs analysis (tEFMA) [15] used this property to filter the infeasible extreme ray in construction during the incremental process of the double description (DD) algorithm [28]. They have interfaced efmTOOL [20] with the solver CPLEX to compute them. Müller et al. [29] presented direct method for computing modules of the thermodynamically constrained optimal flux space of a metabolic network. This method can be used to decompose the set of optimal-yield elementary flux modes in a modular way and to speed up their computation. In the presence of lower and upper bounds on internal metabolite concentration, they have suppressed

80% of infeasible EFMs [18]. If these bounds are unknown, it suppresses at most 50% of infeasible EFMs. The knowledge of thermodynamics can reduce drastically the solution space. The decomposition in feasible subspaces, the Largest Thermodynamically Consistent Sets (LTCSs) proposed by Gerstl et al. [19], have suppressed 90% of the EFMs. The authors then proposed flux tope analysis [30] which is a maximal set of reactions with fixed directions. The flux tope allows one to study the coordination of reaction directions and opens a new way for EFMs computation.

3.4.2 Equilibrium Constants of Reaction

It is possible to compute all the possible thermodynamically feasible EFMs only with equilibrium constant (without knowing the concentrations of the internal metabolites). In [16], the thermodynamic constraint 1 is rewritten as:

$$\prod_i x_i^{S_{ij}} < K_{eq}^j \tag{6}$$

It has been shown in [21] and revisited in [16] by using the Kuhn and Fourier theorem (a consequence of Farkas duality) that this condition is equivalent to the following: a necessary condition of thermodynamic feasibility for any EFM e is

$$e^t \ln(\hat{K}_{eq}) > 0 \tag{7}$$

where $\hat{K}_{eq}^j = \frac{K_{eq}^j}{\prod_i \overline{x_i}^{S_{ij}}}$ is the apparent equilibrium constant of reaction j and $\overline{x_i}$ are the external metabolite. The formula 7 is a linear inequality defining a half space and can thus be directly added as a new constraint in the DD algorithm, which will be processed exactly as the irreversibility constraints. This is done by just adding at initialization to the representation matrix representing the irreversibility constraints a new row whose coefficients are the $\ln(\hat{K}_{eq}^j)$'s [31]. As the inequality in 7 is strict, solutions have then to be checked for equality and those belonging to the boundary hyperplane have to be discarded. If all reactions are reversible, it suppresses at most 50% of infeasible EFMs. Checking the thermodynamic in EFMs allows a better characterization of the solution space but it is limited by the combinatorial explosion of their number in large systems.

References

1. Ataman M, Hatzimanikatis V (2015) Heading in the right direction: thermodynamics-based network analysis and pathway engineering. Curr Opin Biotechnol 36:176–182
2. Dinh M, Fromion V (2017) RBA like problem with thermo-kinetics is non convex. arXiv:1706.01312
3. Schuster S, Hilgetag C (1994) On elementary flux modes in biochemical reaction systems at steady state. J Biol Syst 2(2):165–182
4. Varma A, Palsson B (1994) Metabolic flux balancing: basic concepts, scientific and practical use. Nat Biotechnol 12:994–998
5. Atkins P, de Paula J (2014) Physical chemistry, 10th edn. Freeman, New York

6. Goldberg RN, Bhat TN, Tewari YB (2004) Thermodynamics of enzyme-catalyzed reactions – a database for quantitative biochemistry. Bioinformatics 20:2874–2877
7. Wang P, Neumann DB (1989) A database and retrieval system for the NBS tables of chemical thermodynamic properties. J Chem Inf Comput Sci 29(1):31–38
8. Flamholz A, Noor E, Bar-Even A, Milo R (2012) eQuilibrator – the biochemical thermodynamics calculator. Nucleic Acids Res 40: D770–D775
9. Jankowski M, Henry C, Broadbelt L, Hatzimanikatis V (2008) Group contribution method for thermodynamic analysis of complex metabolic networks. Biophys J 95(3):1487–1499
10. Mavrovouniotis M (1990) Group contributions for estimating standard Gibbs energies of formation of biochemical compounds in aqueous solution. Biotechnol Bioeng 36 (10):1070–1082
11. Beard D, Liang S, Qian H (2002) Energy balance for analysis of complex metabolic networks. Biophys J 83:79–86
12. Henry C, Broadbelt L, Hatzimanikatis V (2007) Thermodynamics-based metabolic flux analysis. Biophys J 92:1792–1805
13. Müller S, Regensburger G, Steuer R (2014) Enzyme allocation problems in kinetic metabolic networks: Optimal solutions are elementary flux modes. J Theor Biol 347:182–190
14. Wortel MT, Peters H, Hulshof J, Teusink B, Bruggeman FJ (2014) Metabolic states with maximal specific rate carry flux through an elementary flux mode. FEBS J 281:1547–1555
15. Gerstl M, Jungreuthmayer C, Zanghellini J (2015) TEFMA: computing thermodynamically feasible elementary flux modes in metabolic networks. Bioinformatics 31:2232–2234
16. Peres S, Jolicoeur M, Moulin C, Dague P, Schuster S (2017) How important is thermodynamics for identifying elementary flux modes? PLoS One 12(2):e0171440
17. Wortel M, Noor E, Ferris M, Bruggeman F, Liebermeister W (2018) Metabolic enzyme cost explains variable trade-offs between microbial growth rate and yield. PLOS Comput Biol 14:1–21
18. Gerstl M, Ruckerbauer D, Mattanovich D, Jungreuthmayer C, Zanghellini J (2015) Metabolomics integrated elementary flux mode analysis in large metabolic networks. Sci Rep 5:8930
19. Gerstl M, Jungreuthmayer C, Muller S, Zanghellini J (2016) Which sets of elementary flux modes form thermodynamically feasible flux distributions? FEBS J 283:782–1794
20. Terzer, M Stelling J (2008) Large-scale computation of elementary flux modes with bit pattern trees. Bioinformatics 24 (19):2229–2235
21. Schuster S, Heinrich R (1991) Minimization of intermediate concentrations as a suggested optimality principle for biochemical networks. J Math Biol 29:425–442
22. Beard D, Babson E, Curtis E, Qian H (2004) Thermodynamic constraints for biochemical networks. J Theor Biol 228(3):327–333
23. Kümmel A, Panke S, Heinemann M (2006) Systematic assignment of thermodynamic constraints in metabolic network models. BMC Bioinf 7:512
24. Zamboni N, Kümmel A, Heinemann M (2008) anNET: a tool for network-embedded thermodynamic analysis of quantitative metabolome data. BMC Bioinf 9:199
25. Jol SJ, Kümmel A, Terzer M, Stelling J, Heinemann M (2012) System-level insights into yeast metabolism by thermodynamic analysis of elementary flux modes. PLoS Comput Biol 8(3):e1002415
26. Vielma J (2015) Mixed integer linear programming formulation techniques. SIAM Rev 57 (1):3–57
27. Salvy P, Fengos G, Ataman M, Pathier T, Soh K, Hatzimanikatis V (2018) pyTFA and matTFA: a Python package and a Matlab toolbox for thermodynamics-based flux analysis. Bioinformatics 35:167–169
28. Motzkin T, Raiffa H, Thompson GL, Thrall RM (1953) The double description method. In: Kuhn HW, Tucker AW (eds) Contributions to the theory of games II. Princeton University Press, Princeton
29. Muller A, Bockmayr A (2014) Flux modules in metabolic networks. J Math Biol 69:1151–1179
30. Gerstl M, Zanghellini J, Müller S, Regensburger G (2018) Flux tope analysis: studying the coordination of reaction directions in metabolic networks. Bioinformatics 35:266–273
31. Peres S, Schuster S, Dague P (2018) Thermodynamic constraints for identifying the elementary flux modes. Biochem Soc Trans 46 (3):641–647. https://doi.org/10.1042/BST20170260

Index

A

B

C

D

E

F

G

H

I

Deepak Nagrath (ed.), *Metabolic Flux Analysis in Eukaryotic Cells: Methods and Protocols*, Methods in Molecular Biology, vol. 2088, https://doi.org/10.1007/978-1-0716-0159-4,